D1217630

Venice, the Tourist Maze

A Cultural Critique
of the World's Most Touristed City

Robert C. Davis and
Garry R. Marvin

UNIVERSITY OF CALIFORNIA PRESS

Berkeley / Los Angeles / London

University of California Press
Berkeley and Los Angeles, California

University of California Press, Ltd.
London, England

© 2004 by the Regents of the University of California

Library of Congress Cataloging-in-Publication Data
Davis, Robert C. (Robert Charles), 1948–
    Venice, the tourist maze : a cultural critique of the world's most
touristed city / Robert C. Davis and Garry R. Marvin.
        p.   cm.
    Includes bibliographical references and index.
    ISBN 0-520-23803-6 (alk. paper)—ISBN 0-520-24120-7
(pbk. : alk. paper)
        1. Venice (Italy)—Description and travel.   2. Travelers—Italy—
Venice—History.   3. Tourism—Italy—Venice—History.   I. Marvin,
Garry R., 1952–   II. Title.
    DG674.2.D34   2004
    945' .31—dc21   2003007854

Manufactured in the United States of America

13   12   11   10   09   08   07   06   05   04
10   9   8   7   6   5   4   3   2   1

The paper used in this publication meets the minimum requirements of
ANSI/NISO Z39.48–1992 (R 1997) *(Permanence of Paper)*.♾

Contents

Illustrations

Figures

Maps

Acknowledgments

We would like to offer our thanks to those many institutions and individuals who have made this work possible. Among the former have been the Stanford University Humanities Seminar, the Cini Foundation, the Vann Seminar of Emory University, the Mediterranean Studies Association, and the Department of Sociology and Social Policy at the University of Surrey, Roehampton. A great many friends and acquaintances in Venice offered us valuable advice and guidance over the course of the several years that it took us to realize this book, and it would be difficult to mention them all here. We would like to especially thank the following Venetians and residents of Venice, however, friends and acquaintances who selflessly contributed their help and time for our benefit, both in giving us advice and in granting us interviews: Cesare Battisti, Sandro Bosato, Matteo Casini, Lady Frances Clarke, Melissa Conn, Vera Costantini, Michela Dal Borgo, Giorgio Del Pedros, Mariangela Du Chaliot, Laurence Lovett, Luca Molà, Carlo Montanaro, Gherardo Ortalli, Fabrizio Reberschegg, Giandomenico Romanelli, Lorenzo "Limba" Russo, Alessandra Sambo, Isabella Scaramuzzi, Jan van der Borg, Franco Vianello Moro, and Guglielmo Zanelli. Closer to home, we have also received considerable assistance from and particularly wish to thank Phil Grabsky, Ana Maria Isaías Nunes de Almeida, Thomas Madden, and Sally McKee. Finally, Stephanie Schwandner-Sievers donated for our benefit her considerable talents as a multilingual interviewer of tourists, Beth Lindsmith provided invaluable help as our manuscript editor and general advisor in matters of prose, and Cindy Davis generously gave of her time, both as our initial mapmaker and in generally putting up with our endless conferences.

Note to Readers

All premodern English texts are transcribed here with their original spelling, punctuation, and capitalization. Some other texts, as indicated in the bibliography, were originally written in other languages (principally French or German) and translated during the seventeenth or eighteenth century into English; these also have been transcribed as they were originally published. All other Italian and French materials, both premodern and modern, including books, journals, and newspapers, were translated by Robert C. Davis, who also transcribed and translated all interviews in Italian and is responsible for their accuracy. All present-day German interviews were conducted, transcribed, and translated by Stephanie Schwandner-Sievers.

The City Built on the Sea

Tourism, as every economist knows, is big business, arguably the biggest in the world today: the latest figures from the Secretariat of the World Tourism Organization suggest that the 664 million international arrivals worldwide in 1999 spent something on the order of $455 billion. Tourism is also the life and life's blood of Venice, Italy, and more than once, while sitting in Piazza San Marco on an August afternoon, we have felt like the better part of those 664 million travelers were passing right before our eyes. Venice may well be, for its size and population, the most touristed city on the planet, but even if it falls somewhat short of this dubious honor in terms of hard numbers, it still seems to enjoy a special status, in the mind of many, as the epitome of the tourist experience. Many of the visitors we spoke with in the course of researching this book have so learned to associate Venice with the touristic ideals of fantasy, pleasure, and—above all—romance that they seem to find it hard to conceive that this place exists or ever existed for any other purpose than bringing enjoyment to themselves or other outsiders. This is hardly a new reaction: Mary McCarthy once observed, "The tourist Venice *is* Venice . . . a folding picture-postcard of itself"; indeed, what we have found pervasively in the world-public mind is that there is no *real* Venice here. Although this is still technically a city (it has a government, after all, as well as city services, some residents, and even a university), visitors often fail to see or intuit such mundanity. Instead, they come here looking for the image that they have already formed of Venice—the vision of the Most Romantic City in the World. Like many another travel writer, Erica Jong advises her readers to hunt this ideal down while in Venice, and in no uncertain terms:

"I recommend that you fall in love in Venice and pursue your love down the narrow *calli,* through the *campi,* under the *sottoportici.* If you can't snare a lover, Venice will surely provide one—even if that love is only the city herself. Venetian love affairs inhabit some ideal realm: they rarely prove durable when reality dawns. Perhaps that's the whole point."[1]

When they actually arrive, do the millions of tourists who come to Venice find Jong's "ideal realm"? They seem to, at least in the movies. At the climax of a recent Danish art-house film, *Italian for Beginners,* the protagonists, a troop of rather drab Danes, end up in Venice, and the city works its predictable magic.[2] One couple, shy and quiet through most of the film, are so overcome with the passion of the moment that they grab an abandoned bed in some back alley and make wild love on the spot. If a viewer should happen to wonder—as we did—how the residents of Copenhagen might respond to finding two Italians doing the same thing in a back alley of their city, it is again enough to remember that for most of the world Venice is not a real city, with a real city's inhabitants and constraints, but a backdrop and a stage for one's gaze, emotions, and passions. Moreover, it is a place seemingly made for intrigue and romance, a labyrinth of narrow streets and waterways that positively invites transgression on the assumption that nobody is watching, or at least nobody who matters.

Becoming the quintessential romantic city did not come naturally to Venice. Until 1797, when it ceased to exist as the *Serenissima Repubblica,* this was a city that existed primarily for practical purposes, not touristic—much less romantic—ones. It was built where and how it was simply to provide its residents with a secure place to live their social, political, and commercial lives. In earlier ages, visitors came here mostly for the same reasons that outsiders always come to cities—for profit, politics, or refuge. Those who admired the place usually did so more for its wealth, so brashly displayed on every side, than for its scenic qualities: many visitors were humbled, even intimidated, when they compared Venice's humming commercial activity, military force, and cosmopolitan crowds to their own, more staid and modest hometowns.

This began to change in the late 1600s, by which time Venice was already well down the slope of its long decline. Visitors from England and France, particularly, began to turn their gimlet eyes on the whole range of products and pomp that had once made the city world famous, and realized that it no longer measured up to what was back home in London or Paris. Yet despite their scorn these foreigners kept coming, right into the 1790s, lured largely by those things that Venice offered that were relatively scarce elsewhere in Europe—a particular sense of style, above all, but also vice: gambling, prostitution, and perversion of whatever sort visitors required. Venice became the continent's brothel, but also, thanks to its limited scale and unique waterways, Europe's first theme park. By the time the collapse of the Serenissima Repubblica came in 1797, Venice was already well on its way to living off tourism alone. As such, we might say, it was perhaps the first postmodern city, selling no product other than itself and its multiple images to the tens or even hundreds of thousands of free-spending foreigners who came there annually.[3]

Much of this was a choice of the Venetians themselves, who during the 1800s were exploited and then abandoned by a series of outsider governments—first Paris, then Vienna, then Rome—cut off by changing patterns of trade from their traditional sources of wealth, and left with little to fall back on beyond their own proven ability to hawk themselves and their city to foreigners. A new form of tourism was soon there to buy their product. Packaged tours, built on the modest but pervasive resources of middle-class Germans, French, and British, and taking advantage of new rail links through the Alps, were soon flooding Venice with foreigners. Special trains might dump as many as ten thousand of these visitors a day into the city for special occasions. Before long, Venetians were beginning to tailor their own holidays, often of ancient vintage, just to suit the expectations and desires of these new visitors.[4]

From these earliest packaged tours to the industrial tourism of the present, visitors have come to this lagoon city looking for Romance with a capital R. In the nineteenth century this was above all a literary venture, as tourists sought (or competed) to plant themselves on the same spot that had once inspired Shakespeare, Goethe, Shelley, Byron, or Browning, and then read aloud the appropriate lines. Such interpretive efforts still continue, though now more through the play of image and representation with which Venice is presented to the world by the mass media in films, travelogues, and above all advertising. Today's Venice is still endlessly created and recreated in a feedback loop of tourist desire and demand, and even if the poetry has by now mostly given way to commercial tropes of gondoliers and Carnival masks, the city still satisfies with remarkable success the needs of those who visit.

The intense affection that Venice arouses in the world public has brought its own price, however. Something like thirteen to fourteen million outsiders currently go there every year, overwhelming a town whose resident population at the start of the twenty-first century was barely sixty-five thousand.[5] Venice is thus a city not just visited but actually *inhabited* by tourists: on average, there are no fewer than eighty-nine foreigners in the city at any given time for every one hundred Venetians who live there: the highest tourist-to-resident ratio in Europe, and nine times that found in, say, Florence.[6] With visitors coming from every corner of the globe, Venice might claim to be one of the great multiethnic cities of the post-modern world. Of course, all these transients do not really live there, at least not in the sense of residing there or paying taxes; yet in terms of human life and public activity, and certainly from the Venetian point of view, they are permanently there, all over town. Although each tourist may take up Venetian space for only a short while, in the aggregate they are there all the time: their transitory occupation might be called a flow, but for the Venetians themselves it appears as a continual, solid mass.

Diverse though they may be and however much each may think that he or she is engaging with the city in a personal way, these tourists are still virtually all engaged in the same fundamental act. They are consuming—not so much Venetian culture, but rather the images of that culture, and consuming them, for the most part, as quickly as possible (over 80 percent of visitors stay less than a day). What Venetians tend to see of them is a homogenous flow: men, women, and children whom they refer to as "the herd," and who in the aggregate seem as undifferentiated as locusts or starlings. Some Venetians (about 40 percent of the total) do make their living tending this herd, selling it souvenirs, ice creams, plates of *fritto misto,* or corn to give the pigeons. Most residents, however, go on living there in spite of, rather than because of, the massive tourist presence. These unfortunates—pensioners, government workers, school children—can do little but suffer what the papers like to call their "days on Calvary," the ever more frequent occasions when sixty and sometimes even one hundred thousand outsiders try to cram themselves at once into what is, after all, a fairly small city.[7]

Tourists and Venetians, then, meet and jostle and try to get by in what is still one of the most beautiful cities in the world. The tourists come, many of them, with an ideal and a cliche battling in their heads, and hope that the ideal, their fantasy of Venetian Romance, will somehow win out, in the city where Henry James already was pouting, over a century ago, that "there is nothing left to discover or describe, and originality is completely impossible."[8] Modern media assure them that this is certainly not so, however, and, if not true love, at least true fantasy (and good shopping) can still be achieved in the World's Most Romantic City. As one recent travelogue on Venice concluded: "Venice is one of those places that surpasses its own publicity. The canvasses of Canaletto and reels of celluloid can't sell it short. Being here is better than the dream."[9]

At the same time, the Venetians, ever older and fewer, keep up their struggle to remain in the city where their ancestors may have been born, but where they themselves are increasingly strangers, superfluous and often unwanted by the tourist monolith that has long since taken over. They write their innumerable litanies of complaint to the papers: about their difficulties in getting around among the crowds, about the pollution and disturbances brought in by the visitors, about the disrespect and ignorance they encounter every day. Those who get by the best seem to have learned simply to look right through the mobs, avoid the "hot" areas of town (although those are increasing in size every year), and live their lives in the early morning and at night.

The development and the interplay between these two parallel realms—one of determined fantasy, the other stuck in a far too real world of overcrowding, decay, and discomfort—are the focus of this book. We have called our investigation a "cultural critique," one that presents the reader with a variety of approaches—sociological, historical, economic, and aesthetic—to a problem that is both deeply rooted in the past and quintessentially postmodern. In the process, we have attempted to set the Venetian experience into what is truly unique about this special city: the particular placeness of Venice. This has required us to engage with the multiple illusions that Venice projects of, and onto, itself, often tempting us away from our own academic disciplines and into the roles of travel writer and sentimental tourist. It has not always been an especially comfortable approach for a pair of social scientists, but we have come to believe that just beneath the pleasing but obscuring surface of this Most Serene City lies a social, topographic, and historical tangle of immense complexity, and that without grappling with the romance and fascination of Venice we could never manage to explain how we think that Venice does or does not work.

To help us in the process of exploring this very complex city, we have taken advantage of Appadurai's notion of "scapes," perspectival constructs through which humans impose meaning on an environment. Unlike those travel writers who have traditionally divided Venice into some number of districts, *sestieri,* or (in one case) gateways, we have seen the experiential city as falling naturally into four of these scapes: Timescape, Landscape, Seascape, and, lastly, Worldscape, and we accordingly have structured this book into these four parts. Each scape implies different, if also simultaneous, ways in which the city has been imagined, constructed, and experienced, both by its visitors and its inhabitants. Each section has been divided in turn into chapters that explore what we take to be the scapes' most salient areas of contact and conflict, between the tourist and the localist vision of the city. Readers may notice a slight disequilibrium in this regard: "Timescape" has only two chapters, while the other sections are allotted three each. At some point even we, with all our enthusiasm for this project, had to admit that this book was getting too long. That, plus a certain refusal to give in to the tyranny of our own self-imposed symmetry, made it easier to drop a third history chapter, where we had originally planned to explore the artistic and aesthetic attractions of Venice for visitors in the eighteenth and nineteenth centuries. This is a topic that already has been well covered elsewhere, by scholars with far better art-historical training than either of us. But perhaps most importantly, we have come to realize that it is image, rather than art, that has long been Venice's primary attraction, whether for the Grand Tourists of two centuries ago or for the mass tourists of today. How this image has been both created and consumed has thus seemed the logical and necessary theme for this book.

Inevitably, in the course of writing this work we have had to grapple with one question that for more than a century has uniquely shadowed this Most Romantic of Cities: is Venice dying? Some have claimed that it most definitely is not, backing this up with such delightfully specious logic as "A city that gives people ideas is not a dead city" (Angkor Wat? Machu Picchu? Atlantis?).[10] Others say that Venice is not only dying but is already dead, killed by its tourist monoculture, pollution, and the inexorably rising tides. Historically, we know that Venice repeatedly has been declared comatose, if not quite deceased: on the ropes, breathing its last, sliding from view beneath the waves. Yet the city, to the extent that it is a city, has kept on, albeit with levels of tourists, rates of physical collapse, and heights of tides that would have been considered horrific there just a couple of decades ago. Somehow Venice keeps going, and every so often some significant improvements are seemingly made, interventions that may hold back the apparently inevitable, final collapse for a few more years. Not wishing to underestimate the flair and skills with which Venetians have approached the salvation of their city (nor the slowness with which they have typically carried out the work), we have tried to avoid snap judgments, especially since more than one apparent social or ecological disaster that we witnessed during the writing of this book seemingly had been resolved by the time of publication. We do believe, however, that while Venice clearly lives from the ceaseless flow of tourists through the arteries and veins of its narrow streets and alleys, it also slowly dies, at least as a Venetian space, through the continual congestion of these lifeways by outsiders.

This project came out of the intersecting interests of a social historian and a social anthropologist. One of us brought to it his specialist knowledge of Venetian social history and his experience of Venetian culture, the other his anthropological interests in ritual practices and symbolic constructions. With this collaboration we hope to open up some new areas where our two disciplines can cross-pollinate; in particular, we hope to demonstrate an ethnographic continuity in the ongoing history of tourism in this most touristed of cities. The historian led the anthropologist through the texts and the topography of the city, in the spaces of which our conversations, observations, speculations, interpretations, agreements, and disagreements that produced this book took place. The anthropologist introduced the historian to the craft of interviews, to symbolic structures, and to the fine art of conducting one's field work in bars. In the end, though, *Venice, the Tourist Maze* was inspired by our own fascination with what is still a very special place—however it is constructed, or deconstructed. Despite the myriad representations of Venice, despite the libraries of books written about it, the city still shimmers—broken, unfixed, and probably unfixable—before all who seek to experience it.

PART ONE

Timescape

CHAPTER I

Pilgrims' Rest

As one of the most scenic and most touristed places in the world, Venice has, perhaps inevitably, inspired a number of scholarly and general studies on its place in the history of tourism. These have been alive to the city's impact on the imaginations and fantasies of foreigners, and aware of the many levels on which it has captivated its admirers. Yet such works for the most part have also been limited in two significant ways. First, they have dealt for the most part with only one, relatively circumscribed aspect of tourism to Venice, the so-called Grand Tour—that leisurely jaunt down to Italy undertaken from the early 1600s to the 1830s by young, mostly British, patrician males, who journeyed there, often with their tutors, as part of what was considered an appropriate education for such elites.[1] Second, as a result, this history has for the most part been told, not as part of the social and economic story of Venice itself, but rather as an element of British (and occasionally French or German) cultural history: specifically, the story of how a cosmopolitan consciousness developed among the ruling elites of these rising states, one that paralleled and abetted the rise of Western European expansion and colonialism during the early modern era.

Such a focus, in the end, has told only half the story, and perhaps not the more interesting half. From it we learn little or nothing about the tourist experience of the city before 1600; nor if, indeed, during these earlier centuries there was anything going on in Venice that in social and cultural terms we reasonably could call "tourism." Moreover, the traditionally xenocentric approach to Venetian tourism, focusing on the visitors rather than the visited, has left the Venetians themselves as oddly passive background figures in what remains fundamentally a northwestern-European story of self-discovery. As such, despite their otherwise well-established fame—already a millennium old in 1600—of being among Europe's shrewdest traders, entrepreneurs, and hustlers, past Venetians have received scant acknowledgment from modern scholars of having had any ability at all when it came to promoting their own city, or even of being especially aware that theirs was a city that could be promoted. This neglect hardly seems justified—especially when one considers how famously the Venetians have proved themselves in this regard since the fall of the Republic in 1797. It must also be kept in mind that in the two centuries before 1600, the European cultural shoe was still very much on the Venetian foot: this was one of the continent's great metropolises—indeed, one of the world cities of the late Middle Ages—sophisticated and alluring to those gaping visitors from such feudal and rural backwaters as the British Isles or even Germany. It is simply hard to imagine that the Venetians, brought up in millennial mercantile traditions, would ignore the steadily increasing tide of visitors that was coming at them from over the Alps or fail to find some way to turn it to their own (and their city's) financial advantage.

Of course, most of the foreigners coming to Venice before 1600 (or, indeed, before 1800) were not there just to see the sights. This was a commercial hub, filled with whole colonies of foreign artisans and merchants pursuing their trades, some in continual rotation back and forth to their home countries and some in a state of almost permanent residency in Venice.[2] One would hesitate to call such professionals "tourists," at least not in any but the most casual sense of the word, any more than it would make sense to apply the term to the flocks of political and religious refugees, visiting or permanent ambassadors, or passing princes that came, went, and stayed in Venice throughout these centuries. Such outsiders, coming for motives of business or politics, might well have enjoyed the ambience in Venice, as well as the food, the business and intellectual connections, and (apparently quite often) the women; some even might have marveled at the sights of a city as unusual, wealthy, and powerful as Venice. It is unlikely that these medieval merchants and artisans spent much of their time exploring the place or seeking out its attractions, simply because they had other business to take care of. We will never know for sure, simply because so few left behind their written reactions to Venice or anywhere else, perhaps because they were illiterate or because they lacked the training in letters that would permit them to elaborate on their impressions about a place so different from their native cities and past experiences.

Central to our modern notions of what it means to be a tourist is the concept of the sightseer: someone, that is, who engages in a socially and culturally constructed way of viewing—what sociologist John Urry has famously and specifically defined as "the tourist gaze." For Urry, this has meant looking at "a set of different scenes, of landscapes or townscapes which are out of the ordinary": at least what is out of the ordinary from the tourist's perspective. Indeed, Urry goes on,

the tourist gaze is directed to features of landscape and townscape which separate them off from everyday experience. Such aspects are views because they are taken to be in some sense out of the ordinary. The viewing of such tourist sights often involves different forms of social patterning, with a much greater sensitivity to visual elements of landscape or townscape than is normally found in [the tourist's] everyday life. People linger over such a gaze.[3]

We do find among the general run of medieval visitors to Venice some good candidates for this sort of self-conscious interacting with the city's extraordinary features. These were pilgrims on their way to the Holy Land. A good many of these worthies kept records—ranging from simple account books to full-fledged diaries—of their trips from Latin Christendom to the shrines and sacred sites of the Levant. Even though their destination lay fifteen hundred miles to the east, many of them treated the stops along the way as worthy of a look and a few lines in their journal. This especially included the Serenissima Repubblica—the Most Serene Republic—of Venice, the jumping-off point for most of them for their sea voyage to the Levant, and a place where they would find themselves with ample free time to have a look around. As they spent their days "visiting the sights of Venice," as one of them put it, many of these pilgrims also behaved in ways—aesthetic, cultural, and recreational—that we have come to associate with what it means to be a tourist.[4] Much like present-day visitors to the city, these pilgrim-tourists found the city as a place poised between the familiar and the exotic. To get ready for their visit they read accounts written by their predecessors; if nothing else was available, then they would make do with gossip and tales about Venice and the Venetians, passed on by other pilgrims or merchants who had already seen it. Once arrived, many were struck by Venice's otherness: not just that of the topography, but also of the local mores, to which they struggled to relate, with greater or lesser success, through their own cultural optic. They developed, for themselves and for those who came after them, a hierarchy of sites to visit, a circuit that could be lengthened or shortened as appropriate to the length of their stay. While in the city they also found one of the tourist essentials common throughout history and the world—an already developed infrastructure of support services ready to aid (or exploit) them during their stay.

At the same time, of course, these were pilgrims before they were tourists: men and women who were on a sacred journey, often in response to a vow, or trying to even the balance sheet of their life's unavoidably sinful nature. They dressed (for the most part) in poverty, journeyed anonymously, and tried their best to avoid the various fleshly and material temptations that Mammon threw into their traveler's path. Yet the fact that these worthies were on one particular kind of (sacred) journey was no bar to them taking interest in the unfamiliar sights and scenes around them. Nor, indeed, did it preclude them from engaging in what (at least to the present-day eye) might seem like secular activities. They were humans, after all, who needed to eat and drink, and who sought, for the most part, to be as comfortable as possible on their long and often dangerous trip to the East and back. Many of them in any case traveled in a group, with a priest or friar to take care of their spiritual needs—leaving them free to explore more worldly concerns, as the occasion warranted.

Ever since 1978, when Victor and Edith Turner had the insight that "a tourist is half a pilgrim, if a pilgrim is half a tourist," many sociologists have grappled with the notion that tourism itself might be some form (or many forms) of a sacred journey.[5] The concept, as it deals with the first half of the Turners' equation, has by now been extended beyond the specifically religious to encompass the Durkheimian notion of the nonordinary, the different, the alien, all of which potentially carry a sacred aura that attracts tourists to them. This encounter with the other, as well as with the more narrowly holy, also served to sanctify the journeys of medieval pilgrims—a combination that especially prevailed in Venice itself. There, having detached themselves—as vacationers have sought to do ever since—from the claims and duties of their ordinary lives, pilgrims usually paused on this first major stop on their sacred itinerary. It was their jumping-off point into an unknown world filled with rumored fantasies, real dangers, and hoped-for spiritual attainment; at the same time, it was rife with attractions and rewards all its own.

Venice's centrality in the overseas pilgrim transport business was assured only late in the fourteenth century, after the city had finally bested a number of other Mediterranean rivals to emerge with a near monopoly over the transit trade between Europe and the Levant. It was a dominance that the city enjoyed and exploited for the next century and a half, from the 1380s until into the 1530s: during these years the Venetians ran regular galleys (and more sporadic merchant cogs) to the Holy Land, attracting a yearly minimum of two to three hundred German, British, French, Spanish, and Italian pilgrims—and some years several times that many.[6] Those holy travelers who wrote about their experiences (or who, being the literate ones, wrote on behalf of others) tended to be clerics. By the end of the fifteenth century, the gossip and written accounts that they had generated about the Pilgrims' Trail in general and Venice in particular had become so extensive that it moved one of them, the Milanese canon Pietro Casola, to become perhaps the first tourist to the city (but hardly the last) to complain that Venice was a place "about which so much has been said and written . . . that it appears to me there is nothing left to say."[7]

By Casola's time Venice clearly had become an attraction in its own right, meriting more pages in most pilgrims' journals than just a few notes on the practical necessities of outfitting and booking one's passage from there to the East.[8] Pilgrim-tourists often wrote of it in rhapsodic terms, much like visitors of later centuries: "Splendidissimo," "grandissimo," and "ornatissimo," Santo Brasca raved about the Ducal Palace, which, Arnold von Harff agreed, "is very fine and is daily being made more beautiful." Casola himself exclaimed, "I declare that it is impossible to tell or write fully of the beauty, the magnificence or the wealth of the city of Venice," although he then went on for some pages trying to do so. As they approached Venice by boat from the mainland, the very sight of this "famous, great, wealthy and noble city . . . standing in wondrous fashion in the midst of the waters, with lofty towers, great churches, splendid houses and palaces," left Friar Felix Fabri and his companions "astonished."[9] These pilgrim-tourists marveled about Venice's cleanliness, its cosmopolitan nature, the power and efficiency of its Arsenal, and, above all, its wealth: even the most widely traveled among them had never seen so many sorts of goods, treasures, and edibles from such a variety of countries, so many "large and beautiful palaces splendidly furnished and decorated," and—especially attractive for pilgrims—so many imposing churches and monasteries. As Casola put it,

I have been to Rome, the chief city of the world, and I have travelled in Italy, and also very much outside of Italy, and I must say—though I do not say it to disparage anyone, but only to tell the truth—that I have not found in any city so many beautiful and ornate churches as there are in Venice. It would take too long to name them all.[10]

Despite such praise, often bordering on adulation, many pilgrims also wrote about their great eagerness to leave Venice and get moving toward the Holy Land, which was, after all, the whole point of their journey. As it happened, however, many found themselves stranded in the city far longer than they had planned. Although the *galee de Zafo* (that is, "galleys of Jaffa") that specialized in trafficking pilgrims to the Holy Land left only twice a year, in late spring and early fall (and, after the 1450s, mostly just in the spring), many pilgrims seem to have been completely unaware of this rather restricted schedule, and could say, like Casola, "I found that I had been in too great a hurry to leave home, and that I must wait several days before the departure of the said galley" (in fact, he was stuck there for nearly three weeks). The Castilian pilgrim Pero Tafur came to Venice so much ahead of time in 1435 that he decided "in the meantime I should travel in Italy, which was well worth while." Even so, on getting back to Venice two months later, Tafur found that he would still have to wait another four weeks.[11]

Such delays could weigh heavily on those who were eager to set out on a difficult journey of discovery and penance—or at least to get it over with.[12] "We were beginning to be exceeding weary of Venice," groused Felix Fabri, and "we were eagerly looking forward to our departure." Indeed, for many pilgrims, Venice on the outbound leg of their voyage turned out to have been the city they stayed in by far the longest—much longer even than Jerusalem. In drawing up their contracts of passage with their galley owner, these travelers insisted that, barring storms, they were not to be kept in any port of call for more than three days; but Venice was their jumping-off point and few pilgrims managed to get moving in less than three or four weeks. Ship captains, grumbled Friar Felix, often "promised that they would begin their voyages directly, which I knew to be a lie." The vagaries of the sea had much to do with this, as the captain-owners waited, sometimes for weeks, for the right weather, for necessary outfitting, or simply for a few more paying passengers. Although pilgrims complained about "the tediousness of waiting" in the city, few would risk a side trip even to Padua, if they, like Casola, were "anxiously waiting for the time of departure, which was put off from day to day": they well knew that their ship could suddenly set off, without warning, leaving them stranded and out the cost of their deposit.[13]

One can detect that there was a certain complicity among Venetians to, if not actively delay pilgrims in their city, then at least not hurry them away too quickly. There were, in fact, many who had an interest in keeping these foreigners around: what Frederic C. Lane has called "the very profitable tourist trade to Palestine" was after all not just a money spinner for those who ran the ships. Ugo Tucci has estimated that in the boom years of pilgrim trafficking these travelers could bring twenty-five thousand or more ducats a year into Venice's economy in what they spent on their tourist needs while in the city—for lodging, food, supplies, entertainment, and gratuities.[14] Although the whole of this activity is not well documented, there are still indications that Venetians of every class tried their best to take advantage of this potential market. Their efforts helped create the social infrastructure that would later put Venice in the forefront of Italian secular tourism, creating a unique relationship with Venice's captive clientele of pilgrim-tourists that made the city more closely resemble Rome—the undisputed queen of tourism in Italy since classical times—than its fellow republic of Florence, despite the Tuscan capital's increasing reputation as the peninsula's artistic center.[15]

Fortunately Venice had a good deal to offer the stranded pilgrim-tourist. Those coming from inland cities were especially taken by the wealth they saw on every side, even more than one might expect in a great port. As Casola put it,

> I see that the special products for which other cities are famous are to be found there [in Venice], and that what is sold elsewhere by the pound and the ounce is sold there by *canthari* [around fifty kilos] and sacks of a *moggio* [333 liters] each. And who could count the many shops so well furnished that they also seem warehouses, with so many cloths of every make . . . and full of spices, groceries and drugs, and so much beautiful white wax! These things stupefy the beholder, and cannot be fully described to those who have not seen them. Though I wished to see everything, I saw only a part, and even that by forcing myself to see all I could.[16]

One could easily find all the necessary provisions here for a sea journey to the Holy Land, and indeed several writers offered a lengthy list of the goods they had purchased there before setting off. Obviously a few pilgrims might decide to load up on some of the finer products that the city had to offer, especially when they again stopped off in Venice on their way home, but Casola's account also makes clear that goods in such variety and abundance were a medieval tourist attraction in their own right, worth admiring even by those who purchased nothing. Indeed, in his account of the city Casola went on for pages, detailing the wonders of Venice's wholesale grain market, its bakeries, butcher shops, fish markets, fruit sellers, and wine outlets.[17]

One could also enjoy Venice for the oddities that were to be found there. Always at the heart of the Serenissima, Piazza San Marco in the fifteenth century was the gathering place of exotic visitors from the Balkans, the Levant, and Africa—soldiers, sailors, merchants, and many others of no particular profession. Such characters made an intriguing sight for these pilgrim-tourists, many of whom themselves came from comparative backwaters in Germany, France, or England and knew they were soon to set off to see the homelands of those same strangers. There were also a variety of exotic animals—rhinoceroses, lions, ostriches, and the like—that were staked about the Piazza and could be viewed for a few coppers. The curious could also find more macabre attractions, as Santo Brasca reported at considerable length: "Then I went to see a Castilian woman born without arms . . . who ate and drank with her feet, sewed, cut, spun, reeled with her feet, and did all the other wifely duties with her feet that another would do with her hands; and it certainly was a prodigious thing to see, and everybody converged on her with many alms."[18]

For the pilgrim-tourists, however, it is not surprising that Venice's greatest attraction was religious, for this was a city that boasted a collection, not so much of holy sites (these were the prerogative of Rome and Jerusalem), but of holy objects that may have been without equal in the western Christian world. Having assiduously collected relics in the Levant for centuries and helped themselves to Constantinople's sacred treasures when they looted that city in 1204, Venice's patrician merchants had over the years brought back an incomparable hoard of saints' relics and other holy objects.[19] The Irish friar Simon Fitz-Simon, who journeyed to the Holy Land in 1323 and was probably one of the earliest pilgrim-tourists to write of Venice, noted that "in the city lie the bodies of saints, which are whole and uncorrupted," and he went on to list, besides Saint Mark himself, Saints Zacharias, Gregory of Nazarenus, Theodore the Martyr, Lucy, Marina, "and many more martyrs, confessors, and holy virgins." By the end of the fifteenth century, in the heyday of pilgrim-tourism, when a visitor could exclaim, "The Relyques at Venyce can not be noumbred," virtually every one of Venice's seventy-six parish churches and eighty monastic houses could boast at least something holy that was suitable as an attraction for pilgrims.[20]

Pilgrim-tourists, especially those who were priests, canons, or friars, inevitably made seeing such relics a top priority of their time in Venice. Some did so because, like Pietro Casola, they had vowed to visit a particular shrine in the city—an indication of the extent to which Venice's holy spots had established a reputation among travelers.[21] Others, like Felix Fabri, tried to take advantage of the protective power of these relics, going in the days just before their departure to pray before the bodies of those "saints [who] are of particular service to those who are about to go on a pilgrimage."[22] Yet, whether clerical or lay, acting on a vow or in simple piety, all pilgrim-tourists seem to have been driven to make contact with holy relics, to see and (if possible) touch as many of them as possible. There was indeed merit to be gained from visiting any relic, but many of them had the additional attraction of also conferring an indulgence if the visit were made during a particular day or festival. The opportunities for such spiritual rewards were not to be neglected, and Lionardo Frescobaldi's rather blunt statement that he "searched for indulgences and for notable places" during his three weeks in Venice in 1384 pretty much sums up the blend of sacred and secular tourism in which most pilgrims on a limited schedule spent their time in Venice.[23] Many pilgrims' descriptions of the city prominently listed all the relics they had seen: the Englishman William Wey and the Milanese Santo Brasca, for example, enumerated (with surprisingly little overlap) fifty-two and forty relics, respectively, which they had managed to see in visiting only a relative handful of the city's many churches. Finding himself stuck for an entire month in Venice, Friar Felix, by contrast, persuaded those in his party for whom he was the spiritual guide to join him in a series of mini-pilgrimages to see the relics in no fewer than thirty-four churches, located all over the city.[24]

Their tales about such single-minded pursuit of the holy creates an impression that these pilgrim-tourists believed that the merit they gleaned from relics was cumulative: each site visited added a bit more to one's overall spiritual accomplishment. When possible, pilgrims sought to intensify their experience by kissing the actual saints' bodies, often "many times," as Felix Fabri noted. The power implicit in the relics could also be harvested by touching them with small objects—especially jewelry— and this the pilgrims might do on their own account or for friends and relations back home, using precious stones or ornaments that had been entrusted to them for the purpose: "whenever [pilgrims] meet with any relics, or come to any holy place, they take those jewels and touch the relics of the holy place with them, that they may perchance derive some sanctity from the touch; and thus they are returned to the friends of the pilgrims dearer and more valuable than before."

Friar Felix, though "the poorest of our company," by his own account seems to have been virtually loaded down with "precious jewels which had been lent me by my friends, patrons and patronesses." After kissing the saints' bodies or relics for his own sake, he then brought out his collection of jewelry from back home; when these had all been applied, he might also dig through his pockets to come up with all sorts of "easily carried trinkets," objects that he knew from past experience he could give as "presents to those who are dear," while at the same time "receiving a reward for doing so." Pilgrim-tourists who were not so well furnished with handy baubles seem to have been able to buy whatever glass beads, paternosters, and little crosses of gold or silver they thought necessary for this purpose: Friar Felix noted that there were obliging Venetian crafts-men hovering nearby, willing to sell whatever items might be desired.[25] Such trinkets might be considered a sort of proto-souvenir, something like the knickknacks sold today by vendors on the Riva degli Schiavoni: mass-produced, easily transported curios. They certainly gained much of their evocative power from having been in contact with the holy; on the other hand, once brought home from Venice, they were also mementos akin to modern-day souvenirs (as we shall see later on), intrinsically car-rying the added significance of a specific Venetian placeness, valued as in-dicative of where they were made and purchased.

Eager to make contact with saints' relics that were spread all over town, many of these late-medieval pilgrims braved the city's tangled streets and canals and sought out their desired objects wherever they might lurk. Still, the sacred attractions held by all the city's parish and conventual churches paled next to those to be found in the Basilica of San Marco, which boasted not only the intact body of the Evangelist Mark himself, but also a host of other treasures. These included both secular objects—crowns, gems, the four bronze Horses of Constantinople, "a large and long unicorn's horn, most highly chased"—and sacred marvels, ranging from standard saints' relics to holy trophies that approached the bizarre. Indeed, fifteenth-century San Marco was something of a medieval Madame Tussaud's of biblical curiosities, offering an enormous range of sacred objects. A visi-tor in the 1600s enumerated a few of the more unusual relics to be found within the treasury:

. . . diverse heads of Saints, enchased in gold; a small ampulla, or glass, with our Saviour's blood; a great morsel of the real cross; one of the nails; a thorn; a frag-ment of the column to which our Lord was bound, when scourged; the standard, or ensign, of Constantine; a piece of St. Luke's arm; a rib of St. Stephen; a finger of Mary Magdalene; numerous other things, which I cannot remember.[26]

San Marco may have attracted many pilgrim-tourists for its relics and cu-rios, but the Basilica, as indeed the entire area, also attracted these visi-tors because it was the ceremonial and architectural focus of the city. Vis-itors more accustomed to the Gothic styles of the north might find the church "beautiful but low," "a small thing," or "not at all in the fashion of those in France," but most of them still waxed enthusiastic about the building's decorations and those of the adjoining Ducal Palace, even if, like von Harff, they had happened to come when the Palace was going through extensive renovations.[27] Pilgrims also came to San Marco when they wanted to pay their respects to the Venetian authorities. For those of high rank this could mean an audience with the doge, for a formal welcome with an embrace and kiss "in the Italian fashion"; it could also be the moment to ask for a ducal intervention—to free personal possessions impounded by an overzealous customs, for example. Even those of more modest standing could arrange a guided tour of the palace with one of the ducal pages.[28] Piazza San Marco also drew pilgrims because it was the place to sign up for passage on the Jaffa galleys: those who were in the market could learn how many galleys were accepting passengers by the number of white flags with red crosses that were on the poles in front of the church. The competition between galley owners could be fierce, according to Friar Felix: although pilgrims might be treated to a free snack of "Cretan wine and comfits from Alexandria" while touring each galley, first they had to weather the assaults of the various touts in the Piazza, who "each invited the pilgrims to sail with their master, and each endeavoured to lead the pilgrims [to their own galley] . . . [and] each abused the other and defamed him to their worships the pilgrims, and each tried to make the other odious to the pilgrims, and suborned men to do so."[29]

Piazza San Marco was also the center of Venetian festive and ceremonial life, home of splendid processions that were world-famous well before 1500. Those who came to Venice in the spring could expect to see as many as three major festivals while they waited for their galleys to depart: Saint Mark's Day on April 25, the Feast of the Ascension in mid-May, and Corpus Christi in early June. Although all three days had a significant place in the Christian calendar, for Venetians they were also and perhaps fundamentally civic festivals, patriotic occasions that brought together in showy processions the entire active citizenry. By the fifteenth century, the three holidays had to some extent melded into one: "a vast spring festival complete with public entertainments," including a fifteen-day fair held in the Piazza and along the adjoining merchants' streets, the "Mercerie."[30]

While the procession celebrating the festival of Saint Mark was a strictly Venetian affair, with foreign visitors relegated to watching from the wings, both the Ascension (or "Sensa," to the Venetians) and Corpus Christi paid special attention and respect to outsiders, and to pilgrims in particular. "They make a great festival for the day of the Ascension," wrote Philippe de Voisins about his visit in 1490, "and all the pilgrims who are [in town] willingly go there."[31] They did so primarily to witness the doge's (and thus the Republic's) Marriage to the Sea, where, together with attendant dignitaries, Venice's ruler was rowed in the enormous and ornate Bucintoro from Piazza San Marco out through the Bocca del Lido to the open Adriatic. Once there, he threw a wedding ring into the water—"as a sign of true and perpetual dominion"—to the accompaniment of fireworks, music, and general aquatic splendor. Pilgrims of the highest rank, like Roberto da Sanseverino, nephew of Francesco Sforza, duke of Milan, got to sail in the Bucintoro with the doge; as Pero Tafur put it, "If strangers or honorable men are present they take them also, carrying crosses and pennons, very richly worked in drawn gold." Those somewhat further down the social scale, such as Santo Brasca or Friar Felix and his four German lords, had to make do with a hired boat, but they may have had even more pleasure, rowing in the thick of "so many boats of citizens and most finely dressed ladies."[32]

For the feast of Corpus Christi, pilgrims of every sort were brought still more directly into the celebration, in particular as part of a procession of confraternities, religious orders, and the seigniory that began with an early mass in the Basilica of San Marco, wound through the Piazza, and then returned to the Basilica as much as four or five hours later. By Casola's day, if not earlier, pilgrims were "expected to assemble in the Church of St. Mark to join the procession"; this meant that they marched toward the end of the ensemble, each paired off with a Venetian noble who accompanied him or her around the Piazza and then up the grand staircase within the Ducal Palace to be personally greeted by the doge.[33] Twelve years later, according to the visiting English pilgrim Richard Guylforde, he and his fellow travelers had been moved from their old position of middling status at the rear of the procession to a point much closer to the center—a good sign that the stature of the pilgrims had increased further in Venetian eyes, as Guylforde himself evidently recognized: "There was greate honoure done to the Pylgrymes for we all moste and leste wente all there nexte the Duke in the sayd processyon, byfore all the Lordes and other Estat."[34]

This central role of foreign pilgrims in one of Venice's key religious and political manifestations is not altogether unexpected: over the course of the 1400s, the departure date for the spring galleys to Jaffa was increasingly set at just after Corpus Christi. Although the city's formal celebration of the holiday dated back to the late thirteenth century, it was only in 1407, about a quarter century after the beginning of Venice's pilgrim boom, that the Senate moved to make the event a grand civic spectacle that would include among the marchers the doge and his councilors, the city's aristocracy, and all the priests and canons attached to the Basilica of San Marco. In 1454, the procession was broadened still further, to include virtually every major corporate body in the city: the *scuole grandi* (great confraternities) and all the regular clergy, as well as the local bishops and abbots. Although the Venetian procession of Corpus Christi, like all such civic-religious celebrations, was multivalent, one of its salient aims had clearly become that of recognizing, honoring, and welcoming pilgrims, to stress their importance to the city—no comparable group of foreigners was ever invited to join in the event on a similar footing. Placing them in the procession, in this highly visible way, can be seen simply as an attempt by the Venetians to do honor to themselves (something they were often prone to do): they had, after all, wrested from competing states a monopoly in the one form of medieval commerce that could bring a city both profit and sanctity, and this was as good a moment as any for Venice to broadcast to the world its centrality in the holy business of pilgrimage. At the same time, however, it was also good politics to treat these men and women with respect and even flattery: it was hoped that, once they returned to their homelands, they would act as good-will ambassadors for the Serenissima, spreading the word about Venetian integrity and courtesy.[35]

Integrating and weaving these special travelers into the core of their sacro-civic manifestations was thus a matter of some importance to the Venetians, and to accomplish this the rulers of the Serenissima were willing to alter and adapt these holidays—Corpus Christi, in particular, but Saint Mark's Day and the Sensa as well—to accommodate this, their first major tourist group. In the process, they created a special festival season, a six-week block of celebrations running from late April through early June, a series of grand spectacles designed to impress, entertain, and enthrall, and to loosen the wallets of pilgrims great and small before they set off for their encounter with the East. One can take this adaptability on the part of Venice's rulers as a sign of their political and mercantile shrewdness, something that historians traditionally have praised about the Serenissima. On the other hand, when sociologists have detected a similar social suppleness in present-day host communities—including a willingness to make a place for foreigners in local festivities—they sometimes lament it as a sign of moral bankruptcy, a willingness to sell out indigenous culture for the tourist dollar.[36] Perhaps there is some truth in this as far as Venice was concerned—too much attention to the good opinions and well-being of foreign visitors may have well been an early indication of the decadence that would overtake the Serenissima in centuries to come. At the same time, however, there is little sign that these re-arrangements were especially detrimental to the Venetian ceremonials themselves. From Venice's point of view, such visitors were being incor-porated into local festivities that were big enough to absorb them and even thrive on their presence. Considering the size of the Venetian polity, its wealth, potency, and (at the time) self-confidence, these civico-religious celebrations could no more be hijacked by the presence of a few hundred non-natives than could the modern-day Saint Patrick's Day Parade in New York City. Indeed, the custom that the Venetians followed on these oc-casions, of pairing every pilgrim in the procession with a senator of the Republic (just as they would later do with indigents and then with newly ransomed slaves), is a good indication of the integrative and inclusion-ary nature of these celebrations.[37]

It was not the state alone that showed a willingness to accommodate and engage these pilgrim-tourists. Ordinary Venetian citizens also rec-ognized the opportunities presented by foreign visitors, and medieval Venice evidently swarmed with would-be guides, touts, agents, and pimps, who sometimes disrupted the tranquillity of Piazza San Marco with their fights over who had first claim on new arrivals. On arriving in town, pilgrims might be set upon by contending packs of *tolomazi* (as such guides were called) offering to escort them about the city, assist in changing money, act as translators, arrange lodgings, and book their pas-sage to the Holy Land. In the process, however, they also might demand extortionate tips from their clients, as well as kickbacks from the money changers, innkeepers, and ship captains whose businesses they fed; they were also reputed to talk their way into foreign hospices (normally off-limits to Venetians), quarrel among themselves over who would get the better class of pilgrim, and book their clients on galleys that were not only not proper pilgrim ships but that were not bound for the Holy Land at all.[38]

Such enthusiasm among enterprising locals for plucking naive and confused pilgrims was a matter of concern for the Venetian authorities, who were interested in maintaining a pilgrimage-transport service that could be both economically and diplomatically profitable to the state.[39] As early as 1229, the Senate therefore place the *tolomazi* under control of the magistrate of the Cattaveri (a board of three nobles charged with chas-ing tax evaders, among other things), while ordering that the guides should be organized into a loosely formed guild, or *universitas,* that would oversee the collection and division of individual booking com-missions and (most importantly) tips. Guides were to be limited in number and officially licensed, so that they might actually help and not merely exploit visitors: pairs of them were ordered to wait every morning and afternoon by the Ducal Palace and the Basilica of San Marco and at the Rialto. Each pair of men were not only meant to know the city's sights and facilities, but they also were supposed to speak, between them, at least two foreign languages.[40]

One of the *tolomazi*'s important functions was to arrange rooms for their charges, and from the beginning of Venice's pilgrim boom in the 1380s these two activities of guiding and lodging were closely intertwined. The city's inns and taverns were brought under the jurisdiction of the Giustizia Nuova, which was committed to establishing an "official" hotel list of the twenty or so *osterie pubbliche,* clustered around San Marco and the Rialto area, each inn offering lodging and board to as many as forty guests. As far as the Venetian state was concerned, its supervision of the *osterie,* like that over the *tolomazi,* was as much in its own interest as for the benefit of foreigners. While the Giustizia Nuova did check to make sure that guests were given (for example) sheets and blankets of appropriate quality and quantity, the magistrate, on behalf of an always rather paranoid state, was also supposed to keep a close eye on who these guests were and where they stayed while in town. At the same time, the Giustizia Nuova wanted to find out how much wine the innkeepers were selling them—this latter to make sure that the state collected as much of its lucrative wine tax (the *dazio del vino a spina*) as possible.[41] Such oversight did not mean that the *osterie* were any less competitive, however, and the rivalry between them for shares of the pilgrim trade could indeed be ferocious. Innkeepers might send their own "garrulous and importuning agents" to the mainland, along the principal inland routes to as far away as Padua, to seek out and hustle any travelers who had not yet decided on their place to stay—or even those who had. Such touts also tried to waylay pilgrims as they were crossing over the Lagoon into the city, as Friar Felix recalled:

[We] were passing the tower which is called the Torre de Malghera [marking the beginning of the Venetian Lagoon], when . . . [a]fter a while there met us another boat with people on board, one of whom asked us what inn we meant to put up in at Venice. When we told him St. George's . . . he began to abuse that inn and its landlord, and stood on the prow of his boat, trying to prevent our going there, and pointing out some other inn to us. As he stood there and noisily tried to persuade us, he suddenly . . . fell from the prow of his boat into the sea. . . . He was dressed in new silk clothes, which received baptism together with him, which caused great laughter on board of our boat.[42]

Although Felix's Saint George was a public inn, it catered almost exclusively to German pilgrims, and "the entire household spoke German . . .
which was a very great comfort to us; for it is very distressing to live with
people without being able to converse with them."[43] Many pilgrims managed to lodge with members of their own "nation": Casola put up at "the
house of the Master Courier of the Milanese merchants"; Santo Brasca
stayed at the house of the Archinti family of Florence; and von Harff was
"taken in by the merchants to the German House . . . the Fondigo Tu-
disco."[44] Many others, however, even some of quite high status, had to
settle for *osterie,* like the White Lion or the Savage Man, even though they
may have agreed with Friar Felix's opinion (which he thought the Venetian state shared) that it was "unbecoming that pilgrims bound on so holy
a pilgrimage should be lodged in public inns . . . [which] are not well
famed."[45] In fact, many pilgrims were lucky to get a room at all: the demand for lodgings grew steadily in Venice, and by the mid-1500s appears
to have far outstripped the capacity of the state's licensed *osterie.* To meet
the shortfall, more than a few private citizens came forward, eager to get
into the act, and fifty or sixty years after Fabri's time the Venetian authorities were complaining that "in 1500 one found only thirteen houses
[that is, *osterie*] that ran hotels and lodged foreigners. Now it is common
knowledge that there are between five and six *thousand* houses that put
up foreigners, something that is almost incredible, but in truth they are
in number very great and infinite, and every day they multiply more."[46]
Such a figure may well seem incredible—it meant that something like a
fifth of all the dwellings in the city had rooms or beds to let—but the men
of the Council of Ten who made this report were not prone to wild ex-
aggeration. Indeed, there survives a much more detailed list from the 1780s
that would seem to back them up, for it provides names and also locates
not only Venice's forty licensed *osterie* of that time, but also the additional
nearly two thousand private *albergatori* (innkeepers) who offered rooms.
It would seem that by then—as indeed is the case today—the Venetian
authorities were resigned to the fact that thousands of individual Venetian home owners (or indeed renters) were determined to ride the rising
tide of tourism in the city by renting out parts of their homes to foreigners;
the best the state could attempt to do was at least monitor and register
this enthusiastic, if amateur, activity.[47]

Just as freewheeling in this regard was public transport in the city. Although a terrestrial network of alleyways and bridges had always existed
for pedestrian traffic, late medieval Venice was far more a boat city than
it is today, especially for local elites, who had themselves rowed in their

own gondolas on even the shortest trips around town. Their parading about is well captured in Carpaccio's *Miracle of the Reliquary of the True Cross at the Rialto Bridge,* painted the same year as Casola's visit to the city and showing the Grand Canal crowded with such private gondolas, each piloted by a liveried gondolier serving his master in very much the same role as a coachman elsewhere. Casola observed that "almost every citizen [that is, patrician] keeps at least one gondola," which would mean, by the most tenable census figures of the time, that there were something like four thousand of these private craft in the city at the end of the 1400s.[48] More significant from the tourist point of view, however, was the emergence in Venice at about this time of gondoliers for hire, men who offered their services by the day to the public. These enterprising individuals were available for any Venetian who did not have his own gondola at hand, but unquestionably they were mostly there to serve foreigners such as these pilgrims, who, as brief visitors, tended to know the city less than anyone and were as likely to get lost as any modern tourist in its maze of alleyways. That pilgrims were timid in confronting the city on foot is confirmed in Felix Fabri's account of the mini-pilgrimages he organized for his group, since the friar differentiated between the times they went on foot or in a gondola. And, indeed, for twenty-nine of the thirty-four churches that the Germans visited, Felix noted that they "rowed," "went in a boat," or "went by water" to get there; judging by his list, Fabri's group appears to have been unwilling to walk to any churches that lay beyond a radius of a mere three hundred meters from their inn.[49]

As with the city's *tolomazi* and *albergatori,* Venetian gondoliers seem to have been more than willing to meet the needs of foreigners like Fabri's Germans. Over the course of the fifteenth and sixteenth centuries, private boatmen would blossom in number, much in the manner of the renters of private rooms. By 1608 Thomas Coryat could write, "Of these Gondolas they say there are around ten thousand about the citie, whereof six thousand are private, serving for the Gentlemen and others, and foure thousand for mercenary men, which get their living by the trade of rowing."[50] If there were indeed four thousand gondoliers hacking for hire in Venice (there are only a tenth that many today), it is no surprise that another seventeenth-century visitor could report that "there are always a World of them standing together at several publick Wharfs; so that you need but cry out, *Gondola,* and you have them launch out presently to you."[51] By the mid-1600s, the trade was becoming so popular among working-class Venetians that some in the ruling elite were worried that it was luring away too many of Venice's shipwrights and house carpenters, and no wonder: an average gondolier could earn somewhat more than a carpenter and perhaps half again as much as a master shipwright at the state Arsenal, mostly for just standing around all day.[52]

By the mid-1500s, Venice's pilgrim trade to the East had largely collapsed. What had for a century and a half been one of the city's great engines of income and prestige was killed off in fairly short order, in synchrony with the epochal changes marking the beginnings of the early modern era. The Ottoman conquest of Palestine and the eastern Mediterranean produced nearly continual warfare between the Turks and Venice, making travel to the Holy Land on Venetian ships too risky for even the doughtiest of knights and the most pious of friars; the English and the Germans—once among the most enthusiastic of pilgrims—largely abandoned the practice of such holy journeys as they embraced Protestantism and its new modes of religious expression; Rome itself, with its jubilee years and Baroque boom, began to replace the Holy Land as an appropriate destination for those who continued in the pilgrimage traditions. Marginalized by the shifts in global trade routes set off by the voyages of Columbus and da Gama, the Venetians eventually shut down their regular galley runs to Jaffa. Their commercial traffic shifted over to less tightly regulated sailing ships, and those few who still aspired to play the role of pilgrim had to book as regular passengers on such vessels: the last properly organized pilgrim voyage from Venice to the Holy Land appears to have taken place some time around 1580.[53]

Yet even as it withered, it is clear that Venice's pilgrim trafficking had put its mark on the city, leaving behind an infrastructure of support services and a wealth of practical experience that would prove vital to the Venetians in handling future tourist waves. When Fynes Moryson and Thomas Coryat came to the city around 1600, they found Venice already well stocked with hotels, guides, eating establishments, transport-for-hire, and (something most pilgrims had avoided mentioning) prostitutes.[54] Brought into life by the varying needs of generations of confused pilgrims and nurtured by the expanding demands of an increasingly wealthy local aristocracy, the service sector of Venice was by the 1600s more than ready to meet the requirements of even the most fastidious and demanding of Grand Tourists. Newcomers like Moryson and Coryat—as well as those who came afterward—could also profit from the directions and advice proffered about Venice by their pilgrim predecessors, in the form of published and manuscript travelers' accounts that stretched back before William Wey and went right up to their near contemporary (and some would say inspiration) Michel Montaigne.[55] Such tales, more or less reliable, were in their way the first tourist guidebooks, soon to be augmented by the real thing—Francesco Sansovino's stalwart guide to the city, *Venetia, città nobilissima,* which first came out in 1581 and went through several editions over the next eighty years, providing visitors (at least those who could read Italian) with a district-by-district, church-by-church disquisition on Venice's many attractions. Such books, whether published in London or Paris, or in Venice itself, proved vital both for helping travelers to overcome their natural nervousness about the oddities they would find while abroad and for spreading a general enthusiasm and interest in places like Venice to an increasingly mobile seventeenth-century public.

Proud of their town and eager to sell it to visitors, Venetians were talented at converting the city's existing attractions to suit changing tastes. The great cycle of spring festivals gave up much of its traditional air of patriotism and civic piety even as it lost the participation of pilgrims. Venetians found it a simple matter, however, with the addition of masques, burlesque performers, gaming parlors (the *ridotti*), legions of prostitutes, and fireworks, to turn these festivals into a second, springtime carnival. Out of the grand festival of Saint Mark's Day–Sensa–Corpus Christi emerged a good-weather bacchanal that was perfectly timed to the opening of the Alpine passes and the arrival of the season's first Grand Tourists from Germany and northwestern Europe. Likewise, churches whose relics had once attracted the attentions (and the donations) of the likes of Friar Felix and William Wey had by the late seventeenth century been reborn as sites of aesthetic pilgrimage. Flocks of connoisseurs were drawn to the city's architecture and sculpture, and above all to the paintings of artists like Titian, Tintoretto, and Veronese. Many expressed much the same reverence before such works that the pilgrim-tourists had once shown to the saintly bodies—relics that now often lay, largely forgotten and neglected, only a few paces away. The sacred has never been limited to the religious alone, and these new aesthetic pilgrims expressed a need, very much like that of medieval pilgrim-tourists, to worship at these same cultural shrines. They were on their own sort of sacred journey, and yet they were also sightseers as much as their predecessors had been. Whether they were the sort of Grand Tourists who devoted themselves to the worship of the masters of Venetian art, or were those who chose to plunge themselves into Venice's special whirl of revelry and debauchery, they would in either case find their way guided and made easier by a host of Venetians, waiting (usually with their hand out for a tip) for a chance to advise, explain, and lead them through this most complex, charming, and seemingly changeless of all of Europe's great cities.

CHAPTER 2

Strumpets and Trumps

Sometime around the end of the 1500s, the tourist sector in Venice—which already had been doing quite well a century earlier—entered a phase of prodigious growth. Changing intellectual attitudes, a spreading desire for useful experience, and more disposable wealth in Germany, Holland, France, and, above all, England made travel fashionable for the wealthy. This especially meant travel to Italy, and over the course of the next two centuries a new species of visitor-for-pleasure came south by the tens of thousands. These so-called Grand Tourists followed an increasingly set route that brought them south in time for the opening of the Alpine passes in the spring, led them to Turin and then Genoa or Milan, thence to Florence and Pisa, down to Rome and possibly Naples, and then home again by way of Florence or Loreto, Padua, Venice, and Verona, and back across the mountains about a year after they had arrived.

The many attractions of Italy were (and are) obvious, but five were often singled out by seasoned travelers. First of these pleasures was, of course, the climate and the foods that were consequently available there—both so much more enjoyable than those of northern Europe, and thought by many as essential curatives for weak or consumptive constitutions. For those who wished it and had the time, there was also the chance, at least through the sixteenth and seventeenth centuries, for a first-class university education in a number of Italian cities. Then, for those interested in comparative politics, there was the practical opportunity that the many Italian states offered to examine close up the workings of different—and often highly contrasting—forms of government. There was also the anthropological attraction of new and unexpected forms of human behavior, expressed in such pleasing ways as dance, music, theater, and modes of dress. Finally, there was a seeming infinity of archaeological and historical treasures, especially attractive to those connoisseurs with a developing taste for the expressions and values of the classical world.[1] For those young Englishmen (and there were many, in the seventeenth century) whose fathers thought that schooling at Cambridge or Oxford was a waste of time or worse, Italy for many decades served indeed as a sort of substitute university. There, green youths of good family could soak up skills in art, rhetoric, poetry, music, and general good behavior, while studying (under the eye of their tutor) the visible ruins of that classical world so increasingly beloved by their countrymen.[2]

It is interesting to note that, although Venice was a hugely popular— even obligatory—stopping point on this Grand Tour circuit of Italy, it really offered very little of any of these sorts of tourist attractions. Certainly the city's climate was nothing special, at least not in comparison to that of Tuscany, Rome, or southern Italy: many found Venetian weather unpleasantly humid, and it was generally considered an impossible place to visit during the summer, when the clinging heat was made much worse by the stench of the canals. Nor was the food customarily eaten there to everyone's liking: too much fish, for one thing, and, as most visitors agreed, "the Venetians are wretched cooks"; in any case, there was scant hope that one could wrangle an invitation to dinner from one of the nobility: " 'Tis seldom, indeed, that he will treat you, yet will accept of a Treat at any time; and nobly entertains his Guests at any publick Feast which costs him nothing."[3] Venice boasted a university at Padua that was admittedly one of the best in the world, but to travel between the two cities took a good ten to twelve hours by boat, "if the Wind be favourable"; many students apparently passed their university time paying only rare visits to the lagoon city.[4] Nor was there anything of the slightest interest to those visitors who were classical mavens or amateur archaeologists, since Venice had been founded only in the fifth century. This left visitors there not only with nothing much to collect, but it even deprived them of one of the Grand Tourists' principal amusements—that of citing (and quoting) whatever classical authority was appropriate to the place they were visiting.[5] By the later eighteenth century the Italo-Byzantine-Levantine styles in which much of the city was built had so fallen out of favor with those of taste that more than a few Grand Tourists were willing to dismiss the whole town as "hav[ing] nothing beautiful or grand, when compared with many other Cities in *Italy*."[6]

In the end, then, Venice would seem to have offered only one of the standard attractions for such youths as hoped to use the Grand Tour as a postgraduate course in civility and culture. The Serenissima was indeed renowned for its special form of government—a painstakingly balanced oligarchic republic where a legally defined and largely closed patriciate ruled on behalf of a largely ceremonial prince while managing to assure itself decent pay for doing so. To British visitors of a Whiggish persuasion, probably no state could have seemed more congenial or expertly crafted, and both they and their French counterparts often wrote at length on the Venetian system of governance in their travel accounts. In the seventeenth century in particular, the Venetian political state was admired enough to merit several works devoted entirely to its study.[7]

Yet even in this regard the city seemed to promise more to the visitor than it could usually deliver. Those who went to Venice to deepen their understanding of practical republican politics generally found that Venetians of the political—which is to say noble—class were maddeningly elusive. "Strangers have so little [social] Commerce with the Natives of this Country," François Misson complained, "that it is difficult to learn their Customs, and Manners of Housekeeping." Only visitors of the highest rank could actually get an audience with the doge or sit in on one of the many council meetings. As for the Broglio, the famous outdoor lobbying of the nobility that took place alongside the Ducal Palace, even the best sort of foreigners found themselves relegated to the sidelines, outside of the nobles' carefully delineated space and off with the masses of ordinary Venetians.[8] The law, visitors were told, made it a crime for men of the ruling class to talk to foreigners, lest they enter into conspiracies against the state, take bribes, or accidentally give away secrets. The government apparently took its own strictures seriously enough: in the 1720s, after a few noblemen, under cover of their Carnival costumes, were discovered talking with a foreigner at a San Marco coffeehouse, the police carried off the establishment's benches and chairs to frustrate any such future casual encounters.[9]

Even if these laws had not been in place (and it seems that those who wished to get around them knew how to do so), many foreign visitors still complained that Venetian noblemen were insufferably arrogant. Buoyed as they were by an overweening pride in their lineages' antiquity and by the knowledge that inside Venice it was a crime to assault them, the Venetian "gentlemen" freely abused visitors and locals alike. Even the meanest among them, it was said (and many were so poor that they had to beg to survive), "often emancipate themselves in giving such offenses, as would meet with proper resentment in any other part of Europe."[10]

On top of this, Venetian patricians also notoriously guarded their wives and daughters from encounters with foreigners: many women were said to be as carefully locked away from public scrutiny as though they were in a convent—which, indeed, many of them actually were.[11]

In spite of lacking the standard Italian attractions, Venice nevertheless remained highly popular with foreigners throughout the seventeenth and eighteenth centuries, for it had different, special qualities that set it off from other tourist haunts in Italy. First among these was certainly the unique topography of the place, which visitors found just as bemusing three hundred years ago they do today. Grand Tourists did not, of course, see Venice as an antimodern fantasy land, as so many tourists do today; nor did they describe it as "dreamlike" and certainly not as "romantic." They did, however, marvel that such a place could exist at all: "A City built in the ocean," exclaimed George Ayscough, "is surely one of the wonders of the world!" To David Jeffereys, Venice was no less than a "Miracle of Nature and Art, which seems to swim on the Superficies of the Water."[12] Unlike modern visitors, most of whom arrive by train or bus, Grand Tourists came to Venice across the Lagoon, usually from the mainland at Fusina, though sometimes up from Chioggia, and they had plenty of time to see the city compose itself out of the mists and broken light above the waters: "The prospect of this city, from the first entrance into the sea, is the most wonderful and extraordinary in the whole world, for the situation is such, that at a distance, which is a full five miles from the nearest land, it appears to the eye, as if floating on the waves."[13]

Once arrived and checked into their lodgings, Grand Tourists of the seventeenth century found a city that was like a great cabinet of curiosities, its oddities made available for their enjoyment by a host of willing Venetians. Visitors toured Venice's many churches, loaded with relics, artwork, and commemorative inscriptions; some paid for guided excursions to the Arsenal; those who had the right introductions could also arrange to see the "Intaglios, curious Works in Wood, Natural Curiosities, and other Things worth Notice" held in the cabinets of various nobles about the city.[14] The prime spot where curiosity seekers went to amaze themselves remained San Marco itself, however, for within the Ducal Palace and the vaunted treasury of San Marco was a virtual freak show of oddities— sacred, profane, and plainly bizarre objects that the Venetians, during the glory years of the Serenissima, had bought or looted all over the world. Here one could view the rock from which Jesus had preached at Tyre; "an image of Our Lady made from the stone which Moses struck with his rod when the Children of Israel were asking him for [water] to drink, and out of it issued a fountain"; two spiral alabaster pillars said to be from King Solomon's temple; images of Saints Francis and Dominic, painted before either of them had lived, by the fabled Abbot Joachim; "one of the large teeth of a giant named Goliath, whom David killed"; and "a wooden crucifix which was struck by a disappointed gambler and which has performed many miracles," in particular, that of shedding blood.[15] Among the most popular (and jealously guarded) attractions was the supposedly original autograph copy of the Gospel of Saint Mark, though as most of those who were favored enough to be allowed a close look at it had to admit, the text was so worn and the pages so decayed that it was impossible to tell even if it were written in Latin or in Greek.[16]

The heyday for such mavens of curiosities was the first half of the seventeenth century. By the 1670s it was Venetian painting that was attracting increasing tourist attention: Misson was one of the first to inform his readers that "they give out, that there are as many fine Paintings at *Venice* as at *Rome,* and we have already seen a good store of them."[17] For the most part, this meant the artwork on display in the San Marco area, especially in the Great Council chamber, the Senate hall, and the various smaller council rooms of the Ducal Palace, spaces that were open to visitors on most days, where the monumental ceiling and wall canvases never failed to provoke comment, if not always complete comprehension.[18] It was not until the early 1700s that Venetian artwork—in particular that of the outlying churches and convents—began to get a systematic treatment in travel narratives and guidebooks: the German visitor Johann Georg Keysler, writing in the 1720s, for example, devoted three full pages to the paintings of the Ducal Palace before moving on to fill nearly another twenty on what was to be found in the surrounding convents and parish churches.[19]

Interestingly, this enthusiasm for Venetian art and the Venetian school of painting was neither universal nor especially enduring among visitors, or at least among those who recorded their reactions to the city. By the later eighteenth century a number of Grand Tourists came to and stayed in Venice—sometimes for several months—without mentioning Tintoretto or Titian in their narratives. Perhaps they were all too aware that such things were already thoroughly covered in the standard guidebooks, and that the chance of their stumbling on a half-forgotten old master in an overlooked parish church was slim indeed. Many visitors were seemingly content just to visit a few of the best known and more obligatory pictures— Veronese's *Feast at Cana* at San Giorgio Maggiore being particularly required—and then largely forget art for the rest of their Venetian stay.[20]

Instead, such travelers sought to amuse themselves with more sensu-

FIGURE 1. Seventeenth-century Carnival in Piazza San Marco. (Image from Museo Correr, Archivio Fotografico, M.9304.)

ous diversions, their aims made obvious by the fact that so many of them showed up in Venice in the dead of winter, when the city would have been at its most unpleasant were it not for Carnival (see figure 1). This was certainly not the only city in Italy with pre-Lenten celebrations—Rome, Florence, and Naples were all quite famous in this regard—and there were also similar festivities in Switzerland, Holland, and Germany; for that matter, a British Grand Tourist could simply stay home in London and make do with the local Shrovetide celebrations. Yet the Venetian Carnival, as almost everyone who experienced it agreed, was different, its amusements so compelling that, as St. Didier noted, "those of other Countries who are desirous to see *Venice,* wait this Opportunity, at which time this City is usually full of Strangers of all Nations." De Blainville, fresh from the 1707 Carnival, agreed:

The Variety and Numbers of Strangers now at *Venice,* almost surpasses what you can imagine; for Foreigners, who are curious of seeing this City, generally reserve their Journey till this Season, when Curiosity makes a prodigious Resort to it of all Nations, Sexes, Ages, and Professions, to gratify their Expectation, which is very high, from what they have heard of the Carnaval Diversions.[21]

Already by the late 1600s observers were recounting how tens of thousands of tourists would jam Venice during the February festivities, with almost as many arriving for the shorter (but warmer) festival of the Sensa, held in mid-May. Lest anyone think that all these visitors were coming for some other reason, de Blainville's near contemporary François Misson went on to assure his readers that "few [foreigners] stay at *Venice* longer than the time of the Carnaval. *Lent* is no sooner come, but all that multitude begins to dislodge: Travellers, Puppet-Players, Bears, Monsters, and Courtesans."[22] A century later, von Archenholtz offered the precise (though completely unsupported) claim that "in the year 1775 the number of those who arrived on the eve of the Ascension day, amounted to 42,480, exclusive of the preceding days." Even half that many foreigners, coming in such a short time, could make news in the Venice of today; in the eighteenth century, before railroads (or any decent roads at all) could begin to realize the full potential of mass tourism, it represented a truly enormous and rapid influx.[23]

One thing that made the Venetian Carnival especially attractive was its length, unusual by the standards of carnivals elsewhere in Italy or in Europe. By the mid-1600s it usually kicked off the day after Christmas or, at the latest, on Epiphany and would therefore have a run of anywhere from six weeks to a good two months before Ash Wednesday closed down the fun. Later, in the spring, Lent had barely run its course when the Venetians were gearing up with their mid-May Ascension Fair, the Festa della Sensa, which was offered to visitors as a combined aquatic spectacle, trade fair, and springtime Carnival. Some travelers suspected that the state had sanctioned this whole mini-Carnival, which lasted around two weeks, specifically to attract back to the city all those—both foreign and Venetian— who would rather be enjoying the spring weather in the countryside. Still, as Keysler advised his readers: "If a Traveller cannot contrive to be at *Venice* in Carnival, the best Way is to order his Route so as to be there about Ascension Day . . . [which] has all the Diversions of the Carnival, as Masquerades and Operas, without any Ridotto, or the dissolute Revellings of the latter."[24]

The Sensa was proceeded by the celebration of Saint Mark's Day and followed in short order by the festivals for Corpus Christi, which essentially meant that from January to June, with only a break for the forty days of Lent, Venice was devoted to some form of carnivalesque activity. Even in the autumn there was the theater season, running from October 5 until Advent; though not a proper carnival, this still had enough carnivalesque features to please those whose travels brought them to Venice in the fall. In addition, the state proclaimed miniature carnivals on many other occasions throughout the year, in particular when one of the nobility was elected to the procurate and when the patriarch or a foreign ambassador arrived in town to take up his office. By the second half of the eighteenth century, the state visit of any foreign notable was considered enough of an excuse to have a full-scale blowout that might last a week or more.[25]

All these festivals had their set course of activities, often involving religious observances, nautical or land processions, and some specific rituals that were seen as defining and appropriate for the event. What set these celebrations apart for ordinary Venetians, however, and certainly for most foreigners—who were otherwise fairly uncomprehending about what was going on around them—was that these were times when masking was permitted. Going about in disguise was an especially Venetian way of celebrating Carnival or, indeed, almost any other festival, perhaps because this was a particularly pedestrian city: "Masquerades," as the Baron Pollnitz observed in 1737, "are more in Fashion here than Elsewhere. People go in Masks to take the Air, as well as to Plays and Balls; and 'tis the favorite Pleasure both of the Grandees and the Commonalty"; in the week leading up to Martedì Grasso (Fat Tuesday) one could see tens of thousands of *maschere* crowding such primary centers as Piazza San Marco or Campo Santo Stefano.[26]

As a result, Venetians and foreigners alike could legally go about the city in some sort of disguise for at least six months out of most years and often even more than that, when it came to private parties and balls. The types of costumes they were permitted to wear varied from season to season, however, as dictated both by law and custom. For Carnival proper, they were allowed to dress up almost any way their fancy dictated, at least up to the final Friday before Fat Tuesday. The earliest tourists to mention Carnival costumes reported a fondness among Venetians for dressing up as men or women from other nations: Jean-Baptiste du Val, writing of the Carnival of 1610, reported seeing locals dressed up "in the clothes of the country" of the Chinese, Abyssinians, Tartars, and Indians; his stressing that such costumes were made of "rare fabrics" leads one to suspect that he was witnessing Venetian merchants showing off their goods to the public as much as dressing up for the festival. A few decades later, Sir John Reresby found that it was popular to dress up as less exotic nationalities: he mentioned French and Dutchmen in particular.[27] Other possibilities, more in line with the basic themes of inversion and transgression that permeated carnivals everywhere, were getting oneself up as a cleric (though this was forbidden) or cross-dressing: Fynes Moryson, writing in the 1590s, noted how one could see "men in wemens and women in mens apparrell at theire pleasure," and Reresby also observed that "if women ever wear the breeches in Italy, it is then, with whom the men change habits."[28]

In later years, costuming for Carnival in Venice underwent some changes. Already by the 1660s, Philip Skippon was noting a great many "extravagant inventions" among the thousands of maskers crowding Piazza San Marco; in particular, he observed groups sporting coordinated costumes, and singled out for mention one "company [of twenty] dressed all in a yellow stuff or coarse silk, having tauny vizards, and huge roses on the shoes, knots in their garters, hat-bands, &c of the same stuff." The fashion for dressing up as other nationalities subsequently appears to have faded somewhat (though cross-dressing remained strong) and was replaced by an enthusiasm for costumes taken from the *commedia dell'arte:* visitors after 1700 reported seeing the Piazza filled with harlequins, pantaloons, and the like.[29] Later in the century, and especially toward the end of the Republic, it became customary among the upper classes to sport just a single outfit, which indeed has become widely associated today with the Venetian Carnival. This was the *bauta,* a hooded cape, and the *tabarro,* a roomy cloak; these, when combined with a three-cornered (usually black) hat and a full-faced white mask (known as a *volto* among Venetians, called a "vizard mask" by tourists), made the wearer completely anonymous and able to move about the city without being recognized.[30]

Some have claimed that the Venetian nobility took up masking as a way to move freely and undetected among those of the lower classes. This probably underestimates the acumen of ordinary Venetians, who were quite able to sense one of the patriciate in their midst. In any case, dressed in the *bauta* and *tabarro,* nobles could move unrecognized among other nobility, especially in parties and in the gaming halls that flourished during Carnival. Not surprisingly, this sort of disguise was widely associated with sexual adventures, and during the last years of the Republic it became the dominant costume for the patriciate during the Carnival weeks and especially for the Sensa, when indeed no other outfit was permitted. Some foreigners found the resulting sameness a bit boring, and their disparaging of the black *bauta* and *tabarro* very much resembles what they had to say about the "funeral air" of the gondolas: "The disguise appears to me to be dismal, inconvenient and destitute of variety," complained Anne du Bocage, for one, going on to suggest that "this habit has been so often worn, that one would think it was high time to fancy one more becoming."[31]

If some foreigners were put off by the costumes, others thought the whole Carnival was both silly and demeaning. Johann Wilhelm von Archenholtz bemoaned "the insufferable crawling and tumult of the crowd" and concluded that "the sight of a numberless crowd of masks . . . can have nothing very attractive." A few years later, the ever-blunt George Ayscough dismissed the entire festival: "Upon the whole, I do not think this city so very full of gaiety, as it is said to be; neither do I find any vast amusement in seeing a number of Christian fools, with varnished faces, as Shylock calls them, with whom I am not acquainted; and to address whom as a stranger, would be looked upon as the height of ill-breeding."[32] Visitors who did wish to join in the fun were advised not to try too hard with their costumes, since those who actually dressed up like something or someone were expected to play the part (which is why Reresby, for one, offered advice on how one was to act the Frenchman or Dutchman). Those who failed to live up to their disguise might well find themselves accosted and abused by low types—drunken gondoliers and "watermen," for example—who during Carnival season could not be punished with a customary thrashing for such impudence. As a result, several writers advised their readers, "Those who are not willing to be Actors on this great Theater, take the Habit of Noblemen [that is, the *bauta* and *tabarro*]; some *Polonian* Dress, or the like, which obliges them to nothing."[33] Even got up in this fashion, as de Blainville complained, "you are extremely jostled as you go along." Things only got much worse on the last day of Carnival, when, it was said, "they grow from merry to stark mad":

All the City is, as it were, quite staring mad, one sees a thousand ridiculous Disguises, young Fellows are driving Bulls [oxen or cows, actually] about the Town, along the Alleys (as most of the Streets are no better) hollowing and making a Noise sufficient to make the Beast as mad as themselves, and indeed for my Part I thought it no safe Pleasure to be in the way of them, and therefore kept within Doors.[34]

Another good reason to stay inside for the festivities of Martedì Grasso was that, with all the bulls, oxen, bears, and large dogs running loose about the city, and with the young men "notoriously abusing and often fighting with those they meet,"

the Shops have been shut, and nothing seen all over the Streets but People armed with Hatchets, Cutlasses, Clubs, and Bludgeons, with other kinds of Weapons, in the whole making a most horrible Noise and Uproar. . . . [E]very Body is at Liberty to arm themselves with what Weapons they think fit, excepting Fire-Arms. . . . But this Indulgence makes People who are at Enmity with one another, be upon their Guard, and very strictly watch one another's Motions.[35]

If so many foreigners appear to have been unwilling to enter into the spirit of the Venetian Carnival, and indeed ended up spending the last days of the festival cringing indoors, one wonders what it was about the whole affair that proved such an attraction. Even the weather itself, so especially wretched in February, seemed to conspire against an enjoyable stay in the city, leaving many a soaked visitor dispiritedly "swimming about like a duck," as Ayscough grumpily described his time being rowed around Venice in a gondola.[36] Yet, though more than one writer warned readers how disappointing the festival could be, visitors continued to come in their thousands: already by 1687, according to François Misson, the Carnival could draw no fewer than "Seven Sovereign Princes and Thirty thousand other Foreigners"; such numbers (of ordinary tourists, if not of princes) seem to have been typical for most of the years in which the Carnival was permitted, from then until the fall of the Republic over a century later.[37]

In reading over the accounts and diaries of these visitors, it soon emerges that, for many foreigners, there were other, more compelling attractions in Carnival besides those that later scholars have traditionally considered "carnivalesque"—that is, masking, social inversion, political license, or gluttony (even if, as Thomas Gailhard once commented, that as far as "glutting themselves with pleasures, Venice goes beyond all other places, if one considers the number, variety and quality of their sports").[38] One of these attractions was the opera, about which almost all visitors who came for the Carnival had something to say. Invented precisely for the Venetian Carnival at the beginning of the seventeenth century, opera was for the first time staged for the paying public in 1637 for the same occasion; by the mid-1600s, as many six or seven houses in the city ran nothing but operas throughout the Carnival session. During its first half century as a public entertainment, opera lured most of those visitors who were in town for Carnival, even though it was an expensive and somewhat rough experience: seats in the "cockpit" could cost eight times as much for an opera as for a regular theatrical comedy and one might still be spit or even urinated on by nobles ensconced in their boxes above. Many English observers during these years seemed primarily fascinated by the sheer spectacle that opera presented them: by the "variety of Seeanes painted & contrived with no lesse art of Perspective, and Machines, for flying in the aire & other wonderfull motions"; even someone as relatively sophisticated as John Evelyn could still feel intrigued by the fact that in one production "they changed scenes thirteen times."[39] Other foreigners praised the music and the singing, which most agreed were "ravishing,"

and it would appear that for many outsiders Venetian opera offered a special chance to enjoy the finest vocal talent in Italy. Evelyn, for one, singled out *"Anna Rencia* a Roman, & reputed the best treble of Women,"
whom he included among the "famous Voices" he had enjoyed; others,
less specific but no less enthusiastic, wrote fondly of, for example, a "famous Girl from *Rome*," "that rare singing woman of *Bologna*," or "These
Men without Beards [who] have delicate Voices."[40] Local impresarios received special praise because they were willing to pay top prices, even for
talent that would sing only for the four or five weeks of Carnival.[41] They
knew that they had to please a broad and discriminating audience, ranging from the Grand Tourists themselves to the socializing Venetian lords
and ladies to the large claque of "Gondoliers or Watermen" who always
got in free and then brought down the house, "clapping their hands and
shouting like Madmen," crying out to their favorite singers, "'Sia tu
benedetta, benedetto el padre che te generò!'"[42]

By the end of the 1600s, however, many observers, especially among the
French, began to complain that Venetian opera had declined to the point
where it could no longer compare to the shows produced in the great European courts.[43] Some continued to praise the precision of the orchestra
and agreed that "the Theatres are Large and Stately, the Decorations Noble." But increasingly often by the late seventeenth century, those who
came to see spectacle left disappointed:

[The theatres] are very badly Illuminated; the Machines are sometimes passable
and as often ridiculous; the number of Actors is very great, they are all very well
in Clothes; but their Actions are most commonly disagreeable. . . . The Ballets
or Dancings between the Acts are generally so pittiful, that they would much better be omitted; for one would imagine these Dancers wore Lead in their Shoes.[44]

This decline in foreign enthusiasm for the Venetian opera (the comedies
had always come in for contempt, as mere buffoonery) may help explain
a rising interest among visitors for the *ridotti*, the licensed gaming houses
that the state allowed to operate only during the weeks of Carnival. As many
as 130 of these salons could pop up in the course of a Carnival, and virtually all those visitors who wrote on the festival seem to have felt obliged
to go to at least one of them—usually the so-called Ridotto Grande, on
Calle Vallaresso, near Piazza San Marco. They, like everyone else who entered, were required to don some sort of disguise, although it was re-
portedly enough to wear "a counterfeit Beard, Nose, or any other thing
that causes an alteration in the Face," in particular the domino, that vestigial mask that was just enough to cover the eyes (as seen in figure 2).[45]

FIGURE 2. Pietro Longhi, *The Ridotto.* (Image from Museo Correr, Archivio Fotografico, V.3421.)

Most of those who wrote on the *ridotto* strongly advised their readers not to gamble: the game may have been simple—in fact, *bassetta,* a simplistic version of blackjack, was the only one allowed—but foreign players were meant to be at a great disadvantage. The cards, for one thing, were of a type unfamiliar to most visitors; moreover, the dealers—who were always Venetian nobles and the only ones allowed into the *ridotto* unmasked— had a pair of watchful advisors ("Ladies in Masques," according to Keysler) to guard them from making mistakes. The dealer also had the right to "dismiss the gamesters when they please, and always come off winners"; at least some of these nobles, if they failed to fleece their adversaries with these advantages, simply cheated: "There are several Gentlemen," commented St. Didier, "so very dextrous in cutting and drawing the Cards, that the most intelligent and quick-sighted Player may be sometimes deceiv'd."[46]

The driving force behind these *ridotti,* and the obvious reason they were licensed in the first place despite their reputation for bankrupting a good many Venetian patricians, was the opportunity they presented the Venetian state to make a profit off its tourists. De Lalande, for one, claimed that foreigners left behind something like fifty thousand Venetian *zecchini* every year in the gaming rooms, and it appears that wealthy tourists genuinely flocked to the *ridotti* during the weeks of Carnival, despite the odds that were stacked against them. As de Blainville observed: "It is surprising to see how madly fond most Strangers at *Venice* are of bringing in their Money to those Nobles, who laugh at them when they have stript them. Sometimes . . . a lucky Hit happens, which makes the Tallyers [that is, the dealers] bite their Fingers' Ends. . . . But this very rarely happens, for one who wins, thousands are ruined."[47] Indeed, the *ridotti* for the most part do not appear to have been especially inviting places: once entering, foreigners found "ten or twelve chambers on a floor, with gaming-tables in them," where "the Crowd is so great that very often one can hardly pass from out of one Room into another"; "but tho' the Throng is so great, yet there is always a profound silence. None are permitted to enter into these places without Masks"—a "Spectacle . . . infinitely more admirable for the Singularity, then diverting in its Pleasures."[48] What the *ridotti* did offer to the enterprising foreigner, however, and at least one reason why many a visitor seems to have ventured into these gloomy, silent halls, was the chance they provided to meet women. Indeed, looking over their travel narratives it soon becomes obvious that this was a primary reason that many of those Grand Tourists who were wealthy and male—as the overwhelming majority of them were, throughout the eighteenth century—put up with Venice and the Carnival at all, even while claiming that they found the whole business demeaning and more than a little frightening.

Chances were that the women in the *ridotti,* or in the public dances staged for Carnival, would be prostitutes. Philip Skippon wrote of visiting one such "publick ball" where "round about on benches sat many whores masked, who expected when any of the company would take them out, and lead them round in the dance. If they were somewhat pleasing in their carriage, then their mates would withdraw and discourse with them a little more privately; and if they liked them when they saw them unmasked, a bargain was struck."[49] Early modern Venice was, of course, very widely known for its courtesans and whores. As early as 1608, the English wanderer Thomas Coryat had decided that there were so many of these women (around twelve thousand, he claimed) and their talents were so unique that no proper discussion of Venice could leave them out. So, after much space devoted to assurances that his information on these women and their skills was definitely not firsthand, he offered what he claimed was the first frank explication of this aspect of Venetian society. Obviously Coryat had lit upon a topic of considerable interest, since in the nearly two centuries between his visit to Venice and the collapse of the Republic in 1797, virtually every visitor to the city made some mention of the whores, even if only to deplore them and their poxy ways. The most widely circulated guidebooks of the era, by Misson, St. Didier, Gailhard, and du Mont, among others, devoted entire chapters to Venetian prostitution, making it a safe assumption that very few Grand Tourists came to Venice without being well informed of the lively sex trade they would find once they got there. Such was in fact Coryat's own assertion: "So infinite are the allurements of these amorous Calypsoes, that the fame of them hath drawn many to Venice from some of the remotest parts of Christendome, to contemplate their beauties, and enjoy their pleasing dalliances."[50]

Venice does not seem to have had an exceptionally large number of prostitutes: Rome was said to have just about as many, and Naples—the largest city in Italy throughout the seventeenth century—was claimed by one traveler to be home to no fewer than sixty thousand.[51] In the same manner as Rome or Florence, prostitutes in Venice had their delineated districts within the city, at least as far as actual brothels went. These were meant to be clustered behind the Rialto fish market, in an area called (and still referred to as) the Carampane, the red-light district in San Matteo parish that grew up around the so-called Castelletto, a group of houses established for licensed prostitution by the government, back in the 1360s. This may well be the zone that foreigners were writing about when they reported that "whole streets of Ladies of Pleasure commonly stand at their Windows and Balconies set out with a mighty profusion of Ribbans, where they expose their best Charms to excite the Inclinations of all passing by them."[52] By the late sixteenth century, however, the Venetian sex trade had spilled out of these narrow confines, and a fair number of prostitutes had set up shop elsewhere in the city. Fynes Moryson, for one, claimed that these women could conduct their business pretty much anywhere in town, "free to dwell in any house they can hyre, and in any streete whatsoever, and to weare what they list." Lady Ann Miller, over two centuries later, observed that it was all the state could do to keep the streetwalkers from invading Piazza San Marco, even outside the Carnival season. During Carnival and especially for the Sensa, they virtually owed the Piazza, where they "live by preying on the Vulgar and the Innocent."[53]

Even for those foreigners who came to Venice specifically for the sex— which one suspects many did—such a profusion of prostitutes was not always such a good thing. It meant that, wherever a traveler went, he was constantly being propositioned, called to, having his clothes tugged. As Misson complained, "drest in red and yellow, like Tulips, with their breasts open, and their faces painted a foot deep, always a nosegay above their ears: you may see them standing by the Dozens at the Doors and Windows, and Passersby seldom 'scape without torn sleeves."[54] Under the circumstances, the long-standing reluctance of tourists to go around Venice on foot seems less peculiar: at least when one went out to see the sights by gondola, he could hope to be accosted by only the higher class of courtesan—those who could afford their own boats. Even so, hiring a gondola had its own risks, as both Moryson and Coryat warned their readers: "If you call for a boate, and say you will go *a spasso,* that is for recreation, howsoever you meane to take the ayre upon the water, he will presently carry you to some Curtezans house, who will best pay him for bringing her Customers, as if there were no other recreation but only with wemen."[55]

As if their simple inescapability were not problematic enough, the many prostitutes of Venice also had a reputation for poxiness. More than a few writers warned their readers that "it is not to be admir'd that the Distemper which usually follows this Vice should be generally spread" among the city's harlots. Considering that these centuries encompassed a European-wide epidemic of syphilis, it seems odd that writers should single out whores of Venice as opposed to, say, those of Rome or Naples. One possibility for this emphasis, however, is that these travelers may have wished to steer their readers clear of the street prostitutes, many of whom came to Venice especially for the festival occasion, often having been thrown out of some neighboring state and carrying the brand of the convicted whore on their shoulder or breast.[56] Instead, young foreign men of good family were directed (implicitly, at least) toward the city's much more exclusive *cortezane,* who could offer them a high-quality, cultured, skilled, and presumably non-poxy sexual experience. Indeed, some writers let it be known that few foreigners could reasonably expect to resist "the amorous woundes of the Venetian Cortezans" and "the attractive enchauntments of their plausible speeches."[57] In the lengthy encomia offered them by a whole successions of Grand Tourists, these professionals were endowed with all the refinements of any geisha in Meiji Japan—talented in singing, versifying, erudite conversation, makeup, and, of course, the arts of love. Just what these arts might be, no writer seems to have been willing to speculate—or to admit of any personal experience—but St. Didier, for one, alluded to "those Base and Extravagant Artifices" (others spoke of "their wanton toyes") that the courtesans knew how to employ on even "the most Venerable Heads of the *Venetian* Nobility . . . as Incitements to those Pleasures which the natural weakness of their Ages yet deprive them from tasting."[58]

Equally alluring—if equally vague—claims about Venetian courtesans turn up in many travelogues and guidebooks, from that of Misson and Jean du Mont, writing in the late 1600s, up to Lalande's *Voyage d'un François en Italie,* published in 1770. For many potential visitors, these women must have seemed dazzlingly different from the common street whores they were familiar with back home in London or Paris, but Venice had another sexual attraction as well. This was the possibility of striking up an affair with an actual Venetian noblewoman. The belief was widespread among foreigners (and spread, of course, by themselves) that the women of the patriciate, generally kept indoors and forced by custom and their families to go about the city heavily veiled and guarded, were in fact incurably licentious. Bottled up the rest of the year, their female salaciousness was believed to break free in a flood every Carnival, under the cover offered them by their masks and disguises. As a result (so it was said), "The *Venetian* Ladies are impatient for these Occasions, and their Husbands equally watchful to preserve the Honour of the Marriage Bed."[59]

Many writers who raised this beguiling possibility did indeed also stress that such women were likely to be tailed by their husbands or some hired thug who would lurk in the background at the *ridotto,* waiting for the lustful visitor to make his move. Throughout the seventeenth century, it was indeed standard practice among travel writers to warn readers of the serious risks of running off with some masked beauty. Yet as the Republic waned and the Venetians themselves moved ever further away from their traditional practices of female seclusion and male obsessions with honor, many visitors claimed that there was no longer much danger of being set upon by assassins in some dark alley, whether for having had a mild flirtation in the *ridotto* or arranging a serious assignation for later. As de Blainville put it,

The greatest Pleasure for Foreigners who are Rich and make a figure, lies in the Ladies of Quality, or Wives of the Noble *Venetians,* who in Carnaval Time, are likewise at Liberty to wear a Mask; and those who have any Dispositions to intriguing, find during this Season, a thousand Ways of deceiving their vigilant Husbands and Keepers, spite of all their Precautions, there being scarcely any Place into which a Mask may not find Admittance.[60]

Those Grand Tourists who braved the bad weather and huge crowds of the season could thus hope for an "easy access to the Ladies [which] contributes not a little to make [one's] Stay in this City agreeable."[61] Widely diffused in various editions and translations, such siren songs no doubt tweaked the imagination of many a reader and probably persuaded more than one Grand Tourist to include Venice in his itinerary.[62] Yet these tales were more than a sort of premodern advertising for libertines: they also made up what might be dubbed an Erotic Myth of Venice—an idyllic city of sexual liberty where mysteriously skilled courtesans traveled about in "*Gondolo's* cover'd with Roses" and employed artful blandishments that only the most morally resilient tourist could (or would want to) resist; where there flourished "whole streets full [of prostitutes] who receive all comers . . . and [who] dress in the gayest colours, with their breasts open and their faces all bedaubed with paint, standing by dozens at the doors and windows to invite their Customers"; and where high-born ladies, with lineages that stretched back for centuries, battled with one another for the chance to seduce a foreigner of quality.[63]

In one of its many possible forms, then, sex appears to have served as a primary motivation in bringing foreigners to the Venetian Carnival, where "young persons, indeed, who only delight in Debauchery and Licentiousness, may tire themselves, if not satiate their desires."[64] Since Carnival (including the Sensa) was in turn the main reason that travelers came to Venice at all, one might well say that sex—or at least the Fair Sex—was a fundamental inducement for visiting the city during these years. Admittedly, many of those who wrote on Venice, identifying themselves with the era's special quality of Reason, assured their readers that they found such licentious frolics distasteful and that "those who find any real Pleasure in them, must have abandoned all Regard to Virtue."[65] Still, whether they offered a rationale or a critique for this erotic paradise, virtually all visitors tended to hold the Venetian state and social order as somehow responsible for what went on. Some believed that the city's sex industry had originally come into being to satisfy the lusts of the many idle young nobles who were prevented by law or tradition from engaging in any truly useful activities, such as trade or governance. Others blamed the social custom of arranged marriages, or no marriage at all, that was imposed on many such youths, leaving them with no legitimate outlet for their sexual desires. As a mirror to this, many pointed to the extreme "severity and rigidness of parents and husbands to their daughters and wives, who are deprived of an honest liberty" and thus end up "greedy of that which is forbidden." Still others held that "so much Libertism" was encouraged by the Republic itself, either to wean young nobles off their tendencies toward rape and homosexuality or for the more devious reason that it "stupefies the boiling Blood of their Youth . . . enervating the Vigour of them whose Impetuous Natures might be dangerous . . . being educated in Softness, [the People] are not desirous of any alterations in Government."[66]

Whichever of these various causes they reported to their readers, almost all writers who dealt with the topic agreed that the result was a city where (during Carnival, at least) intercourse between the sexes—in all senses of the word—was freer than anywhere else in their known world. Many travelogues and guides went further, however, closely linking this erotic paradise with the fabled Liberty of Venice—that freedom of thought, speech, and action so fundamental to what might be termed the legitimate Myth of Venice. Especially after 1700, when Venetian women finally emerged from the convents and virtual seraglios that had characterized their home lives until then, travelers took to linking the Liberty of Venice specifically with the liberty of the city's women. One might indeed question whether most Grand Tourists really understood the Liberty of the Myth in any other sense than that of the libertine; as one guidebook denounced them, local Carnival goers were indeed "guilty of the most scandalous Practices that can be imagined, and this is look'd upon as a Testimony of *Venetian* Liberty."[67]

At the same time, by fobbing the responsibility of the city's booming sex trade off onto the Venetian aristocracy or its social traditions, these travelers were well in line with a second, well-established trope that treated Venice as an alluring meeting place of East and West. For the generality of Grand Tourists, Venice represented the most exotic place they were likely to see, and indeed, "they whom Curiosity prevails upon to undertake long Voyages to learn the Customs, Manners, and Morals of different Nations may in part answer their End, without going farther than *Venice*"; the "barbarous Ethnickes" that one could find there in such abundance—the "Jews, Turks, Armenians, Persians, Moors, Greeks, Sclavonians [Dalmatians], some with their targets and bucklers, and all in their native fashions"—were for many a tourist an attraction in their own right.[68] This inclination to Orientalize Venice included a tendency among some Grand Tourists to project what were evidently their own vices and desires onto their hosts, allowing themselves, as mere visitors, to pass through this exotic/erotically charged "contact zone" as passive, innocent dabblers, rather than as the debauchers, voyeurs, and libertines that many of them evidently were.[69]

Thus we find St. Didier, in beginning a little section of his guide that he titled "Of the Dances of the Girls," stressing the fact that "especially all Strangers [are] so mightily pleas'd with [the sight]" of the local Venetian girls holding their informal, almost spontaneous dancing parties on "the most spacious Places of the Streets or Keys, but most commonly in the little Squares of their Quarters." He then goes on to note that Venetian noblemen (but obviously also more than a few Grand Tourists) liked to "stop their *Gondolos* to behold these Dances upon the sides of the *Canals.*" Their motives for doing so were far from just aesthetic or folkloristic, however, representing instead "the most easy Opportunities of being Familiar with these young Lasses, and afterwards to choose from among 'em such as are agreeable to their Fancies: Insomuch, that these Dancings of the young Women are a sort of Market, where the Beauty of the Merchandises expos'd, occasions oftentimes such to Purchase, as had no Thoughts of it, untill they came thither."[70] Likewise, Jean du Mont, in offering his account of the Sensa, went into some detail on how Venetian mothers—often but not always prostitutes themselves—sold their daughters to the highest bidder, right in Piazza San Marco:

Every Mother that is willing to be rid of her Daughter, carries her thither every Day as to a Market, with all the alluring Ornaments of an advantageous Dress. They have large and very fine Top-knots; their Neck and Breasts are bare; and their Habit consists chiefly of a little Cloak or Mantle of white Muslin, adorn'd with red Ribbons, a single Petticoat of the same Cloth, a neat Pair of Shooes and Silk-Stockins, exactly fitted to their Legs. There is nothing more proper to inspire Love than the sight of those young Creatures, who for the most part are charming Beauties; nor is a Man forc'd to purchase his Happiness at the Rate of a tedious Courtship; for as soon as he feels Nature begin to work, he may immediately address himself to the Mother, who is always ready to make the Bargin. . . . Nor are you oblig'd to buy a Pig in a Poke; for you may view and handle her as much as you will.[71]

Although du Mont made some small effort to pretend that he was writing about sexual and commercial relations between Venetian women and the city's noblemen, it is also clear that he was addressing foreign readers who might wish to explore some of these local customs for themselves. Indeed, he allowed that "this way of buying Girls" might strike some of his audience as "very odd and extravagant," but then went on to admit that just hearing of the practices often "made a very strong impression on the . . . Mind" of certain sorts of foreigners, men who would as a result "spare no Cost to indulge [their] Inclination to Pleasure."[72]

Du Mont must have been correct in his assumption that this practice would titillate and amaze foreigners, since Misson also noted, "You would be surprized to see a Mother deliver up her Daughter for a certain Summ of Money." Most likely such a show was put on as much to attract the specific interest of foreigners as it was for the local nobles. After all, these debauched mothers timed their sales for the Sensa, the period more favored by Grand Tourists to Venice than any except Carnival, when the city was overrun by prostitutes flooding in from all over Italy and Europe.[73] Indeed, readers of these travelogues were told that they had better look sharp if they planned on arriving in town during the last days of Carnival, since the higher-quality courtesans tended to be booked up during these weeks.[74] Coryat explained this extraordinary demand when he asserted that "the name of a Cortezan of Venice is famoused over all Christendome," and that he himself "often heard before both in England, France, Savoy, Italy and also in Venice it selfe concerning these famous women."[75] Peter Mundy, a short time later, allowed how "theis baits drawe many hither, some for Curiositie, others for Luxurie"; while Misson in the 1680s noted that during Carnival "the place of *St. Mark* is fill'd with a Thousand sorts of Jack-Puddings, Strangers and Courtesans come in shoals from all parts of *Europe*."[76]

As they did with the *ridotti*, some foreign visitors stressed prostitution's role in helping the state enrich itself with tourist wealth. St. Didier, for one, noted how "Libertinism . . . empties the Purses of such Strangers, who come in great numbers to *Venice* for the love of Her Lascivious Pleasures," but not much of this foreign exchange, seemingly, stayed in the hands of the women who had earned it. The Venetian state taxed courtesans and whores quite effectively, so much so that the resulting income had by the 1600s become essential to the fisc. As Fynes Moryson pointed out, "The tribute to the State from Cortizans was thought to exceede three hundreth thousand Crownes yearely," enough, according to Coryat, "[to] maintaine a dozen of their galleys . . . and so save them a great charge."[77]

Over the next two centuries, as the Republic lost its empire and a good deal of its trading wealth, it grew more dependent on tourist spending, and in particular on tourist vice. As early as 1688, in commenting on the great crowds that came to the Carnival the year before, François Misson asked his readers to "consider how much Money all this Multitude must bring to *Venice*."[78] A century later, according to von Archenholtz, the government was not able to cancel the Carnival, with all its attendant gambling and whoring, even though these were systematically wasting and bankrupting its own nobility, since without the possibility of finding their expected debaucheries tourists would stay away in droves. As von Archenholtz put it: "After the great decline of trade at Venice, the visits of travellers became the greatest resource of the nation; it was therefore necessary to adopt milder maxims [against vice], in order not to deter them from visiting a country which can by no means do without them."[79]

Venice's role as the Fleshpot of Europe, the Vice Capital of the Continent, lasted as long as the Serenissima itself, only to then be swept away in short order, along with the Republic itself, by the arrival of first French and then Austrian occupiers. On prostitution in the city, nineteenth-century visitors wrote not a word, although it is evident that the Carnival resurrected in the 1880s did at least feature a *ridotto* where tourists could hope to gamble. With the rise of Lido toward the end of the century as an alternative beach vacation for visitors to the city, and the opening of its Casinò soon thereafter, gambling again became a permanent institution associated with Venice. Eventually a second Casinò, for wintertime, was opened right on the Grand Canal, at Palazzo Vendramin-Calegri. By the end of the twentieth century, as Venice's resident population declined sharply, the baccarat roulette tables were crowded almost exclusively by wealthy foreigners, as once again this particular vice began to play an important role in the local economy.

Prostitution, on the other hand, never made much of a comeback in this city, which for some generations now has attempted to promote itself as clean, safe, and family-friendly. Sailors, as we discovered when we first came to Venice over thirty years ago, tend to disparage Venice as a port-of-call, much preferring Genoa or Naples for their ready availability of bars and whores. Admittedly, Venice maintained something of its ancient tradition of brothels—*case di tolleranza,* as the Italians call them—through most of the first part of the twentieth century. A dozen or so bordellos were scattered about the city's historic center, with names like Oriental House and The Mirrors evoking something of the old exotic fame of the Venetian sex trade. The last one of them, the Rosina, located on the Frezzeria, closed down in 1958, leaving the town to the ministrations of just two middle-aged professionals, who went by the *noms d'amour* of Johnson and Carapellese (this latter apparently after a well-known soccer player, whom she was said to closely resemble). But in a town that was becoming as small as Venice, as one aficionado of the topic put it, "where everybody knows everybody, to solicit a prostitute under the eyes of [other] people was truly impossible." So the situation stood until October 2000, when, according to the papers, three street prostitutes—two Nigerian women and a South American transsexual—showed up, looking somewhat confused, in Piazzale Roma: the first professionals to turn up in historic Venice (albeit on the edge) in nearly forty years, they were hastily ushered back to the mainland.[80] In the city that for centuries made itself almost synonymous with sex for sale, it would seem that prostitutes, at least, have become as unknown as cars and trucks.

PART TWO

Landscape

CHAPTER 3

The Heart of the Matter

The heart of tourist Venice is Saint Mark's Square, the Piazza San Marco (see map 1). "Tourists flock here in the thousands every year," boasts a recent guidebook to the city, and far from exaggerating, this greatly understates the case. It is a safe assumption that, at the time of this writing, at least twelve million visitors pass through and around this two-acre spot every year: "a tidal wave that no one is capable of quantifying, because, despite the flourishing of organizations and groups that concern themselves with tourism, nobody . . . has updated data on who comes."[1] To accommodate such a flood of humanity San Marco must be filled and emptied as if it were a tourist tidal basin, as indeed in some ways it is. Even before dawn, on any given morning between April and October, a few visitors have made their appearance; by breakfast time the first tour groups show up, some still rumpled and confused from their all-night bus trips from Munich, Vienna, or Bratislava. By eleven the Piazza is awash with foreigners and will remain so for the next six or seven hours. At peak periods, one might easily encounter ten to fifteen thousand visitors making their way around this space at any one time, filling in the crevices that wrap around the Basilica of San Marco and the Ducal Palace, crowding the Piazzetta dei Leoncini on the north side of the church, and milling around the Piazza itself, as well as the Piazzetta and the Lagoon frontage, or Molo, that runs from the Ponte della Paglia up to the souvenir stands by the Giardinetti Reali (see figure 3).

Taking the elevator up to the observation deck of the Campanile of San Marco and then looking straight down at the open space below us, we have often been struck by the multicolored carpet formed by all the tourists

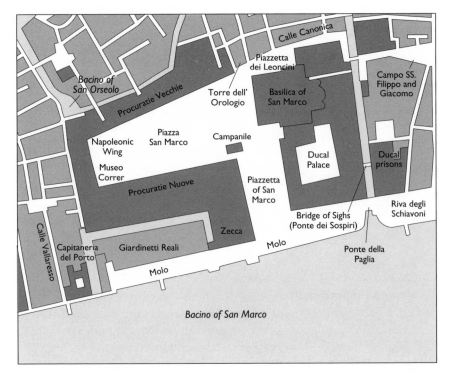

MAP 1. Piazza San Marco.

as they mill about. This mass of humanity has a certain beauty of its own, forming contrapuntal nodes around different tour guides' raised umbrellas or arranging itself in long, patient lines to get into the Basilica or up the Campanile itself. By noon, in the warmer months, there are also hundreds of visitors resting and eating, squatting on the still-shaded steps of the Procuratie Nuove or along the raised center of the Piazzetta dei Leoncini. Only with the setting of the sun do the thousands begin to disperse, and even well after dark some groups are still strolling between the tables of the closed bars and the overflowing trash containers.

It is, indeed, a truism about Venice that "everyone goes to San Marco." This is not the only place that has filled the minds of those who have dreamed of the city—the Grand Canal has a good claim on the imaginations of tourists worldwide as well—but we could say that the Piazza, comprising as it does a set of culturally related and topographically connected places and spaces, does constitute the key touristic signifier of ter-

FIGURE 3. Tourist crowds in front of the Basilica of San Marco. (Photo by authors.)

restrial Venice. San Marco is indeed so powerful and essential an image of Venice that it would seem unthinkable for a tourist to visit the city without entering this space. "It remains unmissable," the guidebooks say, but the opposite is also true: it would be pointless for a visitor to deliberately avoid the Piazza, with the intention of thereby claiming to have experienced some more significant or "authentic" Venice that lay outside this tourist vortex—in the way that, for example, a visitor to London might purposefully stay away from such "tourist traps" as the Tower or Piccadilly, seeking "real" London life.[2] Instead, what we have seen in Venice convinces us that the Piazza is so predominant in both the perceived and expected tourist experience of Venice that not only every tour group, but also virtually all tourists visiting on their own, would never dream of leaving it off their agenda.

San Marco thus offers an excellent starting point for unpacking the culture of tourism in the city, where some basic questions can be asked about what Venice itself has come to mean: What do tourists go there to see? Do they find what they expected? Are they satisfied by the experience they have? The reverse of the question is, of course, this: What do the Vene-

tians now think about this part of their city, and what do they think tourists
see there? Piazza San Marco was for centuries the symbolic and ceremonial core of Venetian self-identity, but in the last fifty years it has become
touristed to the point where Venetians themselves find moving through
it literally difficult. There are only limited ways they can respond to the
masses of tourists who clog the Piazza, robbing them of both use of and
access to what was once called "the most beautiful drawing-room in Europe" (supposedly by Napoleon, who was himself certainly the archetype
of the invasive foreigner). The choices open to them have essentially come
down to three, all difficult: acquiescence, resistance, or abandonment.

The problems of perception and possession of the Piazza are old ones:
they could be said to go back to the first tourists who ever set foot there.
In good part, this is because visitors themselves have always privileged
San Marco over the rest of Venice. Yet, though the Piazza is striking in
its grandeur, foreigners have not always embraced it willingly. It is, after
all, simply the largest stretch of dry land in a city built on water, whose
mass of winding alleys and canals traditionally never seemed especially
inviting to most outsiders. As Wilhelm von Archenholtz noted in the
1770s:

The peculiar situation of Venice forces every body that is in want of bodily exercise, to make St. Mark's Place subservient to his purpose, unless he had rather
paddle through the most narrow and stinking streets. St. Mark's Place being, therefore, the general rendezvous at all times of the day, and frequented from the highest character down to the meanest plebeian, it naturally follows that this place,
magnificent indeed, must in a little time appear to a stranger the most tedious
spot on earth.[3]

Certainly, for almost as long as visitors have come to Venice, they have
found a Piazza San Marco to visit. The Basilica of San Marco, keystone
of the Piazza and one of its great attractions, was finished and consecrated
in 1094 (although its decoration went on for centuries thereafter); eighty
years later neighboring buildings were demolished and several canals filled
in to increase the open space in front of the new church, to the extent that
would roughly delineate the present-day Piazza. The Ducal Palace got
most of its present form and its distinctive facade around 1422; less than
twenty years later, the Spanish pilgrim Pero Tafur toured through it, commenting that, as a whole, "the palace is indeed very noble."[4] Yet what Tafur
and other pilgrim-tourists saw of the Piazza was much different than it
looks today, as Gentile Bellini's *Procession of Corpus Christi* of 1496 quite
clearly shows: although the Basilica and the portions of the Ducal Palace

and Campanile that are visible look quite familiar, the Procuratie Vecchie (along the north side) and the clock tower have yet to be built (though they soon would be—around 1500), and the Procuratie Nuove (on the south side) would have to wait until the mid-seventeenth century for completion. Even though, at 175 by 75 meters, Piazza San Marco has always been among the largest of the city's many squares, until it received the formal architecture of the Procuratie, with their regular colonnades and arcades, the whole area lacked the formal and monumental feel that has characterized it ever since.

This may explain why, while early visitors understood the Piazza's importance as a civic and religious site, they were not apparently so impressed with the visual impact of the space itself. It was "a great square, greater than the Medina del Campo," according to Tafur, but beyond that he had nothing much to say about the Piazza, except that it was "paved with bricks, and surrounded by many-storied houses with porticoes." Pietro Casola, visiting the city at just about the time of Bellini's painting, seemed to allow San Marco only the status of first among equals of the city's squares: "[Venice] has so many beautiful piazzas, beginning with that of St. Mark, that they would suffice for any great city placed on the mainland."[5] A century later, however, it was a different story: when Thomas Coryat passed through the city in 1608, he could admire the finished Procuratie Vecchie, while noting that "the South side [i.e., the Procuratie Nuove] . . . is but little more than halfe ended. For it was but lately begunne." He went on to admire the brand-new mint building, or Zecca, and the tasteful Loggetta ("little loggia") at the base of the Campanile—two important contributions in defining the Piazza, both designed by Jacopo Sansovino—before gushing that, taken altogether, this was "the fairest place in all the citie (which is indeed of that admirable and incomparable beauty, that I thinke no place whatsoever may compare with it)."[6]

Other major changes in the locale were instituted after the Republic fell in 1797. Under the order of Napoleon, the church of San Giminiano, which had originally closed off the western end of the Piazza, was torn down and replaced by the so-called Napoleonic Wing, built in the style of Sansovino's Zecca. The French emperor was also behind the order to tear down the Republic's enormous grain storehouse—the fourteenth-century Granai di Terranova—just up the Molo from the Piazzetta, supposedly because the ancient building was blocking the view from his imperial apartments in the Procuratie Nuove. In its place were installed the so-called Giardinetti Reali, the rather cramped Royal Gardens, on which

Victorian-era visitors often looked with scorn and which are now ignored by virtually all tourists except for those who are looking for a toilet.

As the area around San Marco has evolved over the centuries, so too have there been changes in the ways this large open space has been put to use. During the time of the Serenissima, this quintessentially Venetian space was charged with meanings both spiritual and political, giving historical geographers like Dennis Cosgrove a palimpsest on which to read the import of Venice's republican values:

By the late sixteenth century the central node of Venice expressed a complex symbolic structure, understandable in terms of the humanist ideas shared by the Venetian patriciate and in terms of the Venetian myth. San Marco was a concrete representation of the perfection of Venetian institutions. The Doge's Palace in its original location and external form represented monarchy, while the aristocracy was housed in the two wings of the *Procuratie* which defined the boundaries of the "place of eloquence, of Minerva": the Piazza San Marco. Here republican freedom was celebrated in the daily discourse of Venetian citizens. It opened towards the sacred legitimation of Venice, the Basilica di San Marco. Opposite the Doge's Palace stood the Marcian Library, the seat of humanist wisdom, lateral to the axis which united the lagoon, the "theatre of nature" gesturing towards Venice's *stato da mar* with the "theatre of man" and, via the symbol of time, the Rialto and ultimately the *terra firma*. Nature's beauty and perfection were thus linked to human wisdom and human perfection.[7]

A strong historiographic tradition long vaunted the Piazza and the Piazzetta as central to ritual and ceremonial activities in the Republic. Gentile Bellini's depiction of the Corpus Christi procession of 1496 shows just one of the many ceremonial observances that were regularly staged in San Marco throughout the festal year: that of the city's assembled *scuole grandi,* or great confraternities. Other grand processions—detailed in eyewitness accounts and in a variety of etchings, paintings, and woodcuts—were held to honor a newly invested doge or procurator, naval commanders who returned victorious, or visiting foreign princes. For regular religious observances or simply to commemorate an important event in the Republic's long history, patricians and notables, prelates and musicians were mustered to march about the Piazza. The Piazza, or at least the Piazzetta, was also a purely political arena. The procurators of San Marco convened within the Loggetta of Sansovino, at the base of the Campanile, to oversee Sunday reunions of the Great Council within the Ducal Palace. More significantly, the Venetian nobles themselves met before Great Council and Senate sessions, in that part of the Piazzetta running alongside the Ducal Palace, an area (and occasion) known as the Broglio, where the Republic's patriciate gathered to promenade, make legislative deals, and sell their votes to the highest bidder.[8]

But promenading in San Marco was not the prerogative of just the ruling class. In the Piazza itself, everyone traditionally took part in this amusement, an especially Mediterranean activity that on the surface was (and is) apparently so simple but that in fact has always masked complex rituals of sociability and display. The evening promenade has come to be known as the *passeggiata* elsewhere in Italy, but in Venice it was customarily called the *listòn* and served as a sort of dry-land counterpart to the *fresca*—taking the air in one's gondola on the Grand Canal. In the open-air societies south of the Alps, the *passeggiata* was (and often still is) a prized moment at the end of the day for socializing with one's townsfolk, seeing and being seen. Some of this is visible even in Bellini's majestic *Procession of Corpus Christi,* where the painter evidently did not think it at all odd to include various knots of patricians, merchants, and senators strolling unconcernedly together in the middle background of his work, even as the solemn procession unfolds all about them. Indeed, still further back one can also make out a few women, including one whose extraordinary height probably indicates a noblewoman, a *zentildonna,* wearing the very tall platform shoes, or *zoccoli,* favored by her class. By the late seventeenth century the Venetians had begun to combine their *listòn* with the recently arrived craze for coffee ("which they drink all the day long in *Venice*"), aided by coffee shops like Florian's, which opened in 1720 in the arcade of the Procuratie Nuove and soon set up its tables out in the Piazza proper. During the last years of the Republic, many foreign tourists had themselves rowed to San Marco as soon after they had arrived as possible, to enjoy the spectacle and to allow themselves to be seen. As Lady Ann Miller put it in 1776:

The first orders we gave our gondolier, were to conduct us to the *Place St. Mark.* . . . We were soon there, and found it answers all its descriptions. This is the center of Venetian amusement; here you see everybody; hear all the news of the day, and every point discussed: here are the senators, nobles, merchants, fine ladies, and the meanest of the people: Jews, Turks, Puppets, Greeks, mountebanks, all sorts of jugglers and sights. . . . The Place St. Mark is particularly agreeable to walk in by night; the lights in the coffee-houses illuminating the piazza render it extremely cheerful: a concourse of people resorting here to breathe the cool evening air, is so considerable as to fill the whole square.[9]

Both the social and the ceremonial usages of the Piazza San Marco also had to coexist—in the uneasy manner typical of the early modern world—

with all sorts of private entrepreneurs who sought to invade this public space. Much of the year, the Piazza was very much a marketplace, its broad expanse cluttered by the stalls of money changers, food sellers, and every sort of huckster. As Anne du Bocage put it: "On one side are to be seen puppet-shows, rope-dancers and jugglers. On the other fortune-tellers . . . Mountebanks . . . [and] Story-tellers."[10] At the same time, the Piazza was the regular haunt of preachers, who harangued passersby from a platform chair or portable pulpit; George Asycough wrote of seeing one friar who was "preaching away to several masks [that is, maskers] with great earnestness."[11] One can also see from some of the panoramas of Canaletto that a dentist once practiced in the Piazza, setting up his shop under his large model tooth, right next to the Campanile. Such commercial activities seem to have transpired quite freely in this communal space, even though they could occasionally block or disrupt the ceremonial activities of the state and prevent visitors from visually gaining a full appreciation of the area.[12]

The Piazza also had its function as the main food market for this particular part of Venice, and although fish vendors were required to stay by the public granary on the Molo, those selling other foods could set up in the Piazza itself—Pero Tafur reported that "they hold a market here every Thursday, which is greater than that of the Torre del Campo."[13] In the summertime, the food sellers and their garbage filled the Piazza with varying stenches, which could offend the more fastidious tourists: Hester Lynch Piozzi, for example, complained that "St. Mark's Place is all covered over in a morning with chicken-coops, which stink one to death, as nobody, I believe, thinks of changing their baskets." Perhaps because of the casual attitudes that prevailed around the Piazza, humans seemed to have had little compunction about adding their own contribution to the stench, as de Blainville noted with some outrage:

What is very unsuitable to this Magnificence, and greatly disgusts Strangers, is to see the Street, the Pillars, and the Steps of those fine Porticos perpetually deluged with an Inundation of Urine, with *Sir-reverences* swimming in it like so many floating Islands. This nasty Sight is so far from being disagreeable to the Inhabitants, that they account it Part of their boasted Liberty, to evacuate those Superfluities of Nature, when, where, and before whom they please.[14]

In the nineteenth century, under the French and Austrian occupations, most of this sort of commercial activity and its attendant trash and disorder were swept out of the Piazza and Piazzetta, along with the various festive and ritual manifestations staged there by the republican regime.[15] No longer would it have been possible for Venetians, through their unfolding this space ceremonially, to "read" San Marco in the symbolic or mythic terms proposed by Cosgrove; only the educated elite among future visitors would have this understanding of what had once been the allegorical structuring and significance of the Piazza. Instead, San Marco was turned almost exclusively into a hub for socializing. The Austrians, who had garrisoned the city with upward of fifteen thousand troops, were apparently the ones who introduced the custom of staging regular evening band concerts ("accounted the best in the world") in the middle of the Piazza. Even though the better sort of Venetians tended to boycott these events in protest over the Austrian domination, the concerts and the *listòn* became quite popular with foreigners and "the common classes," an unlikely mix that together helped realize Napoleon's notion of making the Piazza "the most beautiful drawing-room in Europe."[16]

After Venice was finally liberated and united with Italy in 1866, such musical evenings became a social focus of Venetian life. By any given late afternoon in the spring or fall, six or seven thousand strollers had typically poured into the Piazza to promenade to selected patriotic tunes and airs from Verdi and Rossini.[17] Both Mark Twain and Henry James described the attraction of the *listòn* for tourists and Venetians alike, while also making it clear that the two groups did not especially interact in the process. Although both Venetians and tourists disembarked from their gondolas at the Molo, each group encountered the Piazza in a different way. Foreigners, it seems, tended to head straight for the arcade of the Procuratie to gaze at the shop windows while they strolled and looked for other foreigners. Meanwhile, Venetian males would walk around the band, in the central space, while "the ladies are seated on the Piazza . . . where they meet the men of their acquaintance and interchange notes." Only at the tables of the cafés might Venetians and foreigners physically mingle, though even here there was a separation, with each set watching the other as they sat gossiping and eating their flavored ices, known as *granite.*[18]

Some Venetians evidently found the cafés of the Piazza so agreeable that they spent the whole day there, ordering what were, in the days just after unification, the quite inexpensive coffees and ices on offer and nursing them along to last for hours. With its mild climate, low prices, and pedestrian ambience, one American observer commented, "Venice must be a paradise" for the *flâneur,* of whom

some beautiful specimens of the genus invite attention daily at Florian's. I single out one for observation, and, as I must catch a glimpse of him whenever I cross the Piazza—for he appears to be on duty 14 hours a day—the task is an easy one. He is attired in a complete suit of white flannel. His hat is of light straw, and his shoes are of patent leather, cut just sufficiently low to disclose a bit of blue stocking. A high collar and a blue silk cravat, tied in an enormous bow, complete his attire. His appearance is picturesque, but it is cheap. . . . the elements of his visible *déjeuner* at Florian's—a half-cup of coffee . . . and a penny paper do not suggest great opulence on [his] part. . . . I cannot imagine that he, or his fellows, of whom a score are visible at a time, can possibly expend more than a thousand dollars a year apiece. . . . And yet he is one of the dandies of the town.[19]

The relaxed social topography of Piazza San Marco would seem to have held up into the 1950s and 1960s, although, according to some older Venetians we spoke with, the *listòn* had shifted by then from the center of the Piazza toward the old location of the Broglio, along the Piazzetta. There were other changes afoot as well: one Venetian friend told us that he remembered taking part in the promenade as a young man with his friends mostly as a way to meet foreign girls. It does appear that socializing and strolling in San Marco were becoming more the custom for youths than for adults or family groups, and the rising foreign presence may have played a role in this. By the mid to late '70s, in any case, the whole tradition of the *listòn* was breaking down, even among the young. Some blame the acute political antagonisms of the times, which Italians refer to as the *Anni di piombo* ("the leaden years"): we were told that those with strong right- or left-wing loyalties felt too uncomfortable in close contact with their political opponents in the Piazzetta, and so both groups eventually deserted the area altogether, to hold their own *passeggiate* in the *campi* of San Luca and San Bartolomeo. We also have heard others blame the rising tourist presence in San Marco, as it intersected with a steadily falling local population. These taken together made it increasingly difficult and then impossible for Venetians to maintain what is meant to be both an intimate encounter and a public display before the gaze of so many strangers. Instead it was simply easier to move the *listòn* to ever more remote neighborhood *campi:* first in nearby San Luca and San Bartolomeo, and more recently in Campo Santa Margherita and the via Garibaldi on the far edge of Castello.

It is, in fact, clear to anyone who spends much time around the Piazza that, with the exception of those who actually work there—around a thousand employees working for the various public institutions and perhaps eighty companies that mostly deal with tourists—the whole San Marco area has been largely abandoned by Venetians to the foreigners.[20] Part of this may be due to the monumental sterility of the place. Trying to keep visual clutter in the Piazza to a minimum, local authorities have maintained nineteenth-century laws against any private individuals looking to exploit the San Marco area for their own ends. All permanent souvenir stalls, along with food sellers and portrait artists, remain banished to about one hundred meters from the Piazza—that is, to the Riva degli Schiavoni, beyond the Ponte della Paglia, and in the area in front of the Giardinetti Reali. Not that these petty entrepreneurs have gone without a fight: especially after sunset, when the police are rarely in evidence, the more enterprising of them, their stalls equipped with wheels, often try creeping closer to the center of tourist activity, setting up for business in the Piazzetta or right in front of the Basilica itself.[21]

The apparent aim here has been to keep the Piazza space empty and monumental, a place of social gathering following the model introduced by the Austrians, rather than the somewhat messier ceremonial-commercial blend favored under the Venetian Republic itself. Unfortunately, the resulting grandeur has been largely wasted on the locals, since very few of them actually frequent the place any longer.[22] For some years now, "the immense cluster of chairs [that] stretches like a promontory into the smooth lake of the piazza," set out by Florian's and Quadri (and by Café Lavena on the Piazzetta), as well as the show tunes that are cranked out by their house orchestras and the expensive shops that line the arcade, are all aimed exclusively at tourists. San Marco has long since lost that unique combination of intimacy and grandeur that once gave it such appeal that Balzac could exclaim that it served as "a stock exchange, a theater foyer, a reading room, a club, and a confessional." Instead, the modern-day cafés at San Marco host a constantly changing tourist clientele and offer little more charm and probably much less social function than a train-station coffee shop. For those few tourists we spoke with who had actually been to them and sat at their tables, these bars seemed less fixed in their memories as "one of the greatest delights to be found in Venice" than as a sort of foreigners' rite of passage: where one could have the extravagant thrill of paying ten dollars for a single cup of coffee ("and it was so small, too!").[23]

Long since stripped of its role as the city's commercial and ceremonial center, the Piazza has thus only recently also lost its function as a center of local sociability, as "the general *rendezvous* of the promenaders and . . . the fashionable lounge of Venice."[24] Nor did the passing of the *listòn* in Venice merely signify that another quaint Italian tradition was being abandoned by a modernizing society, for the tradition of the public promenade still has social significance in many other small and mid-sized cities throughout Italy, and every evening it continues to draw participants not only from the towns themselves, but also from their suburbs and nearby villages. As Giovanna Del Negro has commented about the practice elsewhere in Italy, "*La passeggiata* is not merely a single, framed event, but a crucial enactment of the . . . [townsfolk's] world view. In this lively atmosphere, complex greetings, glances, gestures, and conversations intertwine to create a richly textured and highly aesthetic canvas of meanings."[25]

The custom of townsfolk gathering together on a daily basis, to see, show off, and be seen, is in many small to middle-sized Italian towns still a key ritual not only of sociability, but also of surveillance. Strollers comment on, gossip about, scrutinize, and judge those they see, thus making the *passeggiata* into a means of expressing a moral community. It can be profoundly important for giving citizens a sense of community, as Del Negro has pointed out: "The *passeggiata* is primarily concerned with connecting with the generalized self and feeling part of a larger whole. Ostensibly, the goal is not to be private in public, to seek anonymity in the crowd, but rather to share a physical and social co-presence with others."[26] Significantly, a participant cannot and should not be anonymous in the *passeggiata,* because any individual will be known, or at least known of, by someone, even in many medium-sized towns. The observers and the observed—and everyone is both at the same time—are not so much simply watching or being watched by others, as they are interacting with those they have knowledge of and with whom they have complex social relations. Townsfolk who do keep up a lively and active *passeggiata* proudly see it as proof of local *civiltà,* that is, civility—or better yet, the civilizing power of their community. In this sense, participants in the traditional Venetian *listòn* were very different from Baudelaire's nineteenth-century *flâneur,* who by definition kept an ironic distance from the social and cultural world through which he strolled, "to take part in the bustle of the city in the security of his anonymous status."[27] When the *passeggiata* is abandoned or reduced to a strictly neighborhood affair, as has by now happened in Venice, the central piazza no longer performs its customary social function of turning mere residents into genuine *cittadini,* or citizens of the whole civic community. What happens for those Venetians who remain in the city is more than just the disappearance of a public arena for socializing (though it is that): they also experience the loss of a primary form of education and socialization in what they themselves have come to term *venezianità,* or "Venetianness."

Even before the Piazza was abandoned by the *passeggiatori,* however, it already had been set off limits for another great Italian social activity:

politics. The site of public gatherings during the uprising of 1848, San Marco was also a popular venue for clashes in the early 1920s between Communist-led workers from the Giudecca factories and the Fascist *squadristi*. All political activity ended after the Second World War, however, when the various parties in Italy's postwar political landscape pledged not to hold either rallies or demonstrations in this space. Some have hailed this decision, saying that it made San Marco into a Piazza della Pace, a "piazza of peace," but there was an important, if unstated subtext here as well: stripped of its place in the local political life, the locale had few social options beyond becoming the city's own Piazza dei Turisti. What political activity was left in Venice was effectively banished from this, its obvious focal point, and sent to the peripheral neighborhoods, leaving the town literally without a political center. It was a situation that made Venice quite unusual among Italian cities: the town has no core where residents can publicly demonstrate their political—and hence social—affiliations. This willingness to eviscerate the city's political landscape for the sake of giving tourists a tranquil experience (something unimaginable in, say, Rome) says a great deal about how Venetians saw their own priorities in the postwar years.

Piazza San Marco is of course still filled with people, but only because in its most recent incarnation it has become a sacred site of tourism. The promenaders who continually crowd it are now the millions of tourists, drawn to a place that enjoys a worldwide fame both as a topographical entity in its own right and as a symbol for the larger cultural monument known as Venice. Speaking of the strength of this attractive pull, Dean MacCannell has noted: "Modern sightseeing has its own moral structure, a collective sense that certain sights must be seen. . . . There are quite literally millions of tourists who have spent their savings to make the pilgrimage to see these sights. Some who have not been 'there' have reported to me that they want to see these sights 'with all their hearts.'"[28] Guides who lead tours through the crowded space of San Marco report that most of the visitors they escort want little more than to simply be in this place. Indeed there is little else that modern tourists can do but gaze around themselves. There is no call for tourists to interact with other foreigners beyond their immediate group; nor is there any Venetian life here to observe, nor a chance of mingling with locals in the cafés of the Piazza. Nor can they adopt the pose of the *flâneur* and gaze upon those around them with bemused detachment, for all they will see in San Marco is a supersaturation of tourists all very much like themselves, equally there to see this "must see" place.

What is left of San Marco as the lived Venice is the architectural shell that bounds it. For those able to detach themselves from the seething mundane life around them, the Piazza can embrace the visitor architecturally, inviting the imagination to roam freely over its variety of spaces and exotic design elements. The Moorish or Turkish features that one sees scattered about the city seem to many visitors to be concentrated here, giving the Piazza and (especially) the Piazzetta the feel of a vast Oriental fantasy land. The lineage of such foreign reactions to San Marco can be traced back to the late eighteenth and early nineteenth centuries, when an Oriental craze took cultured Europe by storm. The landmarks of San Marco went through a rapid aesthetic reevaluation on the part of tourists, and while in 1776 Lady Ann Miller could disparage the Basilica as "in the old absurd Gothic style," and in 1778 George Ayscough could dismiss the Ducal Palace as "a large old pile, gloomy, and not very beautiful," just a few years later a visitor like William Beckford could see these monuments through an entirely different optic, as the ideal stone-and-mortar repositories for Orientalist fantasies: "I cannot help thinking St. Mark's a mosque, and the neighboring palace some vast seraglio, full of arabesque saloons, embroidered sofas, and voluptuous Circassians." Forty years later, Frances Shelley would also rhapsodize of the whole area, "I never saw anything so lovely as the Piazza. I seem to have been transported to Constantinople! Everything is so Eastern and Moresque."[29] Even for those who are familiar with modern-day Disneylands and theme parks, the "fairy scene" of Venice seems capable of sustaining interest on a visual level. No amusement park yet created can surpass the view down the Piazzetta, past the Basilica and the Ducal Palace to the waters of the Bacino, all subtended by the island of San Giorgio Maggiore and the church of the Salute.

Having watched the dynamics of the Piazza, we would question whether the general experience of the locale goes much beyond this strictly visual level of interaction. Although guided groups, ranging from ten to fifty people and receiving instruction from their umbrella-toting guides, seem at first to be everywhere, it soon becomes obvious that the great majority of those who throng to San Marco in the summer months are there on their own. Nor are guidebooks particularly in evidence in the hands of these self-guided tourists, whose interests would appear to lie elsewhere than in understanding the often complex story behind the monuments surrounding them. It is worth remembering that Venice's history actually offers little with which most modern-day visitors can identify personally. Unlike Florence, Rome, or Athens, the city fits rather poorly in the general narrative of Western Civilization. Its congeries of institutions was designed to suppress rather than promote those sorts of exceptional individuals around whom so much of the West's historical master narrative traditionally has been constructed. There are no Medici or Borgias here, only the semimythical Marco Polo, the libertine Casanova (increasingly popular lately), and the seemingly endless line (seventy-six in all) of dour and anonymous doges whose portraits circumscribe the Great Council chamber. Linked less to the great Western tradition than to the Byzantine and Muslim East, the origins of the city's great wealth and the inspirations for much of its architecture and design remain out of sight for the average tourist, over the horizon in Constantinople and the Levant.[30] One guide told us that the only visitors she expects to feel or express a personal historical linkage with Venice, beyond the city's artistic heritage, are German Protestants, who often ask her about Paolo Sarpi and the papal interdicts directed against the Republic, and Jewish Americans or Israelis, who may be knowledgeable about the Venetian Ghetto. For most other tourists, the experience of San Marco remains predominantly a sensual one, with little in the way of historic or moral relevance.

It is also true that a great many visitors to Venice have little idea what the city is all about because they are quite literally there, as indeed in all the places they visit, as accidental tourists. Incentive tourism, increasingly common in societies as disparate as Latin America and East Asia, often rewards the members of work teams or company departments with a coach tour or cruise that may well include a few hours in Venice (that is, at San Marco) as just one of a dozen or more stops in a packaged, ten-day holiday that participants have no role in planning.[31] Likewise, major cruise lines also are increasingly making Venice one of their stops (and several originate or end their tours there) on summer cruises around the Mediterranean. Their ships, as we shall see, can disgorge up to several thousand passengers at a time for a hurried afternoon or morning in the city—which again means San Marco.

When queried, many of these visitors have only the vaguest ideas about Venice's history, how the place came to be, or what the original purpose was of the monuments they are looking at around the Piazza: not surprising, since they have often ended up there through no choice of their own. Their often expressed inability to take Venice completely seriously or to be particularly awed by the great span of its history is undoubtedly abetted by the sensual delights presented by Piazza San Marco itself, its riotous color and playful design set against a compelling background that four hundred years ago Coryat could say "did even amaze or rather ravish my senses." Certainly in comparison to stolid Florence, the much more earnest home of Michelangelo, Brunelleschi, and the Renaissance, Venice positively invites "a state of dreamlike fantasy" as opposed to the active pursuit of culture and history.[32]

This is born out by the relatively low percentage of the millions of tourists coming to Piazza San Marco annually who actually do visit the formal cultural attractions in the area. By general agreement among guides, the most visited site on the Piazza is certainly the Basilica itself, although exactly how many actually enter it has always been unclear because getting into the church proper is free, leaving no record of ticket sales. One can gauge the popularity of the place only by the lines that form to get in: on many afternoons between April and September the queue can stretch a hundred meters, and has been known to reach the entire length of the Piazza, becoming a tourist sight in its own right.[33] It is also a safe assumption that many visitors, discouraged by the thought of waiting an hour or more in the sun to get in, give up on the Basilica altogether. The only certain statistic available as an indicator of how many actually do make it through the ornate bronze doorway is the number who buy a ticket to see the Pala d'Oro, the golden altarpiece set behind the main altar. These, in 1993 (the most recent year for which statistics are available), numbered just under half a million, men and women who had at least enough curiosity about Venice to take the extra time to see this dazzling sheet of gold and jewels. It is fairly obvious just from a glance around the nave that they represent only a small portion of those entering the Basilica proper, many of whom probably have rather less cultural interest in the place:

There are those who enter the Basilica because outside it's raining or it's too hot. There are those who are tired and want to sit down, even on one of the column bases, and there, getting comfortable, they profit [from the occasion] by pulling a sandwich and soda out of their backpack, or even putting out diapers—after all, what other chance will there be?—to change the babe-in-arms.[34]

The most visited site in San Marco in terms of paid admissions is the Ducal Palace. In 1993, just over a million visitors went through the seat of government of the Venetian Republic, to admire the paintings by Titian, Tintoretto, and Veronese that adorn the walls of its chambers. On peak days the mobs moving through the Palazzo can be impressive—as many as ten thousand paid admissions have been registered in a single holiday weekend—though a high percentage of these are evidently less interested in the high cultural offerings than in the ducal prisons, which are reached by passing through the palace and over the Bridge of Sighs.[35] After this, however, formal exposure to cultural or historic Venice drops off sharply

THE HEART OF THE MATTER

among tourists: the Museo Correr, located nearby in the Procuratie Nuove and representing the repository of much of the city's craft and social history, had a mere sixty-five thousand admissions in 1993. Even the Galleria dell'Accademia, located twenty minutes' walk away from San Marco and considered by many to be a world-class art museum, had barely a quarter million visitors annually during the 1990s.[36]

When taken in the context of the total estimated tourists to Venice, the overwhelming majority of whom end up in Piazza San Marco at some point during their visit, such figures seem fairly paltry.[37] A reasonable estimate indicates that less than one in five of those who come to the Piazza buys a ticket for either the Ducal Palace, the Pala d'Oro, or the Museo Correr. One must conclude that the remaining 80 percent, when not waiting to get into the Basilica, spend their time simply wandering around or sitting in Piazza San Marco, their experience primarily one of the ambience that they find themselves in. As Isabella Scaramuzzi notes, "Venice faces a lack of attractions created to draw the attention of visitors: the city itself is the strongest attraction that can be created, and the attention given to Venice is at the top, always and everywhere."[38] This is nowhere better illustrated than by the special nature of the Piazza's number-two paid tourist attraction, the Campanile of San Marco. The Campanile's attractions are strictly visual and ambiental; it carries for the tourist no cultural or historical baggage—indeed, it is not even the original Campanile—but it does present visitors with a stunning view of the bizarre topography of the city and the surrounding Lagoon. As a result, the structure attracts over a million paying visitors annually, about as many as the Ducal Palace itself, and in the summer, it can easily take nearly an hour of waiting to get to the top. On a clear day, at least, most visitors seem to consider the wait well worthwhile: once they finally reach the viewing platform, they have gained not only the ideal overview that completes the entire spatial experience of the San Marco area, but also the perfect vantage point for that essential touristic activity—taking photographs.[39]

Despite, or perhaps just because of, its apparent accessibility on a strictly visual level, the Piazza San Marco can run into the same sorts of problems of visitor overload that plague major tourist sites worldwide. During the high summer season the tourist gaze upon the Piazza is often frustrated and compromised by the tens of thousands of other foreigners who are competing to position themselves so that they too can have unimpeded visual access to the same sites. The situation has so worsened in recent years that even the guidebooks now sneer at the place: "Only at dawn, dead of night, or in the depths of winter," laments Fodor's, "is Piazza San

Marco somewhere to be enjoyed rather than endured."[40] One might want to sympathize with such beleaguered travelers, especially remembering the complaints of late-nineteenth-century visitors whose attempts to sit in solitude on the Ponte della Paglia, gazing on the Bridge of Sighs while reading the appropriate lines from Shelley or Byron, were frustrated by "a horde of savage Germans" passing nearby. As James commented sardonically:

> The sentimental tourist's sole quarrel with his Venice is that he has too many competitors there. He likes to be alone; to be original; to have (to himself, at least) the air of making discoveries. The Venice of today is a vast museum where the little wicker that admits you is perpetually turning and creaking, and you march through the institution with a herd of fellow-gazers. . . . the Piazza, as I say, has resolved itself into a magnificent tread-mill.[41]

Such poetic site-markers as the bound works of Byron, Shelley, or even Ruskin and Jan Morris are rarely if ever to be seen anymore around the Piazza. Nor does one see many visitors gazing soulfully upon the Bridge of Sighs from the Ponte della Paglia—hardly surprising, since during the summer the latter is packed solid with tour groups and street vendors, to the extent that even the simple task of getting oneself photographed silhouetted against the Bridge of Sighs in the background has become almost impossible. Literally tens of thousands of tourists are crossing or lingering on this span every day and never fewer than a hundred of them during the daylight hours (see figure 4). Besides the slow-moving individual tourists and the multiple tour groups of sometimes several dozen, all trying to get their photographs of the Bridge of Sighs or look at the gondolas passing underneath, there are also the service workers from the luxury hotels up the Riva, typically pushing their way through the crowds as they try to hurry back and forth to the Molo, and a fluctuating (but seemingly always increasing) contingent of sidewalk vendors, tattoo artists, and pickpockets. The result can be a good deal of grappling and sometimes outright fights, but no Shelley.[42]

Unquestionably, the sheer volume of visitors to San Marco has not only cluttered up the tourist experience of the area, but also has led to changes in how that experience is structured. This is nowhere more obvious than in the Basilica itself. Inside this venerable church, covered throughout with complex, seemingly endless mosaic decoration, visitor and Venetian alike traditionally have been presented with new insights into the mixture of piety and chauvinism that defined the medieval Serenissima. Henry James once lovingly evoked its "beauty of surface, of tone, of detail, of things near enough to touch and kneel upon and lean against. . . . In this

FIGURE 4. The Ponte della Paglia during tourist season. (Photo by authors.)

sort of beauty the place is incredibly rich, and you may go there every day
and find afresh some lurking pictorial nook."[43] For most of the twentieth century, the Basilica, like most Italian churches, was open to a public
that was free to wander almost at will, having only to pay for viewing the
Pala d'Oro or for ascending up to the choir loft to see the balconies and
the four gilded bronze Horses of Constantinople. This has all changed in
recent years, however (see figure 5). Tourist demand to enter the church
has become so high that ecclesiastical authorities responsible for the upkeep of the structure have roped off a pathway up the middle of the nave,
looping into the right transept, passing in front of the high altar, and then
exiting out the left aisle. Their aim was ostensibly to protect the mosaic
floor of the Basilica from millions of tourist feet wandering freely about
the church, but squeezing so many visitors into this fairly narrow corri-
dor has also produced a peristaltic effect that keeps visitors continually
moving along in the rhythm of what might be described as a convict
shuffle.[44] Just as well, of course, considering how many more are typically outside in the hot sun waiting to get in, but another result is that
the deeply sensuous quality of the structure's mosaic walls and floor, as
well as its many secret nooks that James extolled, is now just out of the
reach of the thousands of visitors who have to experience the place on

FIGURE 5. Tourists waiting in line to enter the Basilica of San Marco. (Photo by Philip Grabsky.)

this conveyor belt. For them, the Basilica has been reduced to a strictly visual experience, rationed out and attenuated, to be enjoyed only at a distance and in enough of a hurry to let the next batch in. It is a process that will only intensify in the years to come: already there is talk of offering computer reservations to enter the Basilica (and thus avoid the lines) and of limiting all visits by guided tours to just ten minutes.[45]

With what strikes us as a certain postmodern inevitability, it seems increasingly the case that the tourist hordes themselves have become one of San Marco's great touristic features. Especially during the high season, when their sheer numbers block any real possibility of reaching either an intellectual or aesthetic rapport with the area's historical space, one can see many tourists doing little more than simply watching other tourists. One hesitates to call the crowds here an attraction, but in fact many tourists look pleased and exhilarated simply to be in the middle of the Something Big that is this enormous, multinational, but essentially homogeneous crowd, content to just sit on the steps of the Campanile or the Piazza and just watch others like themselves—riding on a wave of shared excitement rather like what is generated at rock concerts or sports events. Few that we spoke with expressed any annoyance or disgust at finding themselves immersed in such crowds, not even to the point of voicing the sort of mild sarcasm implicit in a conversation we once over-heard between a heavily accented Australian husband and wife resting on the Ponte dell'Accademia:

He: Do you want to go back to St. Mark's Square?
She: No. There's nothing there. No. There's not much to do there.
He: You could sit and watch a million tourists.

That the crowds that squeeze into San Marco effectively have blocked out the Piazza's historic and aesthetic impact and thereby canceled their own reason for being there in the first place is not a new realization. Over two hundred years ago, John Moore regularly went to San Marco in the morning hours, because at that time the Piazza used to be fairly empty, and so he could meditate in peace on Venice's past glories; he cautioned his readers, however, against going there for anything but

a morning saunter; for in the evening there generally is, on St. Mark's Place, such a mixed multitude of Jews, Turks, and Christians; lawyers, knaves, and pickpockets; mountebanks, old women, and physicians; women of quality, with masks; strumpets barefaced; and in short, such a jumble of senators, citizens, gondoleers, and people of every character and condition, that your ideas are broken, bruised, and dislocated in the crowd, in such a manner, that you can think, or reflect, on nothing; yet this being a state of mind which many people are fond of, the place never fails to be well attended.[46]

Perhaps inevitably, when people are watching people, some people will begin to act up in ways that guarantee that public attention is directed their way. One could even say that there is something of a tradition in this, going back to the many hawkers, jugglers, and acting troops that filled the Piazza during the Serenissima, or to the custom of competing teams of young men forming human pyramids, the so-called *forze d'Ercole* ("competitions of Hercules"), during Carnival. But modern diversions of this sort are usually more fleeting, if only because the police soon arrive to shut them down. The newspapers tell of bicyclists—or, more recently, scooter riders—wheeling around between the crowds; Frisbee tossing, which newspapers usually blame on American students, is also commonplace in the summer.[47]

One young German tourist recently found another way to amuse himself in the vastness of Piazza San Marco: he pulled out a target pistol and started shooting the pigeons. Arrested, he seemed a bit perplexed, and not just because of language difficulties: What was the problem? There

FIGURE 6. Posing with pigeons in Piazza San Marco. (Photo by Philip Grabsky.)

were so many of these birds in the Piazza, after all.[48] The pigeons are indeed not only among the most visible, but also perhaps the most popular, tourist attractions in San Marco, even more popular than the tourists themselves. It is interesting to notice, in the middle of this most remarkably ornate and historically rich of ambients, how much enthusiasm and energy is put into attracting, observing, and photographing such a mundane, ordinary, and rather dirty bird (see figure 6). Tourists, abetted by travel writers and guidebooks, cherish all sorts of myths about these pigeons: how they are descendants of birds released every Palm Sunday by past doges, or brought to Venice by Queen Caterina of Cyprus, or fed by a special endowment left by this or that countess.[49] Certainly a key attraction of the birds is that one can feed them, giving tourists a chance to do something more interactive in the Piazza than just looking around. The three or four licensed grain vendors, who sell a few cents' worth of corn for almost a dollar, do a steady and presumably highly profitable business, although the pigeons feed, grow fat, and reproduce on more than just this high-priced grain, since the tourists themselves leave behind all manner of edible trash.

Not many visitors are seemingly aware that the pigeons have turned into a real problem in Venice. Like so much else linked to tourism in the city, they have flourished to the point of becoming a burden, and increasingly their nests and guano are defacing the city and its monuments. Estimates in the mid-1990s set their population at over one hundred thousand, a number that was unsustainable in the city, especially in winter, when there were fewer tourists around to feed them, leaving many of them to die of hunger and disease. Finally, in 1997, Venice's mayor signed an edict locally known as the "kill-the-pigeons law," which organized a drive to lower the pigeon population—including extensive roundups of the birds (over thirty thousand in 1999 alone), feeding them with sterilizing grain, and forbidding Venetians (or anyone else) to feed them outside of the San Marco area. For a few years, the number of pigeons apparently dropped to around fifty thousand, but this (relative) success was short-lived. By mid-2003, their numbers were back up to one hundred thousand in summer and sixty thousand in winter, a population that authorities seem to have resigned themselves to as irreducible, at least for the time being. As Paolo Cacciari, councilman for the environment, noted, despite the best efforts of the government to weed out their nests and kill off sick birds, it has been "the vendors of seed and the 'dictatorship' of the tourists, who don't want to renounce the fateful photo of the pigeon who pecks," that have rendered such attempts hopeless. And so the pigeons come back, if not more every year, at least more than is good for the city and its monuments.[50]

Looking around Piazza San Marco, one soon realizes that the real devotees of the pigeons are, of course, small children, who live down at pigeon level and often can be seen pursuing the birds around in circles, with a total and single-minded exclusion of their surroundings. Yet adults too seem to be extravagantly attracted to the pigeons of San Marco, as if they are particularly heartened to come across something so familiar in the midst of all this rather intimidating beauty and grandeur. Children who hold a handful of corn to attract a swarm of pigeons (as they will do) can also end up getting mobbed by a swarm of adult photographers, for the most part total strangers, looking for something more homely at which to point their cameras than the Campanile or the Ducal Palace. For the most part, visitors we spoke with took the side of the birds when we raised the problem of pigeon overpopulation for them to consider. Many foreigners left little question that, as far as they were concerned, the birds made the city "more fun" (as one put it), seeming to imply that the damage and expense they caused should be borne by the Venetians, in the interest of making and keeping San Marco as interesting as possible. At least in feeding the birds, we reckoned, these tourists are getting a chance to engage with native Venetians—perhaps among the very few such that they are likely to encounter, if they restrict their time in Venice to Piazza San Marco.

Since they damage the areas where they roost, and since the only Venetians who have a vested interest in them are the corn sellers and photographers of the Piazza, pigeons face a fairly unified local opposition. Not long ago there was even talk in the city government of importing peregrine falcons into San Marco, which has been tried successfully elsewhere in Italy. This has been promoted as the "natural" solution to the pigeon problem, though it has been blocked in the past by animal-rights groups.[51] Perhaps within a few years the pigeons will finally vanish from the Piazza, which more than any other in Europe has been identified with this ancient attraction/antagonism. Until then, though, the birds tend to beguile tourists of all ages, who, after feeding and posing with pigeons, leave behind their own contributions to the city:

little tiny sacks of transparent nylon . . . that are nothing more than the containers of grain for the pigeons that the various, appropriate kiosks sell throughout the San Marco area . . . [and that,] when there blows a breeze just a bit more sustained than usual . . . lift up in a little whirlwind of trash that spins swirlingly in the air to often finish in the faces of passersby.[52]

CHAPTER 4

Lost in the Labyrinth

Venice does not, of course, begin and end with Piazza San Marco, and sooner or later most tourists leave the big square and venture out into the city at large. What they find when they do is a Venice that is very different than the monumental center grouped around the Piazza, a much more low-key city but also an extremely complex one, a confusing maze made up of twin grids of alleys and waterways seemingly laid down without the slightest regard for sense or reason. Many tourists think Venice outside of San Marco is both baffling and somewhat unnerving, even though the great majority of those who leave the Piazza area actually restrict their activities to a fairly small portion of the city (see map 2).

In particular, foreigners tend to visit two of Venice's secondary tourist zones, one around the Rialto Bridge and the other at the Galleria dell' Accademia. Together with San Marco, these form the triangle (the Bermuda-Shorts Triangle?) of prime attractions around which the great majority of the city's visitors endlessly orbit, and which one might justifiably label Tourist Venice, were it not that the entire city is, in fact, Tourist Venice. Like San Marco, these two secondary nodes over the last decades have been largely emptied of the significance, symbolic or functional, that they once held in Venetian life. The Rialto, for one, was for centuries the commercial heart of Venice's vast mercantile empire: specifically, the Rialto Exchange, located at the Campo San Giacomo di Rialto, once symbolized for Venetians "a whole sector of human activity with an appropriate dress, gestures, and decorum . . . [that] signified the bargaining, buying, selling, speculating, and maximizing of personal profits that we would call capitalism and they called tending to their affairs."[1]

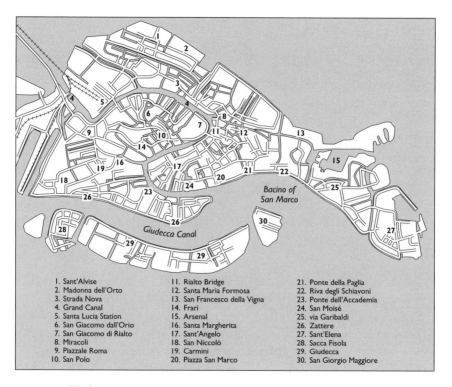

1. Sant'Alvise
2. Madonna dell'Orto
3. Strada Nova
4. Grand Canal
5. Santa Lucia Station
6. San Giacomo dall'Orio
7. San Giacomo di Rialto
8. Miracoli
9. Piazzale Roma
10. San Polo

11. Rialto Bridge
12. Santa Maria Formosa
13. San Francesco della Vigna
14. Frari
15. Arsenal
16. Santa Margherita
17. Sant'Angelo
18. San Niccolò
19. Carmini
20. Piazza San Marco

21. Ponte della Paglia
22. Riva degli Schiavoni
23. Ponte dell'Accademia
24. San Moisé
25. via Garibaldi
26. Zattere
27. Sant'Elena
28. Sacca Fisola
29. Giudecca
30. San Giorgio Maggiore

MAP 2. Venice.

With the fall of the Republic, however, international commerce in Venice withered, and more recently most of the other mercantile activities once associated with this area have likewise faded, such that, for present-day Venetians "tending to their affairs" in this spot has meant selling fish, meat, and vegetables to local customers, and all sorts of tourist schlock to foreigners. Few modern-day tourists, intent on admiring the fresh produce and shopping the stalls for cheap (mostly imported) jewelry and clothing, seem to realize that this *campo,* called San Giacometto, was once Europe's foremost mercantile exchange: even the sixteenth-century statue, the so-called *Gobbo,* or Hunchback, of the Rialto—once the symbol of the Rialto Exchange—is now often half hidden under mounds of discarded vegetable boxes. The world's fading awareness about the area's past significance has been accompanied by a semantic and physical shift in the term "Rialto" itself, which in modern tourist parlance no longer applies to the old neighborhood around San Giacometto, but rather almost ex-

FIGURE 7. View south from the Rialto Bridge. (Photo by authors.)

clusively to the Rialto Bridge nearby.[2] As a prime attraction of tourist Venice, the Rialto Bridge has much in common with Piazza San Marco, combining monumentality with the twin appeals of shopping and simply hanging out. Throughout the high season, the span is jammed with foreigners, some browsing in its expensive shops, and many thousands more just loitering and taking photographs of the Grand Canal from the top of the arch, whose marble balustrades have been polished to a glassy smoothness by millions of tourist hands, elbows, and backsides over the centuries (see figure 7).

The Galleria dell'Accademia, at the third corner of Venice's tourist triangle, has also shifted its place in the Venetian cultural landscape over the years (see figure 8). Originally founded in 1750 as an art school for the city's patricians, the Accademia still enjoys a special status as the principal repository of art from the Venetian school of painting. Opened to the public as an art museum in 1817, twenty years after the fall of the Republic, the gallery got many of its best-known works of art when they were removed from the various religious institutions—convents, churches, and confraternities—that were suppressed under French and Austrian dominion. Despite the undoubted importance of its collection, though, the Accademia is neither as large nor nearly as well attended as Italy's other

FIGURE 8. View of the Galleria dell'Accademia from the Ponte dell'Accademia. (Photo by authors.)

premier galleries: with barely 250,000 paid admissions a year, it hosts fewer than a quarter the number of paying visitors who go to the Uffizi in Florence.[3]

To connect these major nodes with one another, and to link the whole tourist triangle with Venice's major points of entry, the train station and Piazzale Roma, certain primary pedestrian routes have emerged over the years. These tourist thoroughfares are not, for the most part, much to look at or walk along: rarely straight, many of them have sections that are no more than two or three meters wide and sometimes much narrower. It would hardly be obvious that these were primary routes at all if they were not continuously crowded with tourists and lined with shops selling tourist goods. Unimpressive as they may be, these paths still represent the quickest (or at least most direct) route to go from one primary tourist site to another and thence to the outside world. As such, they have become the conduits for a huge share of the city's foot traffic. There are about eight or ten of them, each stitched together from a succession of connecting little streets that the Venetians call *calli*.

One of these routes ties the train station to the Rialto area by passing through Cannaregio; it takes in the Lista di Spagna, just to the east of the station, and then connects with the Strada Nova—arguably two of the most significant "streets" in the city—and thereafter winds through some tortuous and tight passages to the Rialto at Campo San Bartolomeo. Another route, overall more circuitous and much narrower, leaves from Piazzale Roma and arrives at the Rialto by traveling along the opposite bank of the Grand Canal. From this side of the Rialto Bridge, a thoroughfare that hops from *campo* to *campo* sets off along the Ruga Vecchia of San Giovanni, passes Campo Sant'Aponal, skirts the large Campo San Polo, and then swings past the Frari, itself a minor node, with its with altarpieces by Titian and Bellini. At the Frari, this route meets with another, coming from Piazzale Roma by way of San Rocco. These combine and then run along the back of Ca' Foscari; moving some distance back from the Grand Canal, this route finally arrives at the Accademia. From here one can take a short stroll on a busy walkway that leads to the Zattere, with its wide strolling space and ice cream shops, or proceed down Dorsoduro to the Peggy Guggenheim collection at Ca' Venier dei Leoni, but the main route heads over the Ponte dell'Accademia, to Campo Santo Stefano. From there, one can choose to go to the Rialto (via Campo Sant'Angelo) or to San Marco (via Campo San Maurizio). Finally, the whole tourist triangle is itself closed by the routes connecting the Rialto and San Marco. Here there is so much traffic that an entire network of parallel *calli* have been pressed into service: the Calle dei Fabbri, the Merceria San Zulian, and the Salizada di San Lio, among others.

Thanks to the rather ad hoc appearance of these passages, many foreigners who walk them get the impression that they are meandering aimlessly about the city, but of course they are not. Indeed, the city administration has for years tried to assure visitors that these are the town's primary thoroughfares, by having directional signs painted or attached at crucial intersections to guide the unwary on their way. These signs generally feature arrows that indicate the way to San Marco, the Rialto, the train station, and so on: unquestionably they have played a key role in persuading the great mass of tourists to stick to the approved thoroughfares when going about the city. Given Venice's tortuous topography, however, the import of these helpful indicators is not always altogether clear, and many visitors have enjoyed a wry laugh at the Alice-in-Wonderland sight of two adjacent signs indicating "Per S. Marco," with their arrows pointing in opposing directions.[4]

The importance of these thoroughfares has grown steadily along with the tourist influx to the city; they have effectively imposed on Venice a third grid—a tourist grid—on top of the town's more famous double system of canals and alleys. The pattern was already beginning to emerge soon after World War II, as a *New York Times* travel writer noted: "When one gets lost, which invariably happens, it is only necessary to regain a crowded alleyway eventually to end up at a familiar landmark."[5] Another *Times* travel writer, in 1961, apparently felt it necessary to provide his readers with a map and close description of his proposed "Walking Tour in Venice," even though the journey he proposed consisted of mostly these main avenues, allowing him to note that "along this . . . route, one will observe other way-farers bound for St. Mark's and a stranger will not go wrong by following the crowd."[6] By the 1980s, with these primary routes increasingly bracketed by tourist shops and trod along by millions of visitors annually, travel writers stopped bothering to give this sort of advice any longer: these thoroughfares had simply become too obvious.

In theory, funneling tourists along a limited number of major routes has made sense: foreigners know how to get where they want to go and Venetians get to enjoy in privacy those areas of the city that lie outside the thoroughfares. But, as is usually the case with Venice, what is theoretically workable ends up foundering on the twin shoals of the city's ancient quirkiness and the ever-growing numbers of tourists. Tiny at their best and tortuous everywhere else, the main streets of Venice in recent years have been required to carry more pedestrian traffic than they can manage. It recently has been estimated that around 85 percent of all the tourists coming to Venice—which, as of 2003, would amount to well over twelve million annually—spend virtually all their time on land walking in Piazza San Marco or on these little streets.[7] The problem of overusage is compounded by the relaxed attitude of most tourists who come here. Although many visitors briskly move along from one site to another, many others are simply strolling. The shops that line most of these thoroughfares make a determined effort with their window displays to catch the tourist eye—"the Venetians are especially talented at the art of window dressing," as one guidebook puts it—and many visitors are eager to respond to the invitations, treating the city as if it were a gigantic shopping mall.[8] Inveigled by all they see and empowered by the sense of being on the same level as everyone else in this pedestrian world, foreigners do not simply crowd the main thoroughfares; they also browse—continually and abruptly changing direction, stopping, bunching up to chat, or shifting from the shops on one side of the street to those opposite (they often are only a few meters apart) to see what another window has to offer. From such nonchalance frequently come great traffic jams, as other tourists, browsing in the same way, get caught up behind them.

Not surprisingly, this already touchy traffic situation is much exacerbated when tourist levels in the city exceed a certain point—most of those who have studied the situation agree that when over twenty-five thousand visitors are walking around Venice's historic center at one time, there is a good chance that at least some of the main thoroughfares will get clogged. This is especially the case when such routes suddenly turn or narrow—as they almost inevitably do—creating tourist clots that can make even a walk of a few hundred meters every bit as aggravating as driving for an hour in Roman traffic.[9]

For Venetian residents, the forces of convenience and commerce that have focused the tourist flow have not always made their lives easier. It is partly in response to this that locals have made their own back routes off the big pedestrian thoroughfares; when they take them, they use the term *andare per le sconte* ("to go by the hidden [ways]"), or simply *andar sconte*. Some of these are just shortcuts, but others can be quite long. For example, all the way from San Giacometto to Campo San Polo runs a route parallel to the main tourist highway. On it there are no souvenir or fast-food shops—only a few Venetian-style bars, storerooms, and a shop that sells bulk products for house cleaning. Even though this route occasionally comes within sight of the crowds and sparkle on the main thoroughfare, its occupants are for the most part elderly Venetians coming home from shopping or children going to school. Other routes are much shorter—perhaps no more than a few dozen meters, but they can take a person who knows the back alleys off a crowded, slow-moving thoroughfare and whisk him with almost magic ease to his destination. Standing, for example, on the Calle Saoneri (difficult to do, since it is very narrow and highly trafficked: see figure 9), one soon notices that a number of pedestrians are popping through a tunnel called the Sottoportego del Luganegher. They are mostly Venetians or university students, as it turns out, and they know that they can escape the crush on this main thoroughfare by taking a still narrower shortcut to the Frari and (eventually) to Piazzale Roma.

Such alternative routes form part of the essential "backstage" that makes it possible for Venetians and long-term visitors to stay sane in the city: territory that still belongs to those who live there. One long-term resident in Venice, an Italian friend originally from Emilia-Romagna (and still a self-confessed outsider in the city), once went so far as to claim to us that the mark of a "true" Venetian is found less in speaking the dialect (which has in any case faded greatly under the recent impact of television and consumerism) than in knowing the city with the kind of intuitive sense that only a native can have. No one who stays in Venice for the usual two

FIGURE 9. View along the Calle Saoneri. (Photo by authors.)

days or less can expect to pull this off: as Bergeret de Grancourt noted over two centuries ago, "Without [spending] considerable time it is impossible to become knowledgeable about the routes."[10]

As a backstage, however, the *sconte* can only go so far, literally. Sooner or later, all pedestrians, whether resident or foreign, have to get over one particular bridge or through the same narrow passageway if they are to make it from one *campo* or one islet to the next. In such places, Venice has become a city of many pedestrian collisions, none of them particularly painful, but, in the aggregate, still extremely annoying, especially for those who actually live in the city and are trying to go about their normal business. Worried that the Venetian tourist grid is not up to the job of handling so many bodies, many of those whose job it is to deal with mass tourism in the city have been trying for years to relieve the pressure on these limited routes. One popular approach has been to reverse the whole logic of the traffic system and encourage tourists to get off what is referred to as the beaten track of San Marco, the tourist triangle, and the main thoroughfares themselves. Travel writers, always in search of something new, personal, and exclusive to offer their readers, have been playing this key with special insistence for years, encouraging those who think they want to "turn [their] back on the herd, get off the merry-go-round of frenetic sightseeing and museum hopping," and see that evanescent "Venice of the Venetians" to seek out areas that are considered relative backwaters of little interest, even for foreigners with extra time: Cannaregio, for example, or Dorsoduro and much of Castello.[11]

The Venetian authorities largely have subscribed to this goal as well, though apparently more from the fear that too many potential visitors were being scared away from the city by horror stories about such traffic snarls. The city thus has adopted various schemes to get foreigners out of the city's inner tourist triangle and out into the streets at large. The Ministero per i Beni Culturali (Ministry for Cultural Patrimony), for instance, has recently set up "Venezia: Dal museo alla città" (Venice: From museum to city), which is known as CHORUS (L'Associazione Chiese di Venezia), to create in administrative terms a kind of art museum writ large, based not in a single building but spread out around Venice in thirteen (currently) of the city's parish and conventual churches. Paying a single admission, a tourist can not only see many great works of Venetian art in the churches for which they were originally intended, but also can wander through much of the city in the process. Other advantages of this ambitious program have been the publication of a series of inexpensive and very detailed guidebooks for each church in the tour, as well as the opening at regular and extensive hours of a number of churches—such as San Sebastiano, home of several important works by Paolo Veronese—that previously had been closed most of the time.[12] Various cultural organizations also have done their best in this regard, staging activities from art shows to foot races in Venice's periphery in the hopes of drawing foreigners into the back corners of the city. In recent years, neighborhood groups also have taken to staging their own little festivals, celebrating Carnival, a summer regatta, or a local saint: most are happy to welcome a few adventurous tourists to join in.[13]

Cesare Battisti, who is on the Azienda di Promozione Turistica di Venezia (Venetian Tourist Promotion Board; APT), goes even further, asserting that one ought to go around the city without any destination at all. Instead, he once told us, he would like to encourage tourists to simply "get lost" while in Venice: in other words, just head off into the unknown tangle of *calli, campi,* and bridges and take the city as it comes. Battisti is recognizing that, although many of these back areas have never enjoyed much fame and do not have much to offer as specific tourist attractions, they do cater to an almost universal tourist desire—to make some symbolic and personal contact with the visited site. This sense of closure that comes from discovering oneself within the "reality" of a generally recognized tourist landscape can be accomplished almost anywhere in Venice. As anyone who has ventured out into the town's back streets can testify, it is almost always easy to frame a camera or video shot—or just a personal memory, for that matter—that will combine some or sometimes all of the elements of "Venetianness" that allow one to say, "I'm here, I'm in Venice." This is a place where, after all, "unlike just about anywhere else on earth, the most important thing is simple physical presence. No matter what the starting point, a trip to Venice is a voyage to another reality. The truly unique form and spirit of the place assault the senses from the onset. . . . Once in Venice it doesn't matter where you are, as long as you are there."[14]

Walking around Venice just for pleasure is nevertheless a relatively recent innovation for foreigners. From the first days of the pilgrim tourists until the time of Henry James, one took a gondola for any jaunt of more than a few hundred meters. The city was not only confusing and complex for outsiders; it could also be dangerous, or at least annoying: anyone who ventured into an unknown alley could expect, at the very least, to be assaulted by beggar children, whores, or even robbers. With the twentieth century, however, came cheap maps, paired with the hundreds of black-and-white street signs, or *nizioleti* (literally, "handkerchiefs"), that clearly identify every last *calle, ponte,* and *sottoportego* (tunnel).[15] The streets in Venice generally are much cleaner than they were a century ago, but more than anything, what has really made ambling about the town a pleasant tourist attraction in its own right is the collapse of the resident population. As a recent edition of Cook's *Guide to Venice* rather bluntly reassured its more timid readers: "In high season the number of tourists exceeds that of the local population, so as a foreigner, you are unlikely to feel ill-at-ease."[16] These days in Venice the high season runs from March to November.

Bushwhacking through Venice can indeed be its own reward, one that requires very little in the way of understanding the city itself or the history of the people who actually built all these *calli, campi,* and bridges: it is enough just to let oneself go, into the sheer topographical pleasure of wandering through this space. Although spokesmen for Venice and Venetian tourism are forever insisting that "Venice is not Disneyland," their urging that visitors "get lost" in the city is an effective invitation to treat this place as a walking ride through a visual fantasy land, with neither intellectual nor historic depth. "Who built this place, anyway?" a visiting American soldier once asked us. For her, plunging into the whimsical landscape lurking just beyond San Marco had ended up proving (if such proof were needed) only that Venice is no city at all, but rather a caprice or a fantasy that someone apparently had created for her amusement.

When one takes the time, Venice reveals itself as a delight of *calli* running between intimate *campi* that usually are ringed by graceful, if often rather dilapidated, old palaces. True, many of these *calli* dead-end without warning, either coming up short at the bank of a canal or leading to a blind and secretive little *campo,* known as a *campiello.* Other walkways may appear to dead-end but in fact do not and instead shift unexpectedly around a corner or over a little humpbacked bridge or carry on right through an intervening building, in one of the often malodorous tunnels known as *sottoporteghi.* Often claustrophobic, the network of *calli* can also suddenly open up, placing one unexpectedly out in the sunshine (or the winter wind) on one of the spacious *fondamente,* or quays, that run alongside the canals.

Indeed, not many visitors pause to reflect on it, but the city's unusual landscape is closely linked to its particular evolution out of the mudflats and fens of the Lagoon. Originally founded on a series of small islets, Venice was formed by centuries of damming and filling, with each nascent community expanding up to the point where only the network of narrow waterways, the *rii,* or canals, were left between developed land. Up to at least the twelfth century, many scholars now believe, Venetians tended to construct most of their dwellings and churches to face onto the water. Communications between houses and between the islands (which eventually would coalesce into Venice's traditional seventy-two parishes) were necessarily and primarily by boat; the land behind the houses, though it was in a sense the community's center, typically was treated as wasteland—a field, or *campo,* in other words—where residents could dump rubbish, keep gardens and poultry, or stretch cloth. Many *calli* appear to have originated as the narrow tracks leading to and from this communal area, squeezed between the houses and offering those who needed it access to the water.

Some of these original *calli* must have led to a neighboring islet by way of a bridge, but until the fifteenth century, when the Venetians began systematically rebuilding them all in stone, most bridges in the city were miserable little affairs—often just some planks nailed onto old boat hulls, too narrow even for two pedestrians to pass by each other. The uncertain connections they offered, reinforced by the near ubiquity of gondolas, *sandoli, peote, caorli,* and many larger craft, meant that virtually all the city's land routes had a secondary importance to those on the water. After the fifteenth century, walking was left for the poor: men and women with pretensions had given up trying to get around on horseback and switched almost exclusively to their own private gondolas. Probably because of this, as much as anything else, the Venetian *calli* have the feeling of being an afterthought; many were indeed originally no more than back entrances to palaces whose main facade was on a canal. As such, and since space in Venice was always at a premium, there was no reason to make these streets any wider than servants, porters, or beggars needed; squeezed by adjoining buildings, some *calli* narrow down in places to barely a meter in width. Even the main shopping streets, known as *salizade,* are not much wider than that; they often have struck visitors as no straighter and a good deal less impressive than ordinary alleys in other cities. As James Fenimore Cooper described it in the late 1820s, the Salizada di San Moisé, leading him to Piazza San Marco from the west, "was lined with shops, and it seemed a great thoroughfare. Its width varied from ten to twenty feet."[17]

Because the city's bridges were constructed (or reconstructed) in stone during the fifteenth century, when much of the Venetian topography was already in place, they too were often squeezed into marginal space by competing houses and palaces, assuming their variety of charming and sometimes distorted shapes: a number of bridges still carry the name Ponte Storto, "Bent Bridge," or Ponte Stretto, "Narrow Bridge."[18] Yet these unassuming little bridges do not merely pass over water, but also carry one from islet to islet. Many visitors fail to realize that, having crossed even one bridge, they have also greatly complicated their experience of Venice's topography, since this casual act necessarily puts them on a new islet. Though the islets are many, most of them can be reached (or left) by only two or three bridges, with the routes involved often quite specific and not especially obvious. In such situations, seemingly minor decisions about turning right or left can rapidly get tourists completely lost, and whole sections of town that on a map or even in person appear to be contiguous can turn out to be unreachable without taking apparently elaborate deviations. Moreover, since a many a *calle* looks completely different depending on which direction one is walking, it can even be quite difficult to beat a retreat simply by retracing one's steps.

Once caught up in this maze, tourists can find it hard to extricate themselves. Asking a Venetian for directions rarely produces much help, partly because there are so few Venetians around anymore, and many that remain have so cultivated their skills in looking right through bewildered foreigners that it takes real boldness to approach them. Tourists who do manage to get advice often end up with vague directions that do not seem to fit the complexity of the city. A Venetian friend once told us a tale that should probably be considered an urban legend (though she swore it was true), about a bartender who was asked for directions to a particular church by a group of lost French tourists. He told them, in the classic Venetian manner, to go down a certain *calle (andè al fondo),* take the last left *(ultima sinistra),* and then keep going straight (the inevitable *sempre drio*) to their destination. The *barista* was completely flabbergasted, though, when the group came back into his bar half an hour later, soaking wet, completely filthy, and quite angry. It seems that they had taken his directions much too literally, and instead of taking the last turn to the left before the street dead-ended in a canal, as so many *calli* do, they had gone right up to the water's edge and then tried to claw and rappel their way along the wall to their left, periodically slipping and falling into the water as they went.

Getting lost in Venice is by no means a new problem for foreigners, and indeed tourists have been complaining about it for centuries. Back in the 1690s, William Bromley noted how "ordinary people here may go up and down the Town by little Back Alleys, which they call here *Calle;* these by winding up and down, and delivering them over several Bridges, hugely puzzle Strangers at first." In the 1770s, Bergeret de Grancourt lamented that, in Venice, "it is extremely difficult to understand the system," and went on to observe that "in the morning I easily lose myself in the streets, which are for the most part three or four feet wide and make up a labyrinth." Those who wished to go sightseeing in Venice, as they were used to doing elsewhere, could be frustrated by the tricky streets:

Cooper, for example, wrote how one of his traveling companions set out on his first evening in town, eager to see San Marco by moonlight. But though his hotel was just a few hundred meters from the Piazza, "after an unsuccessful search in the maze of lanes, he returned disappointed," unable to find a trace of the place. Just a decade or so earlier, the British major W. E. Frye also noted how easy it was to lose track even of something as large as Piazza San Marco, due to its being in the midst of

a series of alleys connected by a series of bridges which form the *tout ensemble* of this city. . . . All this forms such a perfect labyrinth from the multiplicity and similarity of the alleys and bridges, that it is impossible for any stranger to find his way without a guide. I lost my way regularly every time that I went from my inn to the *Piazza di San Marco* . . . and every time I was obliged to hire a boy to conduct me [back] to my inn. On this account, in order to avoid this perplexity and the expense of hiring a gondola every time I wished to go to the *Piazza di San Marco,* I removed to another inn . . . [which] is not twenty yards from the *Piazza.*[19]

Trying to head off overland on one's own was not recommended in the eighteenth century: "I tried to find my way in and out of this labyrinth without asking anyone," Goethe noted, and attempted to make do instead "only directing myself by the points of the compass. Finally one does disentangle oneself, but it is an incredible maze." That evening the man who has been identified as the most intelligent being of his era, if not of all time, gave up and got a map of the city, though days later he was still marveling at the "strangest labyrinth" he had to go through to get anywhere.[20] At least Goethe was smart enough to know when he was beaten: few other tourists were willing to admit they had to buy a map, perhaps because these were not generally and conveniently available, at least not until 1870, when Baedeker's guide came out with an immaculately detailed plan of the city. Admittedly, this handy and handsome addition was of only limited use since it came without the names of any streets, though all the canals were identified—presumably because visitors were still expected to get around by gondola for the most part. Not until the 1913 edition did *calli* and canals get equal treatment.

In more recent years, the municipal authorities have also joined in, handing out free maps by the thousands at the train station, Piazzale Roma, and the information center at San Marco. Originally these were not all that useful, since they not only failed to name most of the streets, but also only offered a angular, highly stylized schematic of the city, distorting the shapes of islands and leaving out a good many *calli* altogether.

Such maps would have been sketchy for any city; for a place as complex and convoluted as Venice, however, they were hopeless, and one often used to see tourists standing about puzzling in frustration over these familiar blue-and-taupe mis-maps in their hands. Lately, however, free maps have gotten much better, at least naming all the main thoroughfares, though we noticed that a good many tourists still try to make do with nothing more sophisticated than the sketchy advertising maps like those handed out as promotional guides to herd potential customers to McDonald's.[21]

One thing that remains striking about the Venetian situation is how much more in evidence maps are in tourist hands than are guidebooks. This may be in part because a number of the modern-day guides show only the tourist core of the city in any detail. Still, it is probably safe to say that nowhere else in Europe (or perhaps the world) can one see so many people in so many places standing around looking at maps. Some guidebooks warn their readers that those who open up their maps to their full extent will not only block traffic, but also end up looking "like a tourist."[22] Worrying about looking like a tourist is, of course, ridiculous in Venice, since for much of the year in the San Marco triangle virtually everyone is a tourist or a tourism worker anyway; most Venetians seem to assume that anyone whom they see whom they do not know must be a tourist, looking at a map or not. Moreover, as suggestions go, this one is a bit cruel, since it takes weeks or even months to learn one's way efficiently around Venice, and the process is certainly not made any easier if one is too shy to stop and look at a map. It is indeed a safe assumption that, since around five million of Venice's annual visitors have never been there before and are staying for less than a day, the great majority of those one runs into (sometimes literally) on the street are either lost or about to be. Venetians are well aware of this crucial difference between themselves and their visitors. As Paolo Rizzi wrote more than twenty years ago:

It is called a casbah. But one could also say labyrinth. All of Venice is a labyrinth: not so much for us Venetians, as for [our] guests, above all for the foreigners, who almost never know how to get their bearings in the maze of canals and *calli*. In reality, the impression of those who arrive in the city is [one] of generalized confusion. Tourists are disoriented, they do not know how to position themselves, where to go, whom to ask.[23]

What must be remembered, however, is that despite so many tourists trying so conscientiously to find their way with maps, guides, or even by asking directions, getting lost in Venice is not necessarily an unpleasant experience. Those who can handle a bit of "generalized confusion" and are not on a tight schedule have told us that wandering around the city and taking every turn that catches their fancy was for them the best thing about their visit. Perhaps it takes them back to the hide-and-seek emotions of childhood, for they find themselves in an oversized, dreamlike playground filled with narrow passageways, dead ends, tunnels, and half-concealed shortcuts. One can easily get caught up in the swirl of a group of youthful visitors—high-school students, usually—who are enthusiastically (and noisily) charging through this labyrinth, taking pleasure in losing and finding each other, as though they were in a vast, elaborate playground. It may seem a bit strange, but visitors meandering around aimlessly in Venice seem to be in some way "empowered" by the experience, despite the fact that many of them are actually lost. In contrast to their experiences in cities like Florence or Rome (or even London or Paris), many tourists claim to have undergone a kind of dreamy liberation while in Venice, a freedom of action that some describe almost as intoxication.

Much of this boldness and empowerment that foreigners feel in Venice can be traced to the total absence of motorized traffic in the streets, with the result that in this city the modern rules of urban space appear to have been suspended. Both first-time and repeat tourists to the city have told us that this, in fact, is what most struck them about Venice: there are no cars. Many visitors say they had expected at least some vestigial traffic; a few apparently drove over the Ponte della Libertà to Piazzale Roma thinking they could drive right on from there to Piazza San Marco, as they might be able to do in other canal towns, such as Amsterdam or Stockholm. Venetians clearly enjoy this sort of misunderstanding, and one occasionally reads in the local papers, under such rubrics as "Curiosities," articles with headlines like "'Excuse Me, Where Do I Catch the Bus for San Marco?'"[24]

Many foreigners find Venice a great relief after other Italian cities, a "balm to tourists whose nerves are overwrought by leaps for safety in motor-charged European streets. One walks in blessed silence and total security, down the center of the narrow streets . . . with no Vespa or Fiat 500 nearer than Mestre, far away on the mainland."[25] The quiet of the city, the sense that one can enjoy of being quite literally on an equal footing with the natives—who walk the streets with tourists instead of trying to flatten them with cars—greatly emboldens many visitors, who may lay aside the spatial distinctions they would normally make in any other city, Italian or otherwise. Public and private, interior and exterior, all seem somehow more similar when cars are removed from the equation. It accounts for the willingness of tourists to simply plop themselves down on bridge or church steps and means that many foreigners will not be afraid to push their way into the most secluded of little *campielli,* something that they might find quite offensive (or at least amusing) if anyone tried it in their hometowns. Venetians have told us that it is necessary to lock up or guard any entrance that might catch the tourist eye. They know from experience that foreigners will try to sneak into buildings both private and public—these latter often palaces that have been converted for the use of the university, local government, or businesses—and into the most remote *campielli,* just to have a look around. We have heard stories about tourists' impressive ability to evade the concierge and other functionaries and end up blundering into offices on even the fourth or fifth floor of some buildings.

Without cars, trucks, and motorbikes, the streets of Venice unquestionably do have an antimodern, dreamlike quality, giving what is already a surreal labyrinth the additional freedom from the noise and speed associated with the mechanized world outside. Aware of the attraction of such unworldliness, travel writers have for years promoted Venice for its "enchanted" qualities and as a place where one can savor "the unusual experience of not being harassed, as in practically every other great city of the world, by motor traffic."[26] It is this silence, found in Venice's ancient buildings, still water, and softly echoing footsteps, that is the key to the city's fame as the most romantic of vacation spots. Indeed, some tourists have told us that they are so used to a background of motorized din and rapid motion that, at first, the slow rhythms and silences of this unique environment are unsettling, almost eerie.

Like so much else about Venice, this sense of being surrounded by silence is not a new reaction, but rather something that has struck tourists to the city since even before there was motor traffic to flee. Grand Tourists in the days of the Serenissima contrasted Venice's quiet with the wheeled racket they were used to back home—iron-banded carriage wheels running on cobbles can be deafening. Already by the mid-seventeenth century, John Evelyn wrote admiringly of a city that "is almost as silent as the middle of a field, there being neither rattling of coaches nor trampling of horses."[27] But it was the Victorians who particularly embraced the city as a haven from all that was aggressively modern and strident about their own hometowns; for them, Venice was where "instead of the foot-fall of horse and the rumble of carriage-wheels, the gondola glides noiselessly by."[28]

Considering how crowded and uncomfortable Venice's tourist core has become in the high season, the labyrinth that seems to beckon just beyond the tour groups and souvenir shops has acquired considerable appeal, even without offering any specific attractions. In the last few decades, wandering the back streets of the city has become steadily more popular among tourists, encouraged by travel writers and guidebooks alike. This activity also has special appeal for those who see their vacations primarily in terms of shopping. Laura McKenzie, in her video series *Laura McKenzie's Travel Tips,* extols her viewers that "half the fun of shopping in Venice is getting lost on a side street: it's like a giant shopping mall and everywhere you look you'll discover another find." The notion that a shopping expedition in the back streets of Venice is something of a treasure hunt is also pitched by the big cruise lines, whose ads invite those customers who see themselves as "traveling, not touristing," to go out into Venice and search for "hidden treasures in an ancient marketplace."[29]

Naturally, not all of the interests are completely unvested when it comes to getting tourists to spread out across the city. Merchants and restaurant owners who have their establishments far away from San Marco or the Rialto—out on the fringes of Castello or Cannaregio, for instance—are forever coming up with new schemes to persuade or coerce foreigners to come to their part of town for shopping and dining. Some of these ploys are well disguised, as with the recent attempts of the shopkeepers and trinket sellers of the via Garibaldi—a still relatively untouristed part of Castello—to have the city government ban the grand yachts of the rich from anchoring at the adjacent Riva dei Sette Martiri. Although they have sometimes couched their complaints in leftist rhetoric against these *nababbi,* as they like to call the wealthy, these merchants are in fact "those [who] see a resource in the herds of tourists in transhumance along the Bacino of San Marco" and hope that, once the yachts are driven off, the great *lancioni di granturismo*—the long, glass-covered boats that carry forty to sixty tourists into town from the big cruise ships—could be persuaded to land there and disgorge their cargo within blinking distance of the local shops. Appeals with a similar subtext have also been advanced over whether to establish a stop for the *lancioni* and for the public *vaporetto* that comes to the city from the Marco Polo airport in the out-of-the-way Fondamenta di Santa Giustina: "And from there, the herds could go across Castello in the same fashion, leaving behind [their] dollars, marks and little lire *[lirette]*."[30]

It is, of course, quite true that, as they themselves spread steadily across the city like oil over water, it becomes steadily more difficult for tourists to have the delightful experience of wandering around Venice and discovering the modest, unexpected pleasures of its backstage. Finding (or losing) oneself within the Venetian maze depends a good deal on getting away from the herds of fellow tourists, something that is becoming ever harder to accomplish: "Where else but in Venice can you see desperation on the face of the tourist who is *not* lost? The visitor, who, held in the grip of the main routes, with their prominent yellow signs ever pointing the way indulgently back PER SAN MARCO, searching in bewilderment for that chink in the wall that will open into the labyrinth of some imagined secret Venice?"[31]

Finding that secret Venice is a steadily receding goal that continues to call each new generation of tourists who believe that, at least within the disorientation peculiar to the city's labyrinth, one can achieve a deeper, nontouristic experience that means one is a traveler and not just a tourist. This too is not a new problem. Henry James noted well over a hundred years ago that in Venice "there is nothing left to discover or describe, and originality of attitude is completely impossible."[32] Indeed, by now the tourist traffic in the city is such that, with the possible exception of parts of the Giudecca, Sacca Fisola, and Sant'Elena, there is probably no place, however seemingly obscure and remote, that is not gazed upon and probably photographed by at least one tourist on any given day. Even those few foreigners with the fortitude or luck to make their way to, say, the remote Campo Ruga, on the far outskirts of Castello, no longer have the place or the experience to themselves. As they aim their cameras at the seemingly authentic and certainly charming working-class housing around the square, quite likely more tourists will be coming up right behind them to repeat the process. Even worse, chances are they will not be taking pictures of Venetians at all, for by now a scattering of foreigners have found cheap housing in many of these outlying districts: an American friend of ours, a young scholar on sabbatical, told us that a few winters ago he and his family had lived right on Campo Ruga, in fact, and he had had the pleasure of watching a steady stream of tourists who, evidently believing that they had found the "Venice of the Venetians," paused to take pictures of his house, his two girls, and even his laundry.

Those who actually live here—as well as many who study the city—not surprisingly tend to see the Venetian landscape in very different terms than it appears to the average day-tripper. Isabella Scaramuzzi, vice-director of Consorzio per lo Sviluppo Economico e Sociale della Provincia di Venezia (COSES), who has concerned herself for over twenty years with the economics of tourism in Venice, once suggested to us that, rather

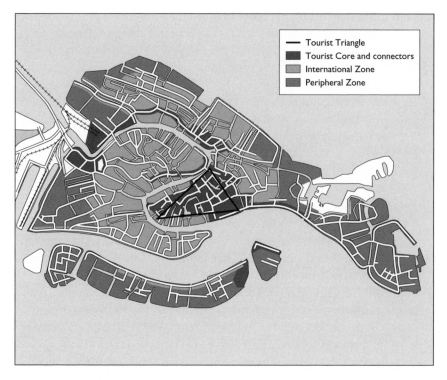

MAP 3. Social zones of Venice.

than see this landscape as an undifferentiated snarl of canals, *campi,* and *calli,* one would do better to view Venice in socioeconomic terms. What emerges is a city that is essentially split into three somewhat concentric circles that radiate out from San Marco (see map 3).[33] Lying at the center, as we have already noted, is the area within the triangle formed by San Marco, the Rialto, and the Accademia: Scaramuzzi calls this the Tourist Core, though less for the monuments and sights that it offers than for the completely touristic nature of life within its bounds. Indeed, searching around in this zone, one is hard pressed to find *any* shop or business that does not cater more or less directly to the needs of foreigners: here is Venice's highest density of hotels, restaurants, fast-food and ice-cream outlets, gift shops, and stores offering "Venetian specialties." Here and around Piazzale Roma and the train station are also mostly where the pickpockets and three-card-monte sharks ply their trades, since this is where they find foreigners most likely to be crowded and bustled to the necessary point of distraction.

Here too are also clustered Italy's best-known, high-end designer shops, brands like Prada, Fendi, and Versace, which have in recent years taken over from local retailers of quality clothing and accessories. It is hard to believe that only a few decades ago Venice had its own purveyors of such fine goods, the last remnants of a thousand-year mercantile tradition that withered away in short order once the city's local population began to crumble.[34] Certainly this current crop of fashionable stores seems to do a slow business: the common belief among many Venetians we know is that such designer shops cannot really expect to make much money in Venice but simply want to include the city's name on their labels. While we have noticed plenty of tourists carrying designer bags, others who cannot afford a five-hundred-dollar purse or pair of shoes seem to at least like having their photo taken while trying such things on—shop attendants are forever having to chase away this sort of bogus customer.

Now that virtually all the stores providing domestic staples in Venice's Tourist Core have shut down and been replaced with tourist services, even the larger institutions, the socioeconomic pillars of the city for the last century or more, have also started to fold. A number of department stores, banks, and insurance companies that had major branches or even a head office there have pulled out of Venice's historic center in the last decade or so.[35] The spaces they left behind are for the most part too big for the merchants of souvenirs and fast foods to take over, and in any case the local authorities finally have begun to pass ordinances against any more tourist shops in this area. Still, in the late 1990s the Benetton family, perhaps inspired by Prada and Louis Vuitton, bought up the massive structure left behind by the departing Banca di Sicilia, near Campo San Bartolomeo, with the intention of setting up a "grand megastore" in its four floors.[36] Other large spaces created by departing corporations are more likely to be turned into hotels or further government offices.

Outside Venice's Tourist Core, Scaramuzzi posits what she calls an International Zone, not so much because there are foreigners there—that would be a given almost anywhere in the city—but rather because this middle area is where many non-Venetians have bought second homes. The heart of this zone, one might say, lies—as it has lain since the days of Browning and Wagner—in the great palaces that front along the Grand Canal, and also in the districts right behind them, where the buildings are rather more spacious than in the Tourist Core and many more of them have gardens. There is generally a more open feeling here, around parishes like San Giacomo dall'Orio, the Carmini, the Fondamenta della Misericordia, and Santa Maria Formosa. This is the band of the city that most attracts outsiders who come to invest in a second home, and it seems to

be that the population of the historic center of Venice, which is shrink-
ing overall, is actually growing here, or at least remaining stable. The reason is, of course, the influx of foreigners: during the year 2000, roughly
three *foresti* (local slang for *forestieri,* or foreigners) moved into the center for every two Venetians who moved out; most of these outsiders are
service workers in the tourist sector—hotels and restaurants, above all—
but a fair number are also buying and renovating houses, giving this area
its particular character, and in the process driving the price of Venetian
housing up to around the highest in Italy.[37]

The Venetians have something of a tradition of referring to these interlopers as "Milanese," presumably because many of the newly rich Italians who came to buy themselves a summer home here a generation or
two ago were from Lombardy. Nowadays, these houses are more likely
to go to non-Italians or even non-Europeans, though for many Venetians
they still all remain *foresti.* Here also is where the best remaining shops
that sell practical merchandise tend to be located. There are also Venetians living in and moving through this zone, of course—this is the area
that houses most of the main municipal and administrative agencies, after all—but the refurbishing of palaces and apartments over the last forty
years has also pushed out many of the poorer locals, who had once been
accustomed to the cheap rentals that abounded in this area. As *La Nuova*
put it:

> The zones that show the most dynamism are those of the center and of prestige:
> Dorsoduro, San Marco, Rialto, Strada Nova, while Castello, Giudecca, and Mu-
> rano show signs of stagnation. And [this] isn't an accident. These are the *sestieri*
> most appreciated by foreigners, by those who realize their dream of a second house
> on the Lagoon, and that by now are a substantial slice of those who are buying . . .
> almost fifty percent [of all purchasers].[38]

Beyond this, mostly as a fringe where the back reaches of the city meet
the waters of the Lagoon, there is an outer ring that makes up the Peripheral Zone. Here are neighborhoods composed of buildings that for
the most part have traditionally been too squalid or too new to attract either investors or tourists, and where, consequently, most of the "real"
Venetians still live. One can find this Periphery in the public housing behind the churches of Sant'Alvise and Madonna dell'Orto, in Cannaregio,
in the working-class houses around the Santa Marta and San Niccolò quarters of Dorsoduro, virtually throughout the Giudecca, on Sacca Fisola,
and in most of Castello and Sant'Elena lying beyond the Rio dell'Arsenale,
if not the Rio della Pietà. Tourists invade these areas only if they are truly

wandering blindly, and when they blunder onto islands like Sant'Elena or Sacca Fisola, they soon hurry out again, after confronting some very unenchanting twentieth-century apartment blocks, shabby little court-yards, and the same sights and smells of poverty that they could just as easily find in the outskirts of Rome or Milan (though there are some ex-cellent and—yes—truly "Venetian" restaurants on Sant'Elena). Even areas such as the Secco Marina in far Castello, with its quaintly old buildings and evidence of a traditional culture, still strike many outsiders as more closed off and resolutely disinterested in accommodating the tourist pres-ence than elsewhere in the city.

The Periphery could be called the last redoubt of traditional Venetian culture, at least if we take that to mean "popular culture." Those who live elsewhere in Venice, even if they can trace their lineage in the city back for generations, might still refer to friends or relatives living on the Giudecca or in Castello's Secco Marina as "true Venetians." This can often be a half-derisive accolade, however, for while residents of the Venetian Periphery do indeed speak the purest dialect,[39] own and operate the most private boats, and still maintain some sense of community life, they are also by and large the oldest, poorest, and least educated of the sixty-five thousand or so Venetians still left in the city. Although the modern apart-ment blocks on Sant'Elena or Sacca Fisola are comparable in their com-forts to similar working-class dwellings in Mestre, many of the dwellings on the Giudecca and far Castello are simply squalid. These are the last re-minder of the often desperate living conditions that prevailed for poor Venetians throughout the city in the 1950s and early 1960s, when critics charged that "most of Venice behind the screen of magnificent palaces on the Grand Canal [is] a rat-infested slum unfit for human habitation . . . [where] thousands of Venetians [are] condemned to live in damp, crum-bling houses just to please the tourists."[40] It is understood by most Vene-tians that the inhabitants of the Venetian Periphery are the most likely in the city to have limited education and be on some form of public assis-tance;[41] the few remaining shops that serve them are often quite primi-tive in terms of their appearance and available selection of goods. Even the *vaporetto* runs that tie such outposts as the Giudecca and Sacca Fisola to the main part of the city have been steadily reduced in recent years.[42]

In the past two decades, the social economy of the Peripheral Zone has been convoluted by the sharp expansion of the University of Venice, many of whose facilities have been deliberately scattered around in a va-riety of abandoned convents and minor palaces all over the western out-skirts of Venice, from San Giobbe in Cannaregio to San Sebastiano in

Dorsoduro. With the university has come a large and highly mobile student population—reported at over 7,750 and growing in 1995—that is willing to crowd into the less desirable apartments of the Periphery, often two or three to a bedroom, so they can pay rents that are astronomical by the standards of the highly inflexible Italian real estate market.[43] Their presence has certainly brought more life to swaths of the city that had been turning steadily more moribund, but students' ability to pay much more than the going rents has also accelerated the exodus of impoverished Venetians from this end of town.

The city's student population has also produced significant social changes in Peripheral Zone centers such as Campo Santa Margherita. As recently as 1992, one travel writer was able to term this "one of [Venice's] quietest, most restful squares, a genuinely Venetian haven away from the Australians and Japanese," but lately Santa Margherita has become the major night-time meeting place for students and young people, who come by the hundreds from all over Venice, the Veneto, and the rest of Europe to gather at the handful of large outdoor café-bars that have sprung up there.[44] This new scene, as we will see further on, has produced a continual clash between the more elderly residents of the area and noisy students who want to hang around the *campo* until nearly dawn; in time it will no doubt persuade still more locals to flee for quieter homes on the mainland.

As though driven by an irresistible centrifuge, hotels are also invading Venice's Peripheral Zone. These outlying areas may be shabby, but they can provide the more discriminating visitors with the alluring possibility of quality hideaways that are removed from the main tourist surge yet still close enough that one can reach the main attractions easily. This is best exemplified by Cipriani's luxury hotel on the Giudecca, where guests are shuttled back and forth from San Marco by private boats, simultaneously avoiding the crowds at the Piazza and the tattered nearby neighborhoods of the Giudecca. The same comfort is offered by the five-star Palazzo dei Dogi, which opened in 1998 in the outer reaches of Cannaregio, near Madonna dell'Orto. Backing as it does on the Lagoon between Venice and Murano, this hotel can provide its well-heeled guests with private and direct connections not only to both cities, but also to the Marco Polo airport on the Terraferma.[45]

As property values have reached astronomical levels in the inner two zones of the city, many foreigners and "Milanesi" have also started looking at the Peripheral Zone more favorably: together with the hoteliers, they are putting this outer district through its own gentrification crisis, as they snap up what available and (relatively) cheap housing they can find there. The waterfront of the Giudecca that faces Venice proper has seen considerable restoration in recent decades, as have the outer reaches of Cannaregio. Particularly in demand are the relatively modest palaces that the less affluent patricians of the Serenissima built in past centuries in these parts of town, which were as unfashionable three centuries ago as they were until the 1990s: now, however, these *palazin* are attracting modern patricians from Milan, Tokyo, and New York. Recent plans to establish a second waterfront along what has always been the forgotten backside of the Giudecca, facing out onto the Lagoon to the south, will no doubt intensify outside pressure on this last redoubt of tradition Venetian culture.[46]

Second homes in Venice have one thing in common with those in any beach or tourist haven: their owners expect to occupy them for only a few months out of the year. In Venice, this generally means May and September, when the weather is at its best and such attractions as the Vogalonga, the Regatta Storica, and the Biennale take place. For the rest of the year, such restored palaces, apartments, and cubbyholes once might have been rented to visiting scholars (we have known a few), but more typically they used to remain vacant—giving Venice the silent wintertime feeling that so many travel writers have mistakenly proclaimed as the moment when "the Venetians have retaken their city," when in fact the place is just as bereft as any other beach resort in the off-season, and for much the same reasons.[47] All this changed after 1999, however, when the state (not city) authorities decided to virtually abolish existing regulations over what sorts of rooms could be rented to outsiders and by whom. The rationale here was partly that Venice was (and is) perennially short of affordable hotel rooms, and that many Venetians had spare rooms and needed the money. We also suspect, however, that a great deal of pressure for liberalizing the rules came from those "Milanesi" who wished to cash in on their otherwise empty second homes. In either case, observers noted that suddenly single, rented rooms and larger bed-and-breakfasts were "popping up like mushrooms in the city." Around two thousand new beds were added to the existing fourteen thousand or so hotel beds already available in Venice (an increase of nearly 15 percent) just between 2001 and 2003, as homeowners—natives or not—rushed to convert attics, storerooms, even corridors into short-term rentals: "On the ground floors everyone is creating tiny apartments *[appartamentini]:* a bit of wallboard, a small bath, and [off] they go." Beyond the fact that such efficiencies are highly vulnerable to flooding from Venice's notorious *acque alte,* this "gold rush" has also proven effective in moving not just second-home owners and long-term renters, but also tourists themselves, to the Peripheral Zone. "In the deserted alleys in the morning," noted the papers, "one sees pop out, from places that had once been uninhabited, tourists with cameras who are searching, disoriented, for the road for San Marco." As even these last neighborhood bastions began to open up, warn some observers, "the danger is that we are taking big steps towards transforming the whole city into a service for tourists."[48]

Though they may disagree on details, most Venetians seem to accept something like Scaramuzzi's socioeconomic "map" as a fair descriptor of their city. They do not see Venice as a labyrinth at all, but rather as the place where they live, socialize, and do business. Admittedly, they sometimes fail to remember their way around town: many we talked to agreed that there might be back corners where even after years they still felt a bit uncertain as to where they were. For them, this represents a minor annoyance but hardly an occasion for wonder or enchantment, any more than a Londoner would say he was thrilled to have gotten lost in the East End. Those who live in Venice, if they need to find a particular street, do just like residents anywhere else: they pull out a map or the *TuttoCittà,* the city guide issued by the Italian phone company. If still more precision is needed—say, a specific address—they can always buy the street index that matches the meandering house numbers to the various *calle* and *campi.* The very idea of calling this, their living space, a labyrinth is, and always has been, an outsider's interpretation, one that carries with it the implication that here is a construct made deliberately complicated and perplexing, designed to thwart easy movement and communication, and intended to amaze and amuse. Such delusions have given rise to vacuities like the one made by a *New York Times* writer in 1969, when he called the city "a stage set where people are actors wandering about between the acts."[49] Very few people, of course, are likely to think that they were born and grew up on a stage set. Foreigners who do experience the city this way, or as a maze waiting to confound them, are really saying less about Venice's topography than about the fact that they experience it the wrong way round, not as it was built over the centuries, as a commercial, industrial, and social site for human use, but as a site of pleasure designed for themselves.

CHAPTER 5

Contested Ground

Recent campaigns to entice tourists into Venice's back streets have been motivated largely by the premise that it is desirable to relieve pressure on the city's overcrowded core around Piazza San Marco and the Rialto, and that this is an obvious way to do it. As it happens, such moves have not been especially successful. Most visitors are in Venice for too short a time to see much of the city anyway and are unwilling to pass up its most famous central monuments to spend their brief stay looking at some lesser-known attractions. In any case, no matter how many tourists are siphoned off from San Marco, there always seems to be plenty more to take their place and keep up the pressure. The flood of foreigners continues to increase at such a pace—from an estimated ten million in 1990 to around twelve to fourteen million at the turn of the century and a projected fifteen million by 2005—that no scheme could lure away enough of them to actually reduce their presence in the center.[1] Indeed, at times it would seem that all the efforts to get tourists out of San Marco have had only one substantial result: by now it is virtually impossible to find any place in Venice that is free of foreigners. This seems to have created a bind for those in the tourist business, as they exhort their readers and clients on the one hand to "get off the beaten track, explore the back streets, and try to meet or at least mingle with the locals," and then promptly turn around to lament that "visitors continue to plague Venice, crowding it to the point of saturation throughout the dog days of summer. Escaping from the throng, contrary to popular belief, is rarely an option. Tourists seem to come at you over bridges and around corners in every quarter of the city. The visitors and their endlessly renewable enthusiasm are the despair of some."[2]

We suspect that when Fodor's writes here of "the despair of some," the reference is not about upsetting the locals. Guidebooks that give advice on "how not to look like a tourist in Venice" generally seem most concerned that their readers not end up offending the sensibilities of other, more refined foreigners, or that they do not give tourism in general a bad name through gross behavior. Still, Venetians do have an old reputation of seldom taking offense from the *foresti,* receiving them benignly, if somewhat avariciously. It has long been a safe bet that one can get away with almost anything in a town that, like Venice, has been "reduced to earning its living as a curiosity-shop," as James put it. Even a century before James's time, Lady Ann Miller could write that the "Venetians are so accustomed to see strangers, as not to be the least surprised at their being dressed in a fashion different to themselves; nor inclined to esteem them objects of ridicule on account of their not speaking the Venetian language."[3]

Perhaps much of this famous Venetian forbearance was based on their having had a goodly part of the city to call their own, however. As we have seen, until the late 1800s visitors to Venice tended to avoid the city's twisting back alleys as much as possible, getting about from sight to sight almost exclusively by gondola. By the early years of the twentieth century, though, the arrival of packaged tours and the increasing cost of renting a gondolier for oneself had begun to put an end to this leisurely, detached way of getting about the city, with the result that areas of Venice that had long been little worlds unto themselves began to experience their first probings by curious foreigners. Something about the special pedestrian ambience of Venice, its tranquillity, may have promoted a certain pushy inquisitiveness on the part of these pioneers: as early as 1930, one *New York Times* travel writer was encouraging readers to simply make themselves at home, even in someone else's house:

Pedestrians are better off provided they know the short cuts, for there are few journeys afoot into the back alleyways of Venice which cannot be shortened by leaving the footway to pass through somebody's dark and mouse-smelling cellar, taking due care to step around the heavy jars of oil standing dimly in the twilight. To the tourist all this is part of the friendliness and intimacy of Venice.[4]

As tourists to Venice have blossomed in number and have pushed themselves ever more insistently into the city's more intimate spaces, this seemingly tranquil world of *campi, calli,* and canals has become increasingly contested territory. Historic Venice is actually quite large—about four kilometers long by two and a half kilometers wide overall—but sometimes

FIGURE 10. Tourists on the steps of Piazza San Marco. (Photo by authors.)

available space in town seems so tight that the weary tourist has to struggle just to find a place to park his rear end. Not everyone is accustomed
to the kind of walking this city demands, where almost all visitors end up
spending most of their time going about on foot, often weighed down
with day packs or cameras, meandering along the maze of stone streets
and over a tiring infinity of bridges. The humid heat that typifies the summer months can turn Venice into an exhausting experience, and tourists
are always looking for a place to sit down and relax for a minute. The city
can appear inviting in this regard: since there is no traffic, no parked cars,
no sidewalks, and none of the uneasy interplay between streets and
pedestrians that characterizes life in other cities, tired foreigners often feel
perfectly at ease simply flopping down anywhere that strikes them as convenient. Yet Venice turns out to be woefully short of public benches, and
sitting in cafés can be expensive, so tourists looking for a place to sit must
take advantage of whatever they find.[5]

This particularly means steps, and the most noted of these—everyone
gets photographed on them—are the long steps running along the south
and west sides of Piazza San Marco (see figure 10). These have the special advantage of being in the shade for much of the day, such that, by

noon on a summer's day, the entire length of the colonnade is often packed tight with tourists sitting shoulder to shoulder, eating snacks, reading guidebooks, or simply resting from the heat. Equally popular with these summer squatters is the raised area of Piazzetta dei Leoncini, north of the Basilica of San Marco, and here, too, picnickers regularly fill in every available meter of sitting space. The stepped plinths of the two columns that grace the Molo also used to be favorite tourist resting zones, but the contact of so many bodies, along with rising air pollution, so damaged the ancient carvings on their bases that these agreeable spots have been surrounded by protective transoms.[6]

Bridges, with their three or four shallow *gradini* (sloping steps) at either end, are somewhat less popular as places to relax, if only because one has to be willing to sit down right in the middle of often rather heavy foot traffic. Still, when the heat is rising, the tourist crush is taking up every other available resting place, and the squatters themselves are perhaps too young or brazen to give any special thought to the inconvenience they may cause others, the bridges too can get filled up. During peak months one can often find a dozen or more such loafers, sometimes staking out extra space with their backpacks, even in the middle of such heavily trafficked nodal bridges as those at the Frari, San Moisè, and Sant'Angelo. There, between the legs of the struggling mass of passersby, these visitors, "disciplined at home, a little less so in Italy," rest, eat, and often leave only a narrow trail for others to cross the bridge, creating human traffic jams that can stretch into the adjoining *calli* (see figure 11).[7]

In fact, almost any steps in Venice will attract foreigners sooner or later. On hot days, for example, visitors can be found sitting on the stepped boat landings, sometimes sticking their feet into the adjoining canal. The comparatively high stoop that leads up to the door of the very central San Salvador also gets its share of squatters all through the morning hours; as soon as the church closes for lunch at noon, the entrance is completely blocked (see figure 12). Even the low step to the main doors of San Giacomo di Rialto (San Giacometto) draws a few weary pilgrims, probably tempted to perch there for the shade offered by the church's ancient portico. To many Venetians such insouciant guests can be annoying, especially when they block the main entrance of a *palazzo* that is an office building or someone's home. Coping with this constant encroachment can call for some complex stratagems just to keep regular access open. One sly bureaucrat from (appropriately enough) the Water Department was recently denounced for setting up a small tube of water that dripped steadily on the steps to his office on the Fondamenta del

FIGURE 11. Tourists crossing the Ponte de le Ostreghe. (Photo by Philip Grabsky.)

Vin, keeping the stones just wet enough that tourists were discouraged from resting there.[8]

With its thirteen million or more annual visitors and a local population of only around sixty-five thousand, historic Venice has the highest ratio of tourists to locals of any city in the world.[9] Moreover, the place's excessive popularity, coupled with its gnarled topography, can lead to human traffic jams, sometimes on a massive scale. According to the local police and to those who have studied the patterns of Venetian tourism, it takes around one hundred thousand visitors coming in a single day to really bring the city to a (literal) standstill. Fortunately, such extraordinary human tidal waves remain fairly rare—there are fewer than ten of them a year, typically occurring at Eastertime, on the Regatta weekend in September, and during the peak (weekend) days of Carnival.[10] When a tourist tsunami hits town, the central areas of Venice are completely swamped. No matter how many *vaporetti* are mustered into service, all these water buses end up packed to capacity; waiting passengers have to be left be-

FIGURE 12. Tourists on the steps at San Salvador. (Photo by authors.)

hind in ever-swelling mobs, as the *vaporetti* themselves necessarily fall further and further behind schedule. People have to push their way physically through the smaller *campi* at San Bartolomeo and San Giacometo; bridges between San Marco and the Rialto are impassable; the capillary *calli* radiating out from the Piazza get completely clogged; and heaven help the visitors when these alleys and bridges are torn apart (as they so often are) for restoration.[11] This is not a static human tide, but one that has its own rhythmic ebb and flow: pouring into the city through just a few points of entry—Piazzale Roma and the station, for the most part—

the flow typically reaches its full force by about 11 A.M., peaking through the early afternoon, and then returning in the opposite direction, away from San Marco (and, to a lesser extent, Lido), toward the points of entry, by around 4 or 5 P.M.

On those holidays when large crowds are predicted, the authorities try to keep track of the traffic influx, tallying up the railway arrivals and counting the number of tour buses that have already disgorged their passengers at Piazzale Roma. When around one hundred thousand tourists have apparently arrived, the highway police then close the Ponte della Libertà to all motorized traffic, effectively banning both private drivers and tour buses from entering the city. Such efforts sound dramatic when announced in the papers, but in fact they often turn out rather like Ethelred ordering the tide to recede: when tourists want to get to Venice badly enough (to enjoy major festivals like the Redentore or Carnival, for example), they are perfectly willing to walk the entire five-kilometer length of the bridge. Moreover, anyone who wants to can usually just push his way onto one of the trains that shuttle between Mestre and Venice every few minutes.

In any case, it often takes far fewer than one hundred thousand visitors to close down Venice's Tourist Core. Since the overwhelming majority of tourists simply want to go back and forth between the station or Piazzale Roma and Piazza San Marco, and since they generally have a limited time to accomplish this, most of them stay on the main thoroughfares. These routes, as we have seen, are really nothing more than a succession of little alleys, stitched together to form the most direct connecting passageways between the station, Piazzale Roma, the Rialto, the Accademia, and San Marco. Along the way there are many pinch points, where even a big road—the Strada Nova in Cannaregio, for instance—abruptly narrows down to just two or three meters, trapping the pedestrian flow and often slowing it to a standstill. Constrictions such as these— the Salizada di San Giovanni Grisostomo, the Calle di Mezzo, and the Calle Saoneri also provide good examples—turn into bottlenecks when as few as forty thousand tourists are in the city, a figure that by some estimates is reached about one day in three.[12]

There is not much that local authorities can do about these abrupt jam-ups, beyond making one-way streets out of some of the especially overused *calli* and bridges between San Marco and the Rialto. As a result, Venetians are constantly lamenting about the "*calli* and *campielli* that explode with tourists, little traffic jams that continually form (caused, always, by groups of tourists)."[13] They can, it is true, sometimes avoid the worst of the crush by taking those back routes and shortcuts that are the heritage and prerogative of true Venetians, but to *andar sconte* has its limits, even for them. Venice, after all, is built on a series of islets, where the maze of *calli* and bridges provide only a limited number of ways to get from one place to another. Sooner or later, Venetians, just like the tourists, end up having to drop back on to the main thoroughfares to get where they need to go, and in such places one often sees local businessmen or students trying desperately to wriggle their way through the stalled crowds. A journalist of our acquaintance, who worked in offices on an extremely busy *salizada* near San Lio, once talked to us with detachment and even humor about the tourist crowds in Venice. But when the conversation turned to how he himself found getting around the city, he suddenly lost his composure: "Look at these crowds right in front of my office," he practically shouted. "Sometimes they are so thick out there [as indeed they were just then] that you have to wait in the doorway for five minutes before you can even step into the street!"

The great tide of visitors generally begins its ebb flow in the midafternoon, and it is running full force by around 5 P.M. on a typical summer's day. At this time, the city's masses of day-trippers—known to the locals as *pendolari* or *gitanti*—are slogging their way back to the station or Piazzale Roma to rejoin their tour bus, catch a train for their next town, or simply leave for the night to set up in a campsite, beach, or hotel on the Terraferma. Sometimes this evening crush is even worse than that of the morning: the routes leading back are often jammed with plodding, tired-out visitors, and the *vaporetti*

are like scenes from the Far West, with punching, shoving, and insults between tourists and residents. To get off, even some kicks are occasionally necessary to dislodge tourists who don't know even a word of Italian and who look on stupefied at the anger of the "indigenes." Only the shouts of the pilots and crew manage to discourage the herd of foreigners, who—as usual—would place themselves with suitcases and backpacks in front of the entrance to the *vaporetto* cabin, creating the traditional human plug that makes the [Venetian] passengers trapped inside fly into a rage.[14]

Eventually almost all of the thirty or forty thousand visitors who come to Venice on an average summer's day find their way out of the city. Left behind are those spending the night in one of the historic center's two hundred hotels (with about twelve thousand beds) or in one of its many new bed-and-breakfasts.[15] There are also a certain number of those the Italians call *saccopelisti,* or "sleeping-baggers," who have long treated quiet, pedestrian Venice as an ideal place for simply dossing down on the pavement for the night, avoiding the cost of a hotel or hostel, and perhaps making some new friends in the process. Though they still turn up in the summer, especially after an evening festival, the era of the *saccopelisti* really had its heyday in the 1970s and 1980s, when Venice was known on the youth circuit worldwide as an ideal place for free lodgings. Back then, we can recall, great crowds of these sleeping-baggers filled the vaults of the Procuratie Vecchie in San Marco and the wide piazza in front of the train station.[16] Other youths have been tempted by the apparent emptiness of the city in August (Venetians, like many Italians, tend to leave their hometown for much of the month) to seek more secluded lodgings—as did the five French youths, for example, surprised in 1980 by a night watchman as they sat around in a private *campiello,* cooking their dinner over a campfire they had started there.[17]

The chorus of Venetians who find their city unlivable simply because it is impassable is growing. Whereas once, not so many years ago, editorialists and those who wrote letters to the press were more likely to wax poetic about rowing on the open Lagoon or the quixotic problems tossed up by their unique city, now they seem to vie with one another for better ways to excoriate the unending tourist crush. A recent example from a letter to the editor of *Nuova Venezia* started off:

How many times, trying hopelessly to open a breach in the impenetrable wall of perspiring tourists that by now clog up even the most remote alleys of Venice, have we dreamed of having available a rapid and efficacious system for getting them out of the way? How many times, shoved at the *vaporetto* landing by herds in halter tops and bathing shorts with armoires [i.e., oversized backpacks] on their shoulders, have we fantasized about being able to throw all of them into the canal?[18]

As if the thoroughfares of Venice were not already crowded enough by the tourists themselves, they are further obstructed by scores of street hawkers seeking to fish their own share of profit from the continuous stream of visitors. Many Venetians refer to these vendors as *vu' cumprà*— a somewhat rude term deriving from the pidgin Italian they supposedly use when they call out, "You want to buy?" to passing pedestrians. Sidewalk vendors from Africa and Asia are to be found all over the European Union; they are especially common in Italy's other major tourist centers, such as Florence, Rome, and Naples, where they offer foreigners and locals such trinkets as carved African statuettes, fashion accessories, knives, and various toys. Especially great numbers of them are attracted to Venice, which offers both a high density of tourists and an interactive pedestrian environment that is highly conducive to their business. Unfortunately, public space is also in short supply in Venice, a situation that has created considerable tension between these outsiders and the many others who wish to make their own use of the city.[19]

The *vu' cumprà* have become one of the standard sights of Venice in recent years. They are almost all men and the majority of them are West African—from Senegal, Ivory Coast, Nigeria, or Ghana—but there are also contingents from China, Bangladesh, Sri Lanka, the Maghreb, and Eastern Europe: the politically correct term for them is the rather unwieldy *venditori ambulanti abusivi extracomunitari,* or "unlicensed, itinerant, non-EU vendors."[20] They usually display their wares on a sheet, which they spread out on the pavement; on certain streets and *fondamente* one sees long lines of them, side by side and sheet by sheet. Generally, each sheet is tended by two sellers, while for every half dozen or so sheets there are also two or more lookouts, stationed at either end of the street. These are there to keep an eye open for the police, because essentially everything about the *vu' cumprà* is illegal. Not only are (virtually) all of them selling without a vendor's license or paying for the right to sell in public space, but most are in Italy without a valid visa or sometimes any papers at all.[21] Moreover, the goods that they sell are usually illegal as well. These include, for example, knives of a forbidden length or switchblades, which are the particular specialty of Chinese vendors. It is especially the Africans, however, who are the most notorious sellers of illicit goods, offering fake designer purses, handbags, belts, and other accessories, as well as designer T-shirts, all of which they sell at a fraction of the price demanded in the exclusive shops that are often right behind them.[22] Consequently, when the police turn up, the lookouts give their warning (usually with a cell phone) and there is a general rout (a *fuggi fuggi,* as the papers call it), as everyone gathers up his goods with lightning speed, stuffs them into his duffel, and dashes off every which way. In the process, they can bowl over the less agile or alert tourists, although some visitors still consider these chaotic moments as sufficiently amusing to make them worth a photograph.[23]

In recent years the increasing activities of the *vu' cumprà* have provoked a good deal of soul searching among Venetians. Many locals take them to be unfortunates of the Third World who deserve a chance to get ahead in the wealthy European world, even if only through selling trinkets on the street.[24] Not all Venetians have much sympathy for the high-fashion shops that sell the genuine handbags and belts and that these immigrants are supposedly defrauding, perhaps because Prada, Versace, and Fendi are

FIGURE 13. *Vu' cumprà* and San Giorgio Maggiore. (Photo by Ana Maria Isaías Nunes de Almeida.)

just as much interlopers in the city as are the Africans standing outside their doors.[25] On the other hand, as with so much else about Venice, the situation clearly has gotten out of control. Once it became obvious to the *vu' cumprà* that this sort of work could pay off, they flooded into town, making it almost impossible for anyone to regulate them or limit their presence to what even the most sympathetic locals would consider a reasonable level. During the high season there are now over five hundred of them on Venice's streets, according to recent police counts, and they tend to concentrate their activities in the city's Tourist Core, which can make them seem even more ubiquitous than they actually are. Their preferred turf covers three major zones: "from the Riva degli Schiavoni to Piazza San Marco, from Strada Nova to the Lista di Spagna, from via XXII Marzo to Campo Santo Stefano" (see figure 13). When the police try to crack down on their activities in these areas, however, they can easily move elsewhere, and groups of them may frequently be found near the Ponte dell'Accademia, in the capillary zone between San Marco and the Rialto, and all around the station and Piazzale Roma.[26]

Despite their apparent fluidity about the city, the *vu' cumprà* seem to have their own internal structure. Each ethnic and racial subgroup sticks to itself and to its own types of merchandise. Furthermore, each nationality has apportioned out the city and its primary vending areas, such that one contingent of Senegalese, for example, is careful not to intrude on the zones reserved for another, and those who arrived in earlier, better years ("in times not yet suspicious") have a claim on the best spots around San Marco and the Riva degli Schiavoni. Individuals who attempt to set up on their own, assuming they could even find a wholesaler to keep them supplied, run the risk of reprisals from those whose territory they invade. The larger Venetian society is not especially aware of the intricacies of territory and hierarchy that have sprung up to make this profitable business operate, except on those (fairly rare) occasions when someone is publicly beaten up for having stepped out of line.[27]

As the *vu' cumprà* have become more numerous and their competition for both space and customers has become more acute, they have also turned more aggressive, at least according to many Venetians and to the newspapers. They have taken to inching their sheets ever closer to the center of main thoroughfares like the Strada Nova or via XXII Marzo, supposedly to slow down the tourists and get their attention.[28] This, by all accounts, they have undoubtedly accomplished, such that on peak days tourists and Venetians find they have been left no space in which to walk, or, rather, "one walks by doing a slalom between the sheets." Woe to anyone who, out of desperation or clumsiness, steps on one of the white sheets or kicks up against the goods on display: *vu' cumprà* have been said to turn on such offenders and attack them verbally and sometimes even physically.[29]

Venice's problems in controlling the *vu' cumprà* are at least partly due to the Italian legal system, which has great difficulty in dealing with undocumented aliens. Unless vendors can be caught in the act of actually fobbing off a counterfeit purse as genuine, they can be brought in only for the minor crime of selling without a license. For such charges, anyone who is unwilling or unable to show the authorities proper documentation (identity card, visa, or work permit) must be released, with an appointment to come back and show the judge his papers. At this point, however, the *vu' cumprà* simply walks out the door and disappears, only to turn up the next day, or even the same afternoon, selling on the street again.[30] "It's like trying to empty the sea with a teaspoon," say the police—

of whom there are surprisingly few in Venice, considering the masses of tourists who flood the place. The authorities rarely have the personnel available even to chase the vendors away from their spots, much less to successfully capture them or confiscate their goods.[31] Moreover, as the vendors become more numerous, there are increasingly frequent stand-offs between them and the police, tense situations in which a patrol of two or three officers (many of whom are unarmed) find themselves sur-rounded by several dozen angry *vu' cumprà* demanding to have their confiscated goods back and to be left in peace.[32]

Still, the real cause of the problem is the tourists. Despite the some-times ludicrous difference between the prices these vendors ask for their goods and those posted in the designer shops, many visitors persist in be-lieving that the *vu' cumprà* are offering real, not counterfeit, products; that they charge less because of their lower overhead; and that, despite all ap-pearances, they are actually working for the designer shops, providing their lucky customers with special locations for "auxiliary sales." Many of the vendors are quite willing to further these delusions, to the extent of pack-aging their sales in counterfeit versions of the same cloth bags that the fine shops use.[33] Other visitors, by contrast, are fully aware that they are buying frauds and are happy to do so, considering that it is possible to buy a handbag for twenty or thirty dollars from a *vu' cumprà* that looks essentially the same as one that would fetch over three hundred in a shop. Whatever their understanding of what they are actually buying, though, tourists are enthusiastic supporters of the *vu' cumprà*. As they wander by in their millions they insure that the business is profitable enough to en-tice ever more vendors to run the risks posed by the police.

Needless to say, those who run the designer shops are furious about this illicit competition, often taking place right in front of their own (high-rent) store windows. On peak afternoons the police switchboards are flooded by calls from merchants complaining that the *vu' cumprà* are not only selling what looks like the same product for a fraction of the price, but that they are even blocking the doors to their shops. As one shop-keeper on the Strada Nova recently moaned: "My store is always sur-rounded by their purses, and by now the customers don't come by any-more. I asked these illegals if they would please move themselves a little further away, and [the only answer] I got from them was to go to that country [i.e., where the sun doesn't shine]."[34] The *vu' cumprà* are not only jostling for space with shopkeepers and tourists in this city, however. As their numbers have flourished, they have also begun to clash with the le-gitimate local street merchants who are also trying to exploit the tourist flow. These vendors have a long if somewhat checkered tradition in Venice. Newspapers in the nineteenth century reported their occasional run-ins with the law when they tried too aggressively to push their trinkets on passing tourists: decorative "shellwork" was evidently popular, as were velvet slippers and collections of souvenir photographs.[35] During the Serenissima, such activities were also widespread, as beautifully attested by Gaetano Zompini's *Le arti che vanno per via,* an exhaustive survey in etching form of all the street crafts, goods, and services offered in mid-eighteenth-century Venice. Indeed, as early as the 1480s, it would appear that trinket sellers in Venice were busy hawking beads and rosaries to pilgrims, as souvenirs and for good luck.[36]

These days such vendors sell mostly souvenir trinkets mass produced in the Third World. They buy their licenses from the city and are kept by law out of Piazza San Marco, so for the most part they gather on the Riva degli Schiavoni, whence they roll their mobile, expandable kiosks to spend a dozen or so hours every day during the tourist months. Here, until a few years ago, they shared what seemed like an ample space (the Riva is in some places a good forty meters wide) with another category of street merchants—the local painters. Of these there are several dozen, some of whom are to be found down on the other side of the Piazzetta, near the Giardinetti Reali. Both painters and trinket sellers have spaces assigned to them by the city, and if they sometimes (some say frequently) exceed the boundaries specified in their licenses, they are at least operating within the limits of the legal system.[37]

This has all changed since the *vu' cumprà* began showing up in increasing numbers, back in the mid-1990s. Seeking to get as close as possible to the tourist flux, they often set themselves up right in the middle of the Riva; according to the more critical Venetians, the densely packed *vu' cumprà* have turned this area, which should be Venice's monumental front door, into "Piazza Grand Bazaar."[38] They have also put themselves virtually on top of the easels of the painters, who now complain that "'by this time we are in a daily battle to win the place we have paid for.'" In their complaints the painters are backed up by the souvenir sellers, as well as by the gondoliers of the nearby Danieli landing, who claim that the *vu' cumprà* have effectively cut off tourist access to their loading dock. Over recent years the animosity (and not a little racism) has escalated into an all-out, daily turf war, with a great deal of posturing, threats, and shouting, punctuated by occasional outbreaks of near or actual violence.[39]

As the various groups jockey for position on the Riva and on the adjacent bridges, their conflicts and arguments have steadily heated up.

Nowadays visitors have to splash their way across puddles as they walk around the area, since the painters have taken to pouring buckets of canal water around where they work, in the hope of keeping the *vu' cumprà* from spreading out their sheets nearby.[40] The adjoining Ponte della Paglia, which is a prime spot for tourists to pause and film the Bridge of Sighs, has in consequence also become a choice spot for trinket sellers; the site is so desirable and so hotly contested that it has recently turned into something of a war zone. The *vu' cumprà* fight among themselves to establish a spot there; once they set up, the licensed vendors, gondoliers, and taxi drivers do whatever they can to drive them away.[41] It all sounds like too many overheated young men with too much time on their hands, but it is also an indication of how valuable a piece of Venetian public space can be, even when it is only the size of a bedsheet. Just how valuable can be inferred, if not from sales figures (these rarely come to light), then at least from the enormous amount of goods the police say they confiscate and destroy. In May 2000, the authorities complained that the five warehouses they had for storing such goods were "exploding"; ten months later the papers noted that the police had just sent two thousand kilos of counterfeit wares "to the pyre"—just a fraction of the "something like ten tonnes of fake purses and leather goods" they would destroy in the first half of 2001.[42]

But the *vu' cumprà* have kept coming back, as tenaciously as ever, and any business that can sustain such losses and keep going must be able to realize immense profits. With potential customers drifting by in their millions, one evidently can make money selling almost any sort of junk in Venice, and it came as small surprise when investigations (paid for in part by Louis Vuitton, as it happened) pointed to connections between the *vu' cumprà* and the Italian underworld. It does indeed seem likely that the Neapolitan mob, the *camorra,* has—or is trying to establish—a hand in every stage of this business, from manufacturing the purses and handbags (in secret factories in Tuscany and Campania) to supplying the vendors, assigning them their territories, and (inevitably) collecting a healthy cut of their take.[43]

Thick on the ground as they often are, the *vu' cumprà,* the painters of the Riva, and all the sundry other street peddlers here still represent only a small, if highly visible, facet of local retail activity. By ancient tradition Venice has always been a highly commercial city, and the hundreds of storefronts around the Piazza, in the Mercerie, and along the main *salizade* testify to the place's mercantile roots. Indeed, half a millennium ago the showrooms and stalls filled with fine fabrics, silks, and spices were a tourist draw in their own right. Even after Venice lost its political freedom and entered its long period of decline, the city at least continued as a provincial center, supplying its residents and those from the adjoining hinterland with virtually everything they needed. Besides the clutch of strictly neighborhood shops—the butcher, baker, barber, pharmacist, stationer-bookshop-tobacconist—there was also a sufficiency of specialty shops scattered around town, of the sort especially useful for residents: plumbers, ironmongers, clothiers, furniture and appliance sellers, and so on. In the past thirty years or so, these prosaic sorts of shops have, one by one, been replaced in Venice by stores that specifically cater to the tourist market, offering goods ranging from the unspeakably tacky to the highest-end designer wares. Clearly, the withering of Venice's traditional retail sector has been closely tied to its overall demographic decline, which ran at around 2 percent per year from the 1970s to the 1990s: upward of fifty butcher shops, for example, closed down between 1970 and 1982, while between 1990 and 2001 the number of bakeries diminished from fifty to thirty-five.[44] At the same time, shopkeepers of all kinds have been increasingly plagued by the difficulty of finding and retaining help, whether sales clerks or apprentices, a factor that may have contributed to the closure of around sixteen hundred artisanal shops in the city since 1984.[45]

Nevertheless, it is also fairly obvious that a goodly percentage of traditional shops in Venice have been closed simply in order to convert their available space to some more profitable use in the tourist sector. Shopkeepers who are doing well can still find themselves evicted when property owners scent the possibility of getting a higher rent by signing a new lease with new merchants interested in catering primarily to mass tourism.[46] Over the years, the specialties of the tourist shops have come in cycles. In the 1960s or 1970s, a foreigner might complain, "All these little shops with the glassware souvenirs make me feel I'm in a trap, a tourist trap."[47] After the Carnival made its comeback in 1979, however, there was a rapid expansion of stores selling masks: whole streets, such as the Fondamenta dell'Osmarin, were suddenly shoulder to shoulder with mask shops.[48] More recently, as the mask market has reached (and probably long surpassed) saturation, one sees an increasing number of fast-food outlets. By the end of the century, Venice could claim the dubious distinction of having more pizzerias than Naples and the highest density of ice-cream shops—virtually all take-away—of any city in Europe.[49]

Interestingly, with the decline of local retail activity, there also is the fading away of one of the distinguishing characteristics of Venetianness, or *venezianità*. This is (or was) the ability of residents to not only know their way around the maze of city streets and canals, but also to know where to find whatever they wanted to buy, in all the different shops scattered about town. As more and more of these stores close down, and everyone, regardless of whether they have lived in the city for a week or for generations, is reduced to taking the shuttle bus to one of the vast "hypermarkets" on the mainland, such arcane knowledge is rapidly becoming pointless. Except in one rather sad context: Venetians now give directions to one another in terms of a retail landscape constructed from memory, as in "Turn at the corner where Salvatini's used to be."

As the storefronts in Venice have been converted to serve tourists, the city's *campi* have undergone a similar conversion. These large and small piazzas served Venetians for centuries as stage settings for their own brand of sociability and neighborliness. During the time of the Serenissima, and indeed until just a few decades ago, the public space of the *campi* was the focus of local community life. This is where fruit, vegetable, and fish vendors set up their stalls during the week and where celebrations were staged on festival days. The home *campo,* rather than Piazza San Marco, was always where Venetians preferred to meet for gossip and play: for groups of mothers to mind their infants, for children to play soccer or ride bikes, for teenagers to hold some variety of a low-rent neighborhood *passeggiata,* and (in the more distant past) for elderly men to play *bocce.* Ranged around the open space, bracketing the parish church, was typically an array of patrician palaces, an indication of the involvement of the gentry in neighborhood affairs; in the ground floors of these buildings were the local taverns, shops, and services that catered to the community. It was, as some social scientists and historians have idealized it, a self-sufficient little world, where one could find almost all the necessities of life, from the material to the social to the spiritual, right at hand.[50]

In the last generation or so, the low-key social dynamics of many *campi* have been overwhelmed by mass tourism, and specifically by the spatial imperatives related to the universal, if seemingly innocuous, tourist activity of dining out. An enormous amount of Venetian commercial space is devoted to restaurants, which is itself somewhat surprising, considering the generally low reputation that Venetian cooking has among those who are fond of Italian cuisine. More than two hundred years ago visitors were already complaining that "the Venetians are wretched cooks," but the real problems with cooking in this town have far more to do with the corrosive effects of mass tourism on taste and service than with anything innate to the Venetians or the dishes they prepare.[51] We have seen many restaurants in Venice that are equipped with large dining areas specially set up to handle tour groups of fifty or sixty, working off a set menu in the fashion of a college dormitory. The results are about what one would expect, but it is also true that many restaurateurs there offer what they think tourists expect to find, which is to say pizza and the ubiquitous spaghetti with clam sauce—both of which originated in southern Italy and are not prepared with much expertise this far north.

Though it is often an indifferent and expensive experience, dining in Venice remains hugely popular among tourists, at least in part for the opportunity it offers them to eat outdoors, *al fresco*. By setting up a portion of their tables under umbrellas out on the neighboring *campo* or along a nearby *fondamenta,* restaurateurs are able to incorporate one of Venice's prime attractions—its antimodern, traffic-free environment—into the ambience of their own dining rooms. This would appear to be a relatively recent development: though neighborhood cafés have long imitated Florian's and Quadri in Piazza San Marco, in setting up tables and chairs on their local *campo,* until the 1970s there was no mention in the tourist literature of restaurants providing this sort of service. In recent years, however, outdoor dining has become the custom even in the most remote corners of the city, such that a walk around Venice during lunchtime has begun to feel distinctly like a stroll through someone else's never-ending dining room.[52]

During the past few decades Venice's *campi* have witnessed a slow erosion of both retail and public space, as stores facing on the piazzas have been converted to restaurants or bars and the *campi* themselves have been colonized for outdoor dining. Many of these establishments, whose interior space actually may be quite small, exist almost entirely outdoors: they are just big enough for a classic stand-up Italian bar and a kitchen, both geared to serve the hundred square meters or so of shaded tables arranged before the front door. This public space, which all the bar-restaurants, from Florian's down to the humblest neighborhood pub, have to rent from the city, is known as the *plateatico* and is the most precious resource of many Venetian restaurants.[53] How big any given establishment's *plateatico* may be—if indeed there may be one at all—is determined by the city government. The rate charged depends on the size allowed and the location: the most expensive, not surprisingly, are those of the grand cafés like Florian's, Quadri, and Lavena, which may run to several hundred thousand dollars per year.[54]

To an extent, the city listens to neighborhoods that want to limit the number of restaurants allowed to set up their linen-draped tables and

FIGURE 14. *Plateatici* at Campo Santa Maria Formosa. (Photo by authors.)

big *ombrelloni* (sun-shades) in the local *campo*. There are, indeed, still a few big or medium-sized *campi* out in back reaches of the city that remain innocent of such embellishments, beyond a few rough tables and chairs right outside the door of the parish pub: at San Francesco della Vigna, for example, at San Pietro in Castello, or San Giacomo dell'Orio. When it comes to the more central *campi,* however, most of them, large or small, have had upward of a quarter of their surfaces converted from public space to *plateatici* for outdoor restaurants: one need only stroll around Santa Maria Formosa, Sant'Angelo, and Santa Margherita to see the results (see figure 14).[55] Elsewhere restaurants have been quite aggressive in taking over a big piece of the broader *fondamente* that run alongside neighborhood canals. The *fondamente* Tre Ponti and Maenad, for example, have been cut back to barely half their original width, often with semipermanent wooden booths that stretch for long distances along the canal side of the quays.

When a *plateatico* is established in this fashion—on the far side of the walkway from the restaurant itself—the result is not only greatly constricted space for pedestrians, but also a great number of collisions, as waiters try to dash back and forth from kitchen to tables through the often considerable press of the crowd. Obviously this is less of a danger in the big *campi,* where, despite the broad expanse of many *plateatici*, there is always a walkway in mid-piazza wide enough to accommodate pedestrians. There is often another sort of spatial conflict in these *campi*, however. To keep dining tourists comfortable in their outdoor ambience, it has been necessary to ban local children from playing—especially from kicking soccer balls around, but essentially from any sorts of games that make noise; in 2003, police even started taking names "of the more lively," threatening them (or their parents) with a fine if they were caught again. Admittedly, this has not bothered all the locals: sometimes older Venetians, "forgetting that, in their turn, they had also been children in a city that offers laughably [small] spaces and equipment for play," have been known to complain to the police about "a treacherous [soccer] ball" or the near miss of a bicycle or skateboard.[56] Still, it is unquestionably true that the brightly colored *ombrelloni* and white linen, though they might make a *campo* seem much more cheerful and inviting to foreigners, also have a chilling effect on local sociability as it has been traditionally expressed in Venice: out in the open, in the face-to-face world of the public piazzas.

There is not much pretense that such outdoor establishments, though they may take up Venetian public space, are really there for the sake of the Venetians themselves. Venetians are not Romans and, though they like café and restaurant socializing, they are not so fond of dining *al fresco*. We have been to many a restaurant in the city—including those famous for their "authentic Venetian cuisine"—and found precious few actual Venetians dining there: but the presence of Venetians drops to near zero when one examines the outdoor *plateatici*. This is admittedly a matter of local taste, but it can also be by design, as one Murano businessman found to his embarrassment in May 2000:

I am the director at a Murano glass-making company, and [one] day, together with a client from Milano, I went to lunch at [a] restaurant . . . in Murano, a place where I go rather frequently for a lunch break. After some minutes, seated at [one of] the outside tables, on the quay, and waiting to have [someone] come and take our orders, we were approached by a waiter who informed us that the tables outside were reserved for tourists, and that therefore, if we wanted to eat, we had to seat ourselves inside. I was left astonished by that assertion, without considering the terrible figure made for the client by my company. . . . To me, it's shameful that the local commercial system should continue to act with arrogance and boorishness not only with regard to tourists, but also, as in my case, to those who live or work in the city.[57]

There is also the constant problem in the *campi* that many if not most restaurants are constantly and stealthily seeking to expand the *plateatico* granted them by the city. The permitted extent of the area (and thus the number of tables that can be set up outdoors) is based on how much the adjoining establishment pays and how much the government decides the *campo* itself will sustain. The police come around only every year or so to check that each restaurant is keeping within its allotted space, however, unless neighboring shops complain that a given *plateatico* has grown to the point that it is interfering with their own customers' access. As a result, if business is good (and the demand is always there for romantic, *al fresco* dining spots), there is generally the temptation to expand bit by bit. Indeed, the nature of the punishment itself has encouraged restaurateurs to stretch out their *plateatici* as much as possible, since the police fine offenders at the rate of around $20 per square meter over the area allotted, up to a maximum of about $450. So, as one observer put it: "Whether you 'cheat' by twenty or by a hundred square meters, the most you can pay is still [the same]." As a result, restaurants have taken to considering the fines simply part of the cost of doing business—a cost well worth enduring, considering the extra trade the larger space brings in; police now regularly come across grossly oversized *plateatici,* some of them three times as large as what the establishment is officially allowed.[58]

Some night spots also encroach on public space in another way. In certain parts of town—around San Marco, along the long canals of Cannaregio, and especially in Campo Santa Margherita in Dorsoduro—a night culture has sprung up in recent years. Initially, the local papers and travel writers wrote glowingly of this awakening of what many had considered one of the deadest cities after dark in all of Italy. Santa Margherita, for example, was praised as "one of the few *campi* in Venice [that] is still awake in the middle of the night, especially in summertime, when there are always miniconcerts and pubs where one can sit outdoors." Indeed, the *campo* has undergone something of a transformation since 1990 or so, with restaurants, but particularly bars, now filling up large swaths of the available open space with their *plateatici*. Although the *campo* is ideally located to form a focal area for university students, many of whose faculties are located nearby, Santa Margherita has become a magnet for young people from all over historic Venice, the islands, and the Terraferma, who flock there to spend hours every evening just hanging out together.[59]

Of course, as with so many things here, anything the Venetians initiate will rapidly attract the attention of foreigners, who promptly start showing up to enjoy whatever new amusements this unique city might have to offer. And, as always, these foreigners, by sheer weight of their numbers, soon change the character of the whole experience. In particular, it seems to be a question of drinking. When Santa Margherita was largely the venue of university students and locals, drinking was fairly subdued: we recall sitting at an outdoor table surrounded by dozens of Italian youths just sipping soft drinks, coffee, and *granate;* those who had alcohol nursed their drinks along, sometimes for hours. With the arrival of masses of foreigners, however, drinking and drunkenness have increased considerably, and with them the noise level has shot up, as has the frequency of fights and arguments. It is now not so uncommon that the police have to be called in to placate some rowdy group, though the presence of the authorities can itself spark near-riots, whether during the Carnival or on warm summer nights.[60]

Such youthful brawls do not themselves impinge on the Santa Margherita ambience as much as other aspects of this new youth culture. By the late 1990s, nearby residents had begun complaining regularly about bars and clubs playing well into the night: "Music and high spirits, chatter until a late hour, laughter in the middle of the night." By the spring of 2001, the *campo* had no fewer than six bars with outdoor facilities, plus several other on nearby *calli* or *fondamente,* including the rather ominously named Round Midnight, a club that was regularly denounced in the neighborhood for blasting the area with high-decibel music until three or four in the morning.[61] Soon local residents had begun to collect petitions of protest, some with three hundred or more signatures, sending them off to their neighborhood assembly, to the city government, and even at one point to the president of Italy, complaining that a night's sleep had become a distant memory: "There are people here that play soccer [in the *campo*] late at night, and the racket goes on until four in the morning." "By now we are at the saturation point . . . there are drums being played in the middle of the night, [and] acts of vandalism." "All in all, more than a *campo,* [it seems] a pit of Dante's [Hell]."[62]

Caught, in this as in many other contestations over Venice's space, between ever-increasing hordes of tourists on the one hand and an apparently helpless government on the other, many Venetians have privately begun to express hostility toward all their city's visitors, insouciant or not. Admittedly, open anger or rudeness toward tourists is still fairly rare, but this may be mostly because those who could not stand the steadily worsening situation have already left. Those who have stayed behind appear to be better at adapting to psychological strains that come with sharing their town with so many millions of foreigners, though many still break under the constant stress and flee to the Terraferma; the older ones who have to stay have little recourse beyond signing an endless series of petitions and protests or writing to the papers.

Many of those who have remained behind have learned to safeguard their sanity in denial. One can see them, as they move together about the *calli* crammed with tourists and *vu' cumprà* or across the *campi* obstructed by enormous *plateatici,* talking right through the strangers all around them and creating a private space of language and recognition that they can share with their friends. This ability to ignore outsiders has become a high art with many older Venetians, who have learned to go about their lives as if the constant flux of tourists around them had no more personal meaning than the weather, the water, or the pigeons in Piazza San Marco. Inevitably, tourists (or at least those who notice anything at all about the Venetians) can detect a certain coldness about many residents of the town, something that the *Gazzettino* had already observed more than twenty years ago: "If once the cordiality and the sympathy of the Venetian were proverbial, now the opposite is true. It seems that the 'real' citizen is becoming scurrilous in his physiological character. He pushes at the *[vaporetto]* piers; when asked for information, he responds 'go wander around somewhere else'; he grumbles at the tie-ups [caused by] impromptu picnics along the street."[63]

Not surprisingly, though unfortunately, it is those whose jobs bring them into the closest, daily contact with tourists—ticket sellers at the main *vaporetto* stops, fast-food or souvenir sellers, waiters, even many gondoliers—who have become the rudest in dealing with them. Mostly these service workers assume (and not without reason) that no one is going to understand them, leaving them free to trash-talk visitors openly, to their faces or between themselves, as if they are standing in front of animals at the zoo. This sort of rudeness is by no means new to Venice—Grand Tourists suspected it and complained about it well over two hundred years ago—and it is unquestionably one of the best benefits of knowing a dialect. It is a handy psychological release valve that may not exist much longer, however, since the Venetian dialect has, in the last generation or so, faded into little more than a strong accent that many Italians can understand.[64] We have seen the results: rude gondoliers or ticket sellers who were understood by an Italian or even a Venetian whom they failed to recognize, followed by a lengthy shouting match, with accusations of *infame!* and *maleducato!* flying back and forth between the insulter and the one who was not supposed to understand the insult.

The general atmosphere of mistrust that such behavior has tended to generate has helped give Venice the reputation of being not only the most expensive tourist city in Italy (which it objectively is), but also the place where one is the most likely to be cheated. There are frequent complaints among tourists about the way they are "plucked like chickens" in Venetian restaurants, especially those close to Piazza San Marco. There is some truth to this: we have occasionally found mysterious (and sloppily written) charges on our bills and have received (already opened) bottles of wine—or better yet, spumante—we never ordered. Still, most Venetian restaurants are more expensive than those elsewhere in Italy only because it costs more to bring in supplies.[65] If waiters and restaurateurs often seem rather brusque or even downright rude, the continuous tide of often blundering and occasionally arrogant tourist-customers at least makes such behavior explicable.

The really notorious offenders when it comes to cheating foreigners in Venice are the taxi operators. Despite the requirement that these boats have the rates clearly posted, many tourists still find themselves reduced "to a haggling worthy of the souk" or, indeed, being forced "to accept any price at all, without being able to check [it]." Thus, though it officially costs about forty dollars for a run from the airport to San Marco, "by now a taxi to Venice [from the airport] costs, in fact, more than a flight to London."[66] The papers regularly run stories about tourists charged hundreds of dollars for a simple run along the Grand Canal—or even just from San Marco to the Rialto!—by taxi pilots who can be caught only if their victims have had enough foresight to photograph the boat to get its registration numbers.[67]

Many tourists have told us that they experienced what they thought was Venetian contempt, but they perceived it not as personally directed rudeness but as a studied neglect demonstrated by Venice toward what is clearly its most precious economic resource: the tourist. There is indeed a sense among Venetians that, even without the slightest effort on their part, mobs of foreigners will keep coming anyway, so the town has been slow to insure the comfort and convenience of its visitors. Tourists are forced to sit in all sorts of inappropriate (and uncomfortable) places because the authorities, for reasons of thrift, aesthetics, or indifference, have never bothered to install benches in places where visitors are most likely to congregate: San Marco, the Rialto, or even along most of the Riva degli Schiavoni.

Local government in Venice has also been notoriously slow in providing visitors with enough toilet facilities. Northern Europeans themselves for time out of mind have held the view that Italian cities are woefully undersupplied with toilets, and those that exist are often either filthy or closed. But Venice, with its very large numbers of walking visitors, often separated for hours at a time from their cars, tour buses, or trains, has struck many as exceptional in this regard even by Italian standards. For decades, the entire tourist flux in and around Piazza San Marco—easily ten to twenty thousand foreigners a day—was served by just four pay toilets, tucked away in the Giardini "Ex-Reali," right around the corner from the Marciana; just one public toilet was available at Piazzale Roma.[68] During the hours when the San Marco facilities were supposed to be open (from 8 A.M. to 8 P.M., May through September), there were always lines of a half hour or so, and desperate visitors often found toilets locked up when the custodians decided to go home early.

We have noticed a great deal of visceral hostility against Venice on the toilet issue, a problem that is only slowly being resolved in the city. For example, thanks in part to (mistaken) anticipation of a heightened demand for the Holy Year of 2000, high-tech pay toilets were installed in various parts of the city—not just around San Marco, in other words. These have proved useful, at least if one is to believe the complaints, from those who live near such facilities, that they are being used at all hours. Nevertheless, many tourists still feel abused by being in a situation where the public facilities are often closed and all they can find are "For Customers Only" signs on the restrooms of restaurants anywhere around the city center. An Australian once joked to us that "these [Venetians] won't even give us a pot to piss in," whereas, by contrast, the locals seem to know perfectly well where to find decent bathrooms any place in the city they happen to be. This knowledge, of facilities so tantalizingly close at hand, perhaps represents for many tourists the ultimate Venetian backstage, from which they are perennially and sometimes painfully excluded. Not surprisingly, more than a few visitors take matters (and a form of revenge) into their own hands—as it were—with the result that in the summer both public gardens and the more out-of-the-way alleys and *sottoporteghi* often reek like open latrines: more public space, one might say, that now carries another of the distinctive marks of tourism.[69]

PART THREE

Seascape

CHAPTER 6

The Floating Signifier

Tourists come to Venice because, above all, the city is built on water—and sometimes they are surprised at just how much water they see. What has by now become the classic entrance into Venice is all water: a drive or a train ride takes one across the Lagoon on the Ponte della Libertà and then the Linea 1 *vaporetto* completes the trip, down the length of the Grand Canal to Piazza San Marco. Passengers on the Linea 1 marvel, as tourists to Venice have done for centuries, at how the buildings of the city "go right down into the water." They also aim their cameras at everything they see that partakes of water: other visitors going by on other *vaporetti,* who photograph them in turn, as well as workboats going about the mundane business of construction, delivery, or garbage collection; they photograph and point out the little branch canals that offer fleeting glimpses into the city's watery soul, shady and suggestive with promise, and they admire the regimental parade of fanciful palaces passing in review, most astraddle water entrances like flooded garages that appear to carry the tides right into their living rooms.

Those who come into Venice along the two-mile backward *S* of the Grand Canal have a chance to see firsthand the unique relationship that the Venetians built up with the water running through their lives. This is why, though they come into nothing but a visual contact with their surroundings, they may still come closer to the city's underlying topographical logic than do the many other new arrivals who, from motives of thrift, speed, or a desire to catch that ineffable prize—"the Venice of the Venetians"—have straightaway plunged into the maze of alleys and attempted to bushwhack their way to San Marco overland. Unfortunately for those among them who might want to go deeper into this fundamental logic, by exploring the scores of smaller canals that lurk, just glimpsed, all over the city, the public *vaporetto* and its somewhat smaller relative, the *motoscafo,* are banned from traveling these smaller waterways. One could, it is true, take one of Venice's water taxis, which can penetrate into all but the smallest of canals, but considering the rather remarkable fares they charge, only the richest tourists would ever consider using such means for just wandering around.

Seen where it can be glimpsed—from neighborhood bridges, along the quays that flank many canals, or at the green-slimed marble steps that terminate the many dead-end alleys—the seascape of Venice can thus seem both disconnected and anonymous, in a vaguely sinister sort of way. Yet it is this water that makes the city unique and explains its tenacity as a tourist draw. Of course, waterways running through a built-up space can exercise a deep attraction, as anyone who has walked through a Dutch town can attest: they bring visual variety, endow the surroundings with secretive and private qualities, and can even (however polluted they often are) convey the sense of something natural penetrating into the midst of dense urban artifice. Many cities, from San Antonio to Stockholm to Singapore, offer the attractions of town fabrics woven of land and water in this fashion. Still, Venice remains the ultimate realization of this particular urban vision: so much so that no one would ever think to call it "the Amsterdam of the Adriatic." Rather, it is other cities that are, or aspire to be, "the Venice of the North," ". . . of the East," ". . . of Asia," and, finally, ". . . of California," which is, of course, simply Venice, California.

Unquestionably, Venice's eponymous claim to be its own archetype derives from the very number of its canals, and the breadth and complexity of the network they form: even after a good many were filled in a hundred years ago and given the whimsically sad designation of *rio terrà,* or "earthed canal," there are still more of them, relative to the city they fragment, than anywhere else—around 170 in all.[1] But certainly as important as their sheer number is the way that canals have structured Venice. These are not the mere ditches for draining swampy land that form the basis for Dutch towns: they are rather the arterial remnants of the sunken delta of the Brenta—traces, as it were, of an underwater river passing through Venice and the Venetian Lagoon and on out to sea at the Bocca del Lido. A millennium ago, the channels were much wider; or rather, the original islands that made up Venice, the cluster known as Rivo Alto, the "high bank," were more widely separated from one another than they are now.

Over the centuries these islands grew together, as their tidal fringes were reclaimed. On many of them, churches and patrician palaces were erected to face out on the canals rather than back toward the interior, a tradition that gave the waterways a central role in Venice for connecting the city's most important centers.

Enclosed and walled in rather than scooped out, canals have primacy in Venice, even when they have been whittled down to the narrowness of alleys by competing palace builders: the waterscape of the city is, paradoxically, its bedrock. Even with a good many of them now filled in, they still represent the most direct way to reach most parts of Venice; unlike the contorted maze of *calli, campi,* and bridges just above them, they generally make sense, at least to those who have a boat handy to take advantage of them. Which is, of course, to say that they make sense to Venetians, and indeed the canals of the city, or more precisely, getting around on those canals, could reasonably be called the last backstage left in Venice, the final spatial possession of the Venetians. Certainly Venetians do use these waterways, as the thousands of large and small boats moored along the more peripheral, wider canals testify.[2] As they move about Venice, they travel in what is a very different city, posited on a different set of principles, than the one occupied by the tourists crowding the quays and bridges above them. Keeping their gaze down on the canal before them, below the level of tourist feet, and calling out only to one another, they see a different city, with a good many more half-collapsed warehouses, sewer outlets, and rather sordid boarded-up storerooms, but also a treasure trove of otherwise secret houses and palaces that show their ornate faces only to these private waters and to those who travel them.

Coming to the city for the short term, without boats or any other way to move about on the canals, tourists in Venice have no direct way to engage with this seascape: they can look at it (indeed, that is why most of them come), but, except for the very few who come with their own craft, they have no way to use it.[3] In this sense, foreigners in Venice are outsiders twice over, effectively locked out of both the city's idiosyncratic culture and the overriding logic of its organization. This double disconnect between Venetians and tourists, complained about by generations of visitors to the city, traditionally has been bridged—if one may use that expression here—by the gondolier. As the interlocutor who has allowed outsiders to exploit the Venetian waterscape and experience the city on the terms in which it was built, the man who rows the gondola long has been held as quintessentially Venetian as the canals themselves: if Venice has the distinction of being the archetype of itself, then the gondolier is the master icon of Venice and, indeed, a key synecdoche of tourism it-self. Even reduced to a few sketchy lines, floating on a neutral background and without landscape references, the gondolier and gondola that haunt restaurant walls, menus, ads, and clip art around the world have the power to conjure up such touristic ideals as graceful servility, relaxed luxury, arcane skills, and unobtrusive knowledge.

Before the fall of the Venetian Republic in 1797, gondoliers and other boatmen plied the waters of Venice in vast numbers: they crowd the canvases of Canaletto, from the 1720s to 1740s, and flocks of them also show up in the great Venice-scapes of Carpaccio, from the 1490s. How many of them there actually were is not certain, but many foreigners who visited the city in these centuries hazarded an estimate—indeed, their sheer number represented something of a tourist attraction in its own right, and the counts ranged anywhere from Fynes Moryson's very modest "some eight hundred, or as others say, a thousand," around 1600, up to Sir John Reresby's whopping "80,000 gundoles, hackney and other" half a century later.[4] Most modern travel writers and guidebooks accept a figure of around ten or twelve thousand gondolas, plus perhaps an equal number of various small transport boats that were also rowed; this is, in fact, fairly typical for the estimates offered by visitors to the city between the 1480s and 1760s.[5]

The implication is that a tremendous portion of Venetian working men made a living by rowing and transport. Since the city censuses that begin in the seventeenth century indicate the presence of around forty or fifty thousand laboring and craftsmen in Venice for most of this period, as many as a quarter of these must have rowed the gondola at least part-time. Indeed, the figure was probably even higher than that, since it was the custom for Venetian nobles and the wealthier citizens who could afford it to have themselves rowed by two gondoliers, one on the raised area at the stern *(da popa)* and one in the well in front *(da mezo);* both rowed, but the latter also saw to the personal needs of the passengers.[6] Most patricians kept two or three of the slim, black boats, or as many as they could afford, both for the sake of status and so that a clan's women and the men of various age groups could go out separately.[7] Starting as early as the fifteenth century, they also had their gondolas rigged out like luxurious coaches, with fine leather seats and a *felze,* or sunshade; the gondoliers themselves often functioned much like coachmen elsewhere, as personal servants dressed up in livery.

Yet not all gondoliers were attached to noble houses or acted as lackeys. Many were in business for themselves, as "hackneys" eking out an existence on paid fares, "hir'd by any man for money." As Thomas Coryat observed as early as 1608, there were a considerable number of such independent oarsmen: "Of these Gondolas they say there are ten thousand about the citie, whereof six thousand are private, serving for the Gentlemen and others, and foure thousand for mercenary men, which get their living by the trade of rowing."[8] Although some of these independent gondoliers worked at transporting goods about the city, most passed their days like modern taxi drivers, either cruising the canals looking for fares or bunched together at stands known as *traghetti* or *stazi,* where potential passengers could seek them out: "There are always a World of them standing together at several publick Wharfs; so that you need but cry out *Gondola,* and you have them lanch *[sic]* out presently to you."[9]

Some among Venice's increasing army of tourists felt positively proud about how their own presence in the city, and the pounds and *livres* they were spending there, helped contribute, in effect, to Venice's social stability, since "the great Number of Gondolas procures a Subsistence to a vast Number of poor People, who would otherwise find great Difficulty to maintain themselves."[10] With so many thousands of oarsmen assiduously on the lookout for fares, and with only limited corporate controls to prevent newcomers from setting up in business, many of these gondoliers-for-hire were evidently in fairly desperate straits, however, and apparently tried everything they could think of to attract customers, especially wealthy foreigners. Tourists expressed pleasure at all the options that were made available for them to enjoy the city:

One may be served with Gondolas in several ways: They who only want them for a Job, need but call *Gondola,* as you do *Coach* at London or Paris, and immediately they come to you. One may also have them at so much by the Voyage, or by the Hour; but the best Way is to have one quite to yourself, to come and wait every Morning at your Gate before you rise out of Bed. . . . [He is] tied down to us during our Stay in the City, to wear our Livery if we please, to come every morning and wait at the hour we set the Night before; to conduct us at any Hour of the Day to any place in the City where we intend to go, and at Night to the Opera or Playhouse.[11]

The sense that gondoliers were a fairly wretched lot — often really no more than beggars offering rides, like rickshaw drivers in the traditional East — continued well past the fall of the Republic, when the number of oarsmen seeking to make a living this way remained in the thousands even as the Venetian nobility was cutting back on expenditures and the city itself, under Austrian dominion, slipped outside the main Italian tourist circuit.[12] Mark Twain, passing through in 1867, just a year after Venice's liberation and unification with the Kingdom of Italy, saved a particular sneer for "the storied gondola of Venice!"—dismissing it as nothing more than "an inky, rusty old canoe with a sable hearse-body clapped on to the middle of it" and mocking his gondolier as "a mangy, barefooted guttersnipe with a portion of his raiment on exhibition which should have been sacred from public scrutiny."[13] Henry James, although far more willing to sentimentalize the man who rowed him about the city (with the somewhat dubious praise of "being obsequious without being, or at least without seeming, abject"), also recognized that his gondolier lived a life of quiet destitution and commented with more apparent sentimentality than irony that "the price he sets on his services is touchingly small."[14]

This was all about to change, however. Freed of its gloomy garrison of fifteen thousand "white-coated, yellow-collared, tight-legged" Austrian soldiers, and connected at last to the growing European rail network, Venice soon found itself bathed in an influx of visitors coming south through the newly opened Mount Cenis tunnel.[15] This new generation of tourists, more middle-class and often more hurried than in years past, usually had neither the interest nor the leisure to take in Venice as James had, "floating about with that delightful sense of being for the moment part of it, which any gentleman in a gondola is free to entertain." Armed with the latest copy of Baedeker (who promised them that with "the aid of the Handbook, coupled with a slight acquaintance of the Italian language," they could *entirely . . . dispense with a guide*"), these visitors were more than willing to tackle the labyrinthine city on their own and on foot, plowing "through churches and galleries in dense irresponsible groups," "fill[ing] the Ducal Palace and the Academy with their uproar."[16] Baedeker thought it enough for his readers, immediately upon arrival, to just take a single two-hour jaunt by gondola up and down the length of the Grand Canal, "to gratify their first curiosity, and obtain a general idea of the peculiarities of Venice." Except for those who decided to stay longer than the four days he recommended as sufficient for the city—and perhaps take in Murano or the Giudecca—Baedeker seems to have considered the gondola as otherwise dispensable for his clients.

Despite such attempts to render them redundant, Venice's "persuasive gondoliers" were quite capable of adapting themselves to this new brand of visitor, just as two centuries earlier they had refashioned themselves from the lackeys of the patriciate to the private escorts of Grand Tourists. Offering their services for shorter periods actually allowed gondoliers to make more on an hourly basis than they had been able to do when hiring out by the day or by the trip: their rates from the 1880s to the First

FIGURE 15. *Gridar gondola* in the nineteenth century. (Image from Museo Correr, Archivio Fotografico, M.570.)

World War were about twenty American cents for the first hour and ten cents for each hour thereafter: perhaps not so much for most visitors to the city, but no longer quite the "touchingly small" sum that James remembered paying in the 1860s (see figure 15).[17] Moreover, the gondoliers soon established themselves in the lucrative business of shunting passengers from the train station to the better hotels, most of which were (and are) located around Piazza San Marco. For this service, which took somewhat less than an hour, they also charged twenty cents, plus extra for the luggage; often they tried to persuade potential clients that two men (at double the rate) were needed, a ploy that Baedeker advised should be firmly rejected with a resolute *Basta uno!*[18] By the 1880s, this crowd of boatmen, while providing foreigners with an essential service, was also becoming a tourist attraction in its own right:

Pass through the railway station . . . and within 20 seconds, you will have experienced your first surprise. Instead of the open square common in European cities, where hacks are drawn up and cabbies strive to catch the passing fare . . . here are the quiet waters of the Canal Grande, unmoved by a ripple. . . . A long line of gondolas is drawn up in front of the station, all equipped with little lanterns. . . . a porter is summoned. He points out a gondola, and the luggage is transferred. . . . A minute afterward we are out in the stream, the gondoliers' oars meeting the wavelets as gently and silently as feathers.[19]

But even as they moved to profit from Venice's modernizing tourist market, gondoliers already had their rivals. The bigger hotels started setting up private squads of boatmen, decking them out in pastiche liveries and sending them with special gondolas to the station. The regular gondoliers, claiming that this competition had "reduced [them] to a state of great poverty," responded with some "serious rioting," beating up a few of the interlopers and sinking their boats.[20] As if this were not enough, a semi-private company gained a license to run a number of *battelli,* or "omnibus-boats," on regular rounds between the station and the major hotels, timed to the train arrivals and departures. Such as they were, the omnibuses were cheap, costing only about a quarter of the price of a gondola; holding up to eight tourists and their luggage, they soon became a mainstay for those coming in on package tours. The gondoliers seem to have been willing to tolerate this sort of competition, perhaps secure in the knowledge that Baedeker himself had condemned these ungainly, rowed vessels as "slow, often crowded, and affording no view," concluding that they were "not recommended."[21] But the boatmen were far less happy when, in the fall of 1881, the first *vaporetto,* or "little steamer," appeared. Initially, there was just one of these motorized versions of the omnibuses, and it ran along the Grand Canal only once every forty-five minutes; within a few months, however, a second was added, and this was enough to bring the gondoliers out on strike (see figure 16). They had a lot of support from the British press, which freely predicted that "these ugly brown steam-boats tearing up the waters of the Grand Canal" would lead directly to their "absolute extinction": "The gondola and the picturesquely-clad gondolier may for a few years longer maintain a languid existence. . . . But who can quote Byron with the smell of train-oil in his nostrils? The news must be a shock

FIGURE 16. First *vaporetto* and landing stage, 1881. (Image from Museo Correr, Archivio Fotografico, V.7758.)

to the aesthetic feelings of every properly constituted mind, while its effect on Mr. Ruskin is a subject too painful to contemplate."[22]

Some American visitors, by contrast, apparently agreed with Mark Twain that by the late nineteenth century the gondolier was little more than a sham anachronism anyway, and that his disappearance with the coming of the *vaporetto* was an inevitable expression of progress in a modernizing world. Even those who claimed to be "as fierce against the steamboats on the canal as any other lovers of the picturesque" soon decided that the *vaporetto* was not so bad when it emerged that it "took us everywhere for 2 cents when the gondoliers asked 50."[23] In the end, the gondoliers' strike folded in just three days, but not without some consequences that were significant in transforming these boatmen yet again, to what would become their present-day incarnation.

This seemingly futile strike in 1881 was the moment when the gondoliers made the first concessions toward giving up their long-held predominance in the city's transport sector and began to assume instead a new role as Venice's icon and protectors.[24] The mayor, with whom they were negotiating, refused to seriously consider abolishing either the *vaporetti* or the omnibus (though that would disappear of its own accord before the First World War). He did agree, however, to limit the number of stops the steam-launches could make—twelve, initially—and to restrict them to the Grand Canal and the hours of daylight. The *vaporetto* in consequence functioned thereafter as a sort of bus—Venice's own form of very cheap mass transit for moving both locals and tourists around on something of a regular schedule; it would not actually penetrate into the city's innards or offer anything like the languid boating experience through shadowy waterways that had once been the delight and prerogative of Venetian patricians and Grand Tourists.[25] This role, in fact if not in law, would stay with the gondoliers, who in this crucial year also won the right to limit their own population. Numbering about two thousand at the time of the strike, they were permitted to close their corporate ranks and henceforth enroll only sons and grandsons of active members; within a few decades, thanks to attrition, there remained barely five hundred gondoliers still active at the city's ten *stazi*.[26]

As a result, and despite the ever-multiplying number of *vaporetti* (they were soon running every fifteen and then every ten minutes), Venice's gondoliers were never again the sorts of abject, boat-wielding beggars that Twain and James had encountered. Even the arrival of the first gasoline-powered motorboats around 1909 failed to spell the doom of the gondola, though these competed much more directly for the sightseeing trade than had the *vaporetti*. Indeed, before long the better hotels not only were using these smaller and highly maneuverable vessels to pick up preferred clients at the station, but also were making them available to their patrons for traveling around Venice's back canals and seeing the Lagoon by moonlight. As in the past, the Anglo-American press promptly predicted in anguished tones that this new competition could result only in complete disaster for the gondoliers:

What will be the result? The utility of the gondola as a conveyance of passengers and luggage has passed away forever—just as with the advent of steam launches in the lagoons the genuine romance and songs of the gondolier passed away. Above all things, the tourist requires rapid transportation in "seeing Venice." This the motor boat provides him. [The gondolier] may still remain for some time a picturesque personage until, like the horse cab in land cities, he gives way to the Frankenstein monster, which modern invention and the demands of modern utility have brought into being. . . . And Venice, the Queen of the Adriatic, with her last unique glory gone . . . and her canals resounding with uncouth noises, no longer the inspiration of poets and artists, will become sordid, commonplace, and unlovely, a sacrifice on the altar of modernism.[27]

As it happened, the duel between gondola and motorboat was not so quickly resolved, and indeed continues to the present day. Despite the steadily increasing number of powerboats in the city and the repeated predictions both in Venice and abroad of the gondoliers' own imminent demise, there remain today about four hundred gondolas in active service— almost as many as there were in the years just after the First World War. The problem, over the course of the twentieth century, was less in finding enough work to support this many oarsmen than in finding enough gondoliers themselves: as the population of Venice nose-dived after 1960, there were no longer enough sons willing to follow their fathers into the business, in the traditional manner. As a result, in 1980 the city intervened and opened up gondoliering to outside entrants by means of a public exam, or *concorso*. Although the descendants of gondolier families would still have a decided edge in the competition, applicants would be accepted based on their dexterity at managing the unwieldy craft, their ability at handling tourists, their linguistic skills, and their knowledge of the city's monuments. The job thus became rather more like a civil-service position, with the tacit understanding, however, that no one was thinking of allowing anyone from outside the immediate region of Venice to join. As Venice, together with Italy, moved closer to the European Union, this semiclosed system was put to the test, most famously in the late 1990s, when one Alessandra Hai, who had the temerity to be not only a German but also a woman, applied twice (unsuccessfully, as it turned out) for a position, under the equal-opportunity statutes of the EU.[28]

The real shift in the nature of the gondola had less to do with who actually rowed it, however, than with the iconic function that both boat and oarsman were beginning to assume in a Venice completely consumed by mass tourism. In the middle of a media debate in the mid-1920s over whether the archaic duo should be replaced once and for all by motorboats and steamers, no less than Mussolini himself weighed in to intone that, even with the Fascist passion for modernization, gondola and gondolier would remain protected, since "there are some things so holy, that no material gain can justify their sacrifice."[29] If gondoliers could no longer compete with motorized vessels in getting around Venice's waterways, they could at least offer a tangible sample of the city that once was, selling their services not so much to actually get anywhere, but rather simply to be on the water in the traditional manner. This they were clearly doing by 1930, when Baedeker began listing rates both by the hour and by the trip, but it was only after the Second World War that gondoliers essentially ceased to offer trips with any particular destination and instead just provided tourists with the experience of a gondola ride for its own sake.[30]

FIGURE 17. Gondola in the seventeenth century. (Image from Museo Correr, Archivio Fotografico, M.24307.)

Appropriately for its new iconic role, the gondola itself completed the last stage in its physical evolution at just about this time. In its earliest incarnations, as the plaything of wealthy Venetians, the craft was somewhat different in form: it was in fact about four or five feet shorter and perhaps 10 percent wider at the beam than it has since become.[31] At both bow and stern it once terminated in massive peaks "which are rais'd up to the full height of a Man," with the one in front emblazoned (or armed, it could be said) with a massive steel multipronged *ferro* that cleaved the space before the vessel like a giant broad-axe.[32] As a final touch, gondolas were outfitted with small cabins, the *felze* that allowed noble passengers protection from the elements (see figure 17). Initially, as is evident in Carpaccio's Venice-scapes, *felze* were little more than spindly tents draped with removable cloth, but by the eighteenth century they had evolved into complex black lacquered boxes with louvered doors that could be opened at the sides and front when the occupants wished to see out or to be admired, or closed tight for occasions that required complete anonymity and privacy.[33]

Although they evidently served well enough for several generations of

Grand Tourists, these early editions of the gondola were too bulky and aggressive for the mass-tourist traffic that invaded Venice after the 1880s. By degrees the craft was lengthened and slimmed down; while maintaining its traditional all-black coloring, it was also flattened at either end and the *ferro* reduced to a mere vestige of its former self. To make it easier to handle by a single oarsman—who, by having to row on just one side, tended to drive the gondola in a circle—a more elaborate *forcola,* or oarlock, was devised, sculpted in three dimensions to give the gondolier more options for placing his oar. For much the same reason, the hull of the gondola was also adjusted to make the left side about nine inches longer than the right. This seemingly quixotic design decision, which gives gondolas their distinctive, almost surreal lilt, very neatly helps compensate for the thrusting power of the gondolier's oar, which always tends to drive the boat to the right. With only a single boatman on board, the gondola's seats could also be rearranged to allow the boat to seat more passengers, by adding a rather uncomfortable miniature chair, known as a *seggiolino,* to take the place of the now-vanished gondolier *da mezo* in the bow.[34] Finally, the *felze,* which evidently had been blocking tourists' views of passing buildings, was shrunken to a vestigial summer awning (known as the *tendalìn*) and then, by the 1930s, removed altogether.[35]

Alexandre St. Didier was thus a bit hasty when he wrote in 1680, "I do much question if Human Industry can add any farther Perfection to the Gondola's they use at *Venice.* . . . Their Figure and Lightness are extraordinary."[36] Still, the craft has at least remained essentially unchanged for nearly a century. It was also perhaps not completely coincidental that just about this same time the gondoliers formalized for themselves their immediately and widely recognized costume: black trousers, a white and wide-lapeled canvas sailor's smock over a black-and-white-striped T-shirt (with a navy-blue overshirt in winter months), a red or blue scarf around the neck, and all of it topped off by the characteristic gondolier's straw boater. Certainly there is no indication in photographs or drawings from before the First World War that gondoliers sported this rather boyish getup with any regularity: their costume appears to have been far more ad hoc and personalized than that, from before the days of Mark Twain until well into the 1920s.[37] Forty years later, however, Clive Barnes, writing for the *New York Times,* could positively sigh with satisfaction and relief when he arrived in Venice and discovered that "yes, the gondoliers do wear striped jerseys and straw hats."[38]

The reason, of course, is that for an icon to function it must be both reliable and instantly recognizable, and for gondoliers to dress in shorts,

FIGURE 18. False gondola. (Photo by authors.)

tank tops, or lambswool jackets (depending on the season) just would not do. It is worth noting that in the 1980s, when Venice's waterscape was invaded by the so-called *gondolieri abusivi abusivi* (that is, *very*, VERY illegal gondoliers)—men with suspected mob ties who sought to waylay tourists arriving by car at Tronchetto or San Giuliano and carry them off, under the pretense of offering a gondola ride, in fake motorized boats— both they and their craft were done up to look even more real than the real thing. The vessel they used was a black *sandolo* with an inboard motor, and while the *sandolo* is very much a traditional Venetian boat, it is also a good 20 percent shorter than a gondola and lacks both the latter's distinctive *ferro* and rightward cant. For about twenty years, before their boats were finally sequestered and they were put out of business, these *abusivi abusivi* putted about the more peripheral canals of the city, with seats, cushions, and especially costumes that were typically more faithful replicas of the iconic gondola and gondolier than those offered by the real gondoliers themselves (figure 18).[39]

Even though their reputation as the interlocutors to the Venetian waterscape has been significantly weakened since the opening of their ranks in 1980, gondoliers generally still seem to believe that they possess, at least in a corporate sense, the authority and the responsibility for bringing tourists into contact with Venice's waterways and thus its very soul. It is an authority they have been willing to fight for, whether with the *abusivi abusivi* or with the legitimate and licensed water-taxi drivers, who in the last century have stolen their transport business and continue to ruin their sight-seeing enterprise by habitually speeding along the canals and throwing up an excessive bow wave (the notorious *moto ondoso*) that can sometimes fling an oarsman (and his passengers) right out of the gondola and into the water.[40]

After observing a good many gondoliers, we have concluded that they tend to base such claims for authority less on any particular knowledge of Venice's culture and history or on their skills in dealing with tourists (that is the business of the professional guides, after all) than on their mastery of rowing, the quintessentially Venetian form of movement. Certainly it is no easy job to propel one of these oversized black canoes— flat-bottomed, 35.65 feet long but only 4.66 feet wide, and slightly bent at the stern. Insofar as the gondola has become the key icon through which tourists can enter Venice's waterscape and thus engage with the true logic of the city itself, it is the gondolier's own great skill at mastering this unwieldy and unique vessel that permits visitors to experience Venice as it should be experienced. This, at least, is a conviction held dear by a good many gondoliers, who make much of their skills with the oar and their adroitness of maneuver, talking of aiming for a "certain elegance of oarsmanship" and enjoying "a natural gift of beautiful rowing." It is not, of course, an effortless talent, and one can sometimes spot younger ones out on the back canals—practicing how to place their feet and twirl their oar, looking to find their own *stile di voga,* or style of rowing.[41]

Gondoliers, as everyone knows, row standing while facing the prow, *alla veneziana,* using just a single oar on one side, supported (but not secured) by the bizarre wooden claw of the *forcola.*[42] For centuries observers have marveled at the "facility [with which] these Fellows avoid one another, and with what Ease, Dexterity and Quickness, they pass by all Obstacles that are continually coming in their Way." Such skills are real, and there is no faking the art of rowing a gondola: visitors who get the opportunity and have the temerity to have a try often end up ruing their rashness.[43] Yet the better sort of gondoliers are also able to do what they have to do with a kind of seeming inattention, that casual *sprezzatura* praised by Castiglione as a necessary quality for any aspiring courtier.

If proof were needed of the success with which gondoliers have sold their product to the tourist public, one need only look to the price curve of what they have charged for their services. Once it became clear, immediately after the Second World War, that gondoliers were no longer offering a trip to anywhere in particular, but rather the *experience* of a gondola ride, and thus of Venice itself, their fees could begin to more accurately reflect the seemingly unquenchable tourist demand for such an experience. The trajectory is impressive: the fifty-minute ride that cost $0.42 in 1930 and $1.00 in 1945 rose to $1.60 in 1952, to $3.00 by 1962, $5.00 in 1965, $24.50 in 1979, nearly $40.00 in 1988, $53.00 in 1992, and finally reached around $70.00 in 1999. Even taking considerable inflation and ongoing currency fluctuations into account, this still represents a roughly tenfold increase in real dollars in barely half a century; moreover, it only reflects the cost of a jaunt during the daylight hours—the tariff traditionally leaps up another 20 percent during the romantic hours just around sunset.[44] And high as these prices may seem, they are only the posted rates. Individual gondoliers do not feel especially bound by such official fares, and in recent years many of those who stand on the main thoroughfares and offer their services (*gridar gondola*) to passing tourists seem to have no compunction about demanding the simple parity of one hour for one hundred dollars.[45] Even visitors whose guidebooks have given them the confidence of knowing the supposed rates sometimes find they have trouble actually getting them—gondoliers may simply refuse to accept such offers, safe in the knowledge that their own *cartello* of posted prices, while legally nailed up, is also usually well out of sight.

To many tourists—and not a few professional travel writers—such demands seem outrageous, way beyond the value of the service that is being offered: in New York or Vienna even the greediest horse-and-buggy driver might hesitate before demanding two dollars a minute for his (minimal) labors. Not a few visitors to Venice point with a certain vindictive glee at the frequent sight of groups of gondoliers sitting around idly in the shade, with no clients at hand and nothing to do (figure 19). Foreigners often take this as a clear sign that in their greed the gondoliers have priced themselves clear out of the market and, foolish Italians that they are, they are left with no work at all. This, however, misreads the situation completely, for gondoliers have not only gauged very accurately what the market will bear, but they are really by all accounts also making a very good living at what they do.

The misunderstanding that many observers form here is that gondoliers are waiting around unsuccessfully for what most tourists consider to be "traditional" fares: that is, honeymooning couples or besotted Venice lovers seeking their apotheotic moment with the Most Romantic City in the World. Certainly, at a hundred dollars a jaunt, more than a few new-

FIGURE 19. Loafing gondoliers. (Photo by authors.)

lyweds do find a sunset gondola ride under the Bridge of Sighs a hard bullet to bite, but one has to remember that what is "traditional" about this coupling of Venice, the gondola, and lovers is really only a late-Victorian concoction, one that was in fact seized on by travel promoters as a romantic ideal only in the 1950s, after the custom already had largely died out.[46] More importantly, however, over 80 percent of the gondoliers' business, according to those we spoke with, does not consist of such couples at all, but rather the members of tour groups, who are loaded up six at a time and sent on their way with almost industrial efficiency.

Gondoliers have indeed adapted themselves well to this particular and most notorious aspect of mass tourism (see figure 20). Individual oarsmen, in sharp distinction to the Jamesian ideal—where "[your] gondolier . . . is your very good friend . . . part of your daily life, your double, your shadow, your complement"—often have almost no personal contact at all with their clients.[47] Instead, the arrangements for the *nolo* (that is, the rental for the ride) are carried out between the head of a *stazio*'s

FIGURE 20. The gondola as mass entertainment. (Photo by authors.)

cooperative and the tour guide. The tourists themselves usually have signed on for a ride (at an additional fee) as part of a general package for their stopover in Venice, chosen from a menu that might also include a trip to the glassworks of Murano, the lace shops of Burano, or the Jewish Ghetto. The additional cost — usually twenty dollars per person these days — does not seem excessive to the individual, and in any case, the price is often hidden in the general charges of their tour: as one gondolier told us, "They pay without knowing they paid" *(Pagono senza sapere di pagare)*. Each gondola filled with six passengers is thus generating an income of $120, as opposed to the "official" $70, producing a tidy profit for the tour guide and (we would imagine) something extra for the gondolier. Chances are, then, that the boatmen one sees lounging around at the Molo, San Moisé, or Santa Chiara are not idle because they have foolishly priced themselves out of business, but because, on the contrary, they are merely between groups, enjoying a break in a schedule that was already parceled out and assigned days or even weeks beforehand. These can be seen written up on the little white pen-boards that each *stazio* maintains: the names of the tour groups, their time of arrival, and their size in numbers of gondolas—all regularly consulted by the waiting gondo-liers throughout the day.

Many gondoliers seem to like this system: the work comes to them and all the arrangements are ironed out by someone else. But there is lit-tle incentive for things to be done differently in any case, thanks to the co-op form of organization that predominates among those who offer *gondole a nolo*. All of these gondoliers who hire out at the city's ten *stazi* are meant to turn over their take at the official tariff to their group leader, who then divides the total among all members of the co-op every month, based on the time each put in and various social factors, such as years of service or personal exigencies.[48] Even those who are (as they say) "crazy for the money" *(matti per i soldi)* thus have little incentive to hustle extra business by buttonholing passing foreigners, since any additional take they earn would still be put in the general pool: the only way to gain a larger monthly share is to put in more hours (whether actively rowing or just sitting around smoking in the shade), and such turns of duty *(volte)* are strictly allocated and regulated.[49]

The increasing dominance of tour groups in the gondola business has also created fundamental differences in the nature of the ride itself. To keep groups together and to increase efficiency, gondoliers these days run their clients in virtual flotillas of boats. Called the *carovana,* this is quite liter-ally a caravan of six, eight, twelve, or up to even thirty gondolas with six passengers each and perhaps a singer for them all, moving in tight, linear formation or spreading out side by side on the open waters of the Grand Canal (see figure 21).[50] Making up a *carovana* is a complex art, calling on the skills and coordination of an entire co-op, including a contingent of gondoliers, the group leader, and a few aged *ganzeri,* or gaffers, to hold the boats, as well as the tour guide. Standing at the *traghetto* of San Orse-olo, behind the Piazza, we had to look on in awe at the assembly-line process with which coachloads of fifty or sixty tourists from northern Eu-rope, America, or Japan were loaded up six at a time in these awkward and tippy boats and sent off down the adjoining Rio delle Procuratie. One of the *stazi* in particular, at the foot of the Calle Vallaresso near Piazza San Marco, has been specially adapted to handle the big groups, boasting it-self the only station in the city that has four loading docks, allowing the gondoliers there to "digest" *(smaltire)* groups of fifty or more in minutes.

Since *carovane* are both big and unwieldy, they hardly can be expected to meander at random about the city, and though tourists who travel this way may think they are enjoying a relaxed, Jamesian wander, their little fleet is in fact moving on as fixed and preset a course as if the boats were

FIGURE 21. Gondolas in *carovana*. (Photo by authors.)

running on rails.[51] Each of the ten *stazi* has actually been allocated a closely defined territory, spelled out in its *mariegole,* or operating charter, and through this domain each gondoliers' co-op has plotted out a set route that allows them to offer their passengers a balance of boating experience on the Grand Canal and on the more secluded inner waterways, while taking in a couple of landmark churches or palaces along the way.[52] Still more important, big clusters of boats are able to return smoothly to their home *stazio* within the required fifty minutes, while also avoiding the sorts of traffic tie-ups on the narrower canals that would inevitably result if gondoliers from one *traghetto* were allowed to wander at will into the territory allocated to another. Tourists thus very likely see only the same restricted series of canals along which their gondola has just passed with a previous contingent and to which their gondolier will return within an hour or so. Such a procedure, though depressingly different from what many may consider the ideal of a gondola ride, really only makes common sense, since the number of foreigners willing to pay for a trip is truly enormous. We were told by Franco Vianello Moro, current head of the Ente Gondola, or gondoliers' co-op, that "no fewer than three million passengers pay for a gondola ride every year."[53] This figure might seem stunning, since it means that around a quarter of all visitors to the city take a jaunt in a gondola, but we calculated that, just as a minimum, if the two-hundred-odd gondoliers on duty at any one time made only a half dozen daily jaunts with five or six passengers, just between May and September—which they obviously do—they would easily carry more than a million.

All this brings in enormous wealth, not all of it completely licit: we saw so many hundred-dollar bills casually passing half-concealed back and forth from guides to the gondoliers or their *ganzeri* that we felt like on-lookers at a race track or an open-air drug market.[54] Clearly such a complex operation must be kept running smoothly if it is to stay so highly profitable, and the *mariegole* of the ten cooperatives are designed to keep the gondolas running on their designated tracks and the gondoliers safely inside their own confines, out of the hair of their neighbors. The result, needless to say, has been to further restrict and replicate the routes available for gondola rides. We were told that if any tourist actually had the audacity to ask for a gondola tour of the entire city, the gondolier he or she approached was required to figure out a route beforehand and then call all the other *stazi* whose territory he would have to cross, in order to gain permission. Though all the boatmen feel themselves justified in demanding a truly enormous fee for such an effort, many find it easier just to refuse this sort of quixotic request on the spot, as no more than a tourist whim.

The fascinating aspect of all this, of course, is how such a shabby, degraded, expensive, and seemingly pointless event continues to attract so many willing customers. The more so considering what a high percentage of those who make the trip do so in the middle of a hot and humid summer's day, with at least a good portion of their jaunt spent out in the full sun of the Grand Canal, with their craft pitching about erratically from the continual waves of scores of passing taxis, supply boats, and *vaporetti,* and the salt air filled with diesel fumes and the roar of engines (and sometimes sirens) of every description. Nothing, it would seem, could be further from a Merchant Ivory–style celluloid fantasy of dawdling along the canals of Venice with a beloved and under the care of an attentive and obsequious gondolier. To add to the discordance of the image, chances are high that the trip will be made along with a half dozen or more gondolas *a carovana,* accompanied by an overripe tenor warbling out hackneyed Neapolitan ballads to tunes provided by a wheezy accordion; the gondoliers, who often barely have to twitch their oars to keep together, may well amuse themselves smoking, talking loudly on their cell phones, or cheerfully swapping insulting comments in dialect about the appearance and behavior of their passengers.

It has thus been surprising but also somehow reassuring to have most of the gondola passengers we spoke with as they were disembarking assure us that they were quite pleased by their ride. If there were any negative reactions at all, they usually had to do with matters external to the gondola and the trip itself: some expressed embarrassment at being put on public display in this fashion—not, of course, on display to the Venetians, whose opinions are irrelevant and unknowable, but rather to the other tourists who peer down at them and take their photo from bridges or passing *vaporetti,* and who (it was imagined) must think them oversentimental rubes for paying so much for such blatantly romantic purposes. Others complained about the stink of the water or the shabbiness of the buildings they passed, but this seems to have represented more a failing on the part of an overly "real" and gritty Venice (or the ineffable "they" who are supposed to be taking care of the place) rather than of the ride itself. Moreover, many tourists are well aware of the water quality in Venice before they ever leave home, and as one young woman told us, "We've traveled from Australia, which has taken us hours and hours to get here. . . . It was so nice just to have the gondola, people we knew are behind us, and you waved at them, and who cares about the smell?" Perhaps the most typical reaction to the ride, which we heard over and over, was that expressed by a group of American girls, who, immediately on returning to shore, enthused with teenage insouciance, "Well, we couldn't come to Venice without going on a gondola ride: you just *have* to do it!" This notion, that going in a gondola is some sort of obligation that must be fulfilled if a visit to Venice is to be complete or successful, was something we heard often, and it resonates well with recent sociological thought on sightseeing as a ritual process. Dean MacCannell, for one, has observed, "Modern international sightseeing possesses its own moral structure, a collective sense that certain sights must be seen," and certainly we picked up the sense from disembarking tourists of a ritual obligation successfully dispensed with.[55]

Moreover, this rite cannot (or should not) be carried out in any other sort of boat than a gondola, rowed by a properly attired gondolier. Traveling through Venice by *vaporetto* or taxi may well expose foreigners to a great deal more of the Venetian waterscape than they would see from a highly restricted gondola ride—a ride that is, after all, a trip to nowhere—but going about by those other means would fail to realize a much more important tourist aim: that of being in the closest possible proximity to—indeed, directly inside of—one of the most potent and universally known of all tourist icons. The all but unquenchable public demand for the *experience* of a gondola ride (as opposed to actually going somewhere in the boat) is clearly well understood by the gondoliers, who have responded not only by raising their fares over 1,000 percent in the space of just two generations, but also by coming up with the concept of the *carovana* cranking around endlessly on its preset route. Those (steadily dwindling) few who can still recall how, in the pre- and immediately post-war era, one could hire a gondola and an attentive, gossipy gondolier for a whole day's ramble for around five dollars, may rail against this rather louche present-day reality, but they tend to forget that now, as then, gondoliers provide their clients not with transportation as such but with a packaged cultural production, a combined souvenir/ritual event that has since 1945 been very much able to set its own price structure. As MacCannell, once again, has noted about cultural productions in general, their "value is not determined by the amount of labor required for their production," but is rather "a function of the quality and quantity of *experience* they promise."[56]

The cultural production that is on offer in a gondola ride, it must be stressed, is definitely not an opportunity to come into closer contact with the "real Venice of the Venetians." Any tourist who steps into a gondola hoping for this is headed for disappointment: if anything, the gondola takes its passengers directly *away* from whatever there still exists that is genuine (that is, "untourist") about Venice. In this regard, traveling the canals by gondola is not at all like, for example, seeing London from the top of a double-decker bus, in company with the heterogeneity of ordinary Londoners going about their business; still less like seeing India through the windows and amid the racket of a third-class railway carriage. To step into a gondola is instead to advance still deeper within the tourist bubble that already englobes the greater part of Venice itself: it is a safe bet—indeed, almost a dead certainty—that anyone seen in a gondola, alone or in a group, is a tourist, part of a by-now fictive Venetian waterscape that is no longer of interest to anyone except other tourists.

Since they are not actually taken anywhere, of course, it also cannot be said that tourists are exposed to the real logic of the Venetian waterways. Nor, it must be admitted, do they seem to learn much about or to have any particular interest in the operation of the gondola itself. Despite all their gondolier's efforts and after-hours practice, few passengers pay much attention to his ability with the oar: "The only thing they really notice and applaud," a gondolier told us, "is one of the most basic [skills]": when the boatman slips gracefully under a low bridge *(schiava il ponte)* with a single oar stroke. We concluded that one could compare gondoliers at work with fine race cars stuck in rush-hour traffic: all talent and no action. The only opportunities most of them have to show off their oarsmanship while crawling along in a dozen-gondola *carovana* is in desperately trying to counteract the pitching waves and whitecaps continually thrown up by passing taxis and *vaporetti*.

It is also true that the great majority of passengers express scant specific interest in the architectural treasures that drift by: "They don't care whether this palace is Gothic or that one is Romanesque," one gondolier told us; rather, if they want to know anything, "they want to know about Venetian life"—*la vita veneziana*—tales that can lend at least a slight validity to this exclusively tourist way of getting around. Buildings are more interesting if they can be attached to known historical personalities, virtually all of whom turn out to have themselves been tourists in the city: finding out where the likes of Mozart, Goethe, Wagner, or Browning stayed seems to be a highlight of any jaunt. We were also told that the Fenice opera house was an extremely popular sight—but only after it burned down in January 1996.

Despite—or, indeed, we would claim because of—the rather limited range of their gondola trip, those visitors (and these are certainly the great majority) who can accept that this is a *ride,* just like a ride on an amusement park Ferris wheel, do often end up with an experience that they find rewarding on its own terms. In searching for an underlying dynamic to this quixotic ritual, we have noticed that a gondola's passengers are less likely to direct their attentions outward, toward the passing city, than inward, toward themselves, as they enjoy this mutual adventure that places them physically within an icon of such renown. This is especially true of the *carovana,* which produces its own brief, shared dynamic, often structured around its sheer size and the interconnectedness of the passengers in the different boats, since they are usually (though not always) members of the same tour group—"people we knew [who] are behind us, and you waved at them." A good many tour groups apparently supply those who choose the gondola ride with bottles of wine or spumante; many other tourists simply bring their own bottles along and sometimes get fairly drunk—"You need a bottle each!" one group of young New Zealanders told us—and when the day is hot, we have seen their group self-absorption lead to splash fights between gondolas. Older passengers, we were told, especially those out after lunch in warm weather, may simply fall asleep. Indeed, though the general ambience out on the water in the midday sun may not be especially comfortable, many passengers told us they found their ride extremely relaxing and that lounging on the padded cushions of their gondola was hugely rewarding because "we just loved having a chance to sit down after so much walking!"

For the run of adults, in any event, the great pleasure is chatting with and photographing others from their tour group, whether in their own gondola or in adjoining boats. Passengers' urge to secure this brief experience with film or video also extends to the gondolier. When tourists photograph their gondolier this way—high up behind them on the *popa,* confidently gripping his thick oar—they further underscore the importance of his status as an icon of Venice, rather than as a transport worker or even as a Venetian per se. In this regard, he is quite unlike the London cabbie, for example—another transport worker, also highly trained and knowledgeable in his craft, and presumably as quintessential a Londoner as the gondolier is a Venetian. While visitors to London are quite eager to photograph the unique black cabs—one of that city's key markers, recognizable everywhere—there is no point in their taking a picture of a cabbie by himself, since away from his vehicle, he is just another figure on the street. The gondolier, by contrast, has an outfit as distinct as that of the Beefeaters or the guards at Buckingham Palace. Capturing him on film creates for the tourist photographer an image delightfully filled with shifting, postmodern attributes. The photographed gondolier is, first of all, like the London taxicab, a marker that indicates a tourist has been to Venice; at the same time, he is so universally familiar that he is also a "sight" that must been seen in Venice, as much as Piazza San Marco or the Ducal Palace; and finally, captured on film or video as the visitor's own boatman, he is also a marker again—the desirable proof that one has been in the presence of, and indeed inside, not Venice itself as much as one of the ultimate icons of the tourist experience.[57]

The *carovana* can provide yet another point of self-reference for its passengers if they decide to pay the extra $150 for a singer and accordionist (who together make up *una musica,* in the gondoliers' parlance), who emphasize their focal function by always traveling in the central gondola. Again, those who lament the inappropriateness of these tenors who warble out "O sole mio," "Torno a Sorrento," or sundry Las Vegas lounge hits—at the expense of some considerably more obscure but supposedly genuine Venetian dialect ballads—miss the point, which is that they are certainly not there to create anything like an "authentic Venetian experience."[58] Indeed, it should be kept in mind that the whole notion of the singing gondolier crooning to a loving pair of passengers by moonlight is at best a fictive tradition. The occasional boatman may still sing (though more with ironic intent than not), but already over a century ago singing had been put on a staged, commercial footing, with the gondoliers rowing semiprofessionals under the balconies of the major hotels after dark to serenade the tourists above; the tunes they chose then were different but no more Venetian than they are today: "Verdi and Bellini at their worst," as one sour British tourist observed in 1896. "'Ah che la Morte,' and its companion absurdities."[59] These days, at least the singer and his accompanist try to provide what their clients know and expect: a pleasurable, unifying, and (one might say) nonthreatening group experience. Passengers may join in the singing, and, if they are feeling lively enough to ask for it, there may even result a form of gondola karaoke, with the accordionist backing them in everything from (we were told) pop songs to the Japanese national anthem.

In one very real sense, of course, the singer, the accordionist, and indeed the modern gondolier and the *carovana* itself cannot be other than "authentic," for they are genuine creations of postwar Venetian mass tourism, with only the most tenuous structural links to the gondola and gondolier of past centuries. As a unique mode of transportation, beautifully designed for getting about a unique city, the gondola really did go extinct some time around the First World War, just as so many anguished Anglo-Americans had been predicting it would. No longer the key that opens up the water-logic of the city, the gondola is now a postmodern experience, a ritual ride to nowhere filled with both irony and obligation. But, though gondola and gondolier may have become a whole bundle of self-referential cliches piled on top of each other, they still provide the ultimate and necessary validation of any tourist's visit to Venice. As long as tourists are able to say that *sono stati in onda*, that "they have been on the water," as a gondolier told us, they can leave the city happy. After all, for centuries foreigners have recognized that gondolas "have made Venice the most individual of all cities in the world," and even at the prices they have had to pay and the rather dismal circumstances under which they often have to go, taking a gondola ride is still worthwhile to possibly millions of tourists annually: the act incorporates them into the myth and image of Venice, making them a part of Venice in a way that only this city offers.[60]

It is doubtful that observations such as these would carry great weight with the Venetians themselves, who (we hope) rarely spend much time arguing over gondola semiotics in their local bars. For them, the gondola is no more a form of transport or means of getting around than it is for tourists, but unlike foreign visitors, a good many native Venetians have never actually taken a proper gondola ride. Of those who have, a number turn out to have had a gondolier for a friend and thus got to go for free; others told us that they had a relative who gave them a *nolo* as a present, to mark a ceremonial occasion such as graduation from high school or the university. But, by and large, most Venetians told us that, for them, going about in a gondola—with the significant exception of taking the *gondola da traghetto,* which ferries standing passengers across the Grand Canal at several fixed points for a nominal charge—is strictly for the tourists. Having lived with this situation for decades, Venetians seem to ignore a priori anyone in a gondola, no longer giving a second glance to such postmodern spectacles as tourists on bridges taking pictures of other tourists in gondolas who are in turn taking pictures of them, or even the occasional sight of a fully outfitted, beflowered, and gold-trimmed wedding gondola setting off to nowhere in particular with a Japanese couple dressed as bride and groom in the most traditional European style.[61]

Curious as to what might happen when a modern-day Venetian and a gondola were brought together, we offered a local friend a jaunt one evening a few years ago. Persuading him was not easy, however: though born and raised in the historic center, he claimed to have never set foot in a *gondola a nolo* his whole life; it was fairly obvious that he found the whole idea of taking part in such blatantly tourist amusements to be rather embarrassing. Rather to our disappointment, moreover, we soon discovered that even having a Venetian along would not persuade our gondolier to lower his asking price—indeed, we seem to have ended up paying somewhat more than the going rate. Once settled into the gondola, we found that our friend was not content to let the boatman take us on the set course for that *mariegola;* this was his city, where he had lived for nearly fifty years, and he did not wish to just wander around: he wanted to go *somewhere.* Specifically, he wanted to pay a visit to his aunt, who lived near the Campo San Polo. Thanks to the overriding logic of the Venetian canal system, our gondolier got us there surprisingly quickly—too quickly for our taste, considering that we were paying him a flat rate—and immediately our friend began to shout up to his aunt, four floors above the canal. After a bit, a tiny old woman actually did stick her head out the window and peer down at us. Our friend, who had been hoping to tease her with this unexpected—yet achingly traditional—means for a Venetian to make a social call, may not have gotten the reaction he expected, but it certainly seemed what he deserved: "Sandrino!" his aunt called down with what sounded to us like mock horror. "What in heaven's name are you doing in *that thing?*"

CHAPTER 7

Behind the Stage

Beyond and all around Venice lie the islands of the Venetian Lagoon, scattered across the 250 square miles, or 55,000 hectares, of marshes and open waters that stretch from Chioggia in the south to the fens and fish farms up north (see map 4). When seen in satellite photos, Venice is the cluster of islands shaped like a fish—a red fish, thanks to its tile roofs— swimming west just above the midpoint of this watery waste. Venice still is, as it has been called for centuries, La Dominante, the dominant fish in this school. The other islands are by comparison just minnows: some in clusters of their own and some just isolated islets, while a good many are mudflats (which the Venetians call *barene*) that spend barely enough time above the tides to grow themselves a few scrubby bushes. Venice may indeed have flowered—first into the world's most beautiful and then into its most touristed city—but it is, underneath all its marble-clad palaces, its history, and its tourists, just one cluster of muddy little islands among the many that dot this estuary.[1] And, although none of the other islands of the Venetian Lagoon has ever been (or could be) as renowned as Venice itself, they, the world to which they belong, and the culture that their communities created have played their own essential role in developing tourism in Venice.

Something of a breed apart in this insular world are the long barrier islands, the *lidi,* that separate and protect the Lagoon from the Adriatic Sea. The Venetians got their start and were based for several centuries on the northernmost of these *lidi,* at a spot called Malamocco. Only in 810, when threatened by an army of Franks, did these proto-Venetians pack up and move to the island cluster known as Rivo Alto, later called Rialto.

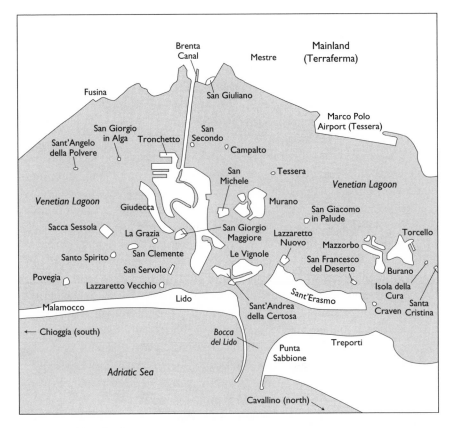

MAP 4. Venetian Lagoon.

Here, according to legend, they founded the city that would be called
Venice, but they were far from alone in this rather unappealing stretch of
marsh and bog. To the south was the fishing town of Chioggia, which
had survived from Roman times; to the north were a number of newer
island communities that were already flourishing when Venice itself was
still unoccupied. These included Burano, Mazzorbo, and, above all, Torcello, which boasted its own cathedral—moved bodily during the seventh
century from the Terraferma to this new home in the swamps to save it
from the rampaging Lombards.

In the beginning, then, Venice was not merely one of the various island clusters in the Lagoon, but also just one of its many communities.
It was better located than most, however, and was probably wealthier from
the outset, thanks to the patrician refugees it had attracted from Padua

and the trade it generated overseas. Before too long, Venice would become the leading city in these waters, as one by one the other island towns fell into its orbit and under its control: the last to go was Chioggia, which held on to its independence until the late fourteenth century. As they submitted to La Dominante, most of these island towns soon stagnated and were left far behind by Venice: Burano, Malamocco, Murano, and Chioggia would never amount to much thereafter, while Torcello and Mazzorbo eventually withered away altogether.

Though it became in time a literal backwater, the Lagoon was far from deserted during the days of the Serenissima. The little towns hung on through crafts, boat building, and, above all, fishing. They shipped their catch to Venice: fish, crustaceans, and shellfish seemingly without limit, as Thomas Coryat noted in 1606: "The abundance of fish, which is twice a day brought into the citie, is so great, that they have not onely exceeding plenty for themselves, but also doe communicate that commodity to their neighbour townes." Those Grand Tourists who were more accustomed to regular meat on their plates might not have welcomed having to eat so much seafood, but the Venetians as a people saw fish as not only fundamental to their cuisine, but also as the basis of an entire estuary culture, one closely identified with life out on the waters of the Lagoon.[2] Many of the larger islands were also planted as market gardens: especially favored were Sant'Erasmo and the aptly named Le Vignole (the Vineyards), which lay only a few minutes northeast of the city.[3]

The Lagoon was also the Venetians' favorite playground. Anyone who had a boat, barge, *sandolo,* or gondola—which is to say, nearly everyone—regularly went out on the open waters to sail, row, and fish, both for the catch and for relaxation. It was an idyllic, timeless world, where "the motion of the men, as they stand to their oars is very graceful, and the other occupants of the boats, as they sit or lie about, look both handsome and happy."[4] Those of the patriciate often just liked to have themselves rowed around for pleasure—*andar a spasso,* as they called it—taking the air away from the city and perhaps concluding with a picnic on one of the outer islands or a mudbath on the Isola della Cura, an islet not far from Torcello, whose minor hot springs still flow.[5] In the reedy backwaters and around the fish farms, or *valli,* were also popular spots for hunting wildfowl: Carpaccio's fifteenth-century depiction of noblemen hunting birds from rowed boats with bow-and-pellet is well known; two hundred years later their descendants, along with many commoners, used muskets for the job.

The Lagoon was also alive with piety from the earliest days of Venetian settlement. The smaller islets were especially ideal places for monasteries and convents, and at one time or another virtually every extant religious order had a house on one of these little mudbanks. There, isolated and serene, yet not too remote or uncomfortable, the sons and daughters of the elites, who made up a goodly part of these communities, could live out their lives still reasonably in touch with—and often in sight of— their kin back in the teeming city. Thanks to family connections, many of these religious houses were well endowed, even sumptuous.[6] Over the last half millennium of the Serenissima, a dozen or more such monasteries and convents flourished around the Lagoon: when one of them died out or went bankrupt, the state itself would take it over and hold it until another order came along looking for similar accommodations.

During the centuries of the Serenissima not many tourists saw this world: far from the glitter on the Grand Canal and Piazza San Marco, it remained Venice's backstage. What most visitors did encounter of the Lagoon and its islands amounted to what passed by their gondola as they were being rowed from the Terraferma to Venice, and then back again; not many of them had either the time or the interest for roaming further afield in any case. One who did so was Jean-Baptiste du Val, who, as secretary to the French ambassador to Venice between 1607 and 1610, was perhaps stuck in the city long enough to want to get out and explore somewhat. Du Val visited a good many of the island monasteries and convents, primarily to see their relics and religious shrines, but also with an interest in the paintings of Venetian masters that were to be found in island churches on Murano, San Giorgio Maggiore, Malamocco, Santo Spirito, Sant'Andrea della Certosa, and Povegia.[7]

Besides Venice itself, there were said to be, all in all, some twenty-five built-up or inhabited islands in the Lagoon, and in time some of these began to figure in the tourist circuit. Alexandre St. Didier, for example, encouraged his readers to visit the southerly barrier island of Pelestrina, which he called Polesin. He especially recommended the town to his readers as "the best Peopl'd and most agreeable" of the Lagoon settlements, with "houses [that are] small, but pretty and neat," and around fourteen thousand inhabitants living from gardening and fishing.[8] Not everyone, of course, had as much time as du Val or St. Didier for such wandering about, although both John Evelyn and Philip Skippon, among others, were willing to make the trip so that they could explore the back fens and visit the monasteries and churches with their "rare Paintings & Carvings, with inlayd work, &c."[9] One place that almost everyone wanted to go was Murano, the closest island town to Venice.[10] Back in the Middle Ages, tourists had journeyed over to "pleasant Murano" primarily to take in the air, which was meant to be softer and healthier than in Venice proper, and to stroll in the town's many patrician gardens.[11] Already by the fifteenth century, however, Murano had also become a tourist attraction for its glass industry, for which it was renowned all over Europe. The chance to see glass production in action and to browse among the blown-glass products on display made Murano a fixture on the Venetian circuit, and by the late 1600s Nicolas de Fer would decree, "Travellers always visit the Glasshouse at *Mirano [sic]*."[12]

Tourist interest in the Lagoon and its communities—never very great to begin with—largely evaporated during the last years of the Venetian Republic. In part this was economic: Murano, for example, went into eclipse around the mid-1700s as competitive glassworks abroad either stole Muranese secrets or invented superior techniques; moreover, as glassblowing became more widely diffused throughout Europe, it was no longer quite as exciting to watch the craftsmen in action.[13] In these same years, most of the island monasteries and convents were suppressed, their adepts dispersed, and their artistic masterpieces pulled out to be displayed at the Accademia. Even going out *a spasso* on the Lagoon's waters, in the manner of du Val and Skippon, seems to have lost its appeal for tourists in the later eighteenth century, perhaps because of the increasing likelihood that one's boat would be pulled over by the ubiquitous and supposedly rapacious customs patrols to be thoroughly searched: any books, edibles, or jewelry on board could be declared contraband, subject to confiscation or heavy fines—to be paid on the spot.[14]

Never rich, the Lagoon world became positively impoverished with the fall of the Republic. When Frances Shelley passed through Mestre in 1816, she found that this once lively town, the primary jumping-off point for Venice from the Brenta Canal, had turned into "the most desolate-looking place possible, [with] every mark of decay," where "[i]nnumerable beggars assailed us as we walked to the gondola." Some visitors, admittedly, quite liked such scenes of desolation and decay: Lord Byron and Percy Shelley scoured the Lagoon's islands and backwaters, rhapsodizing over the faded glories of the Serenissima that they saw as incarnated in the ruins they found there. Others who followed them found the back stretches of the Lagoon filled with romance and "scenes entirely out of the sphere of ordinary life, and unlike any thing else to be found on earth." Better yet, advantageous exchange rates in the early 1800s meant almost anyone could afford a gondola and pass his days "living on the water," as James Fenimore Cooper put it. Cooper felt that, in his jaunts around the backwaters of the Lagoon, he could discern what little still remained of the Venice of legend:

Most of the islands, of which there are many that do not properly belong to the town, are occupied; one containing a convent with its accessories, another a church, and a third a hamlet of the fishermen. The effect of all this is as pleasing as it is novel; and one rows about this place, catching new views of its beauties, as one rows round and through a noble fleet, examining ships.[15]

This ancient social world of the Lagoon was being systemically dismantled, however, and what the Serenissima had begun, the French and Austrians completed in the next century. By the mid-1800s, almost all the island monasteries had been deconsecrated and abandoned to fishermen, squatters, or picnickers.[16] Buildings that had once been extensive and solid—on Santo Spirito, La Grazia, and San Giacomo in Palude, among many others—were left to fall into ruin. Little island communities that had once served local shipping dispersed as boat traffic shifted to take advantage of the Austrians' new deep-water passage linking Mestre with the Bocca del Lido. In 1840, one entire islet was allowed simply to disappear, along with its twelve inhabitants, swallowed up by the shifting currents.[17] The old free-spending tourists, enjoying once generous exchange rates, were rarely to be seen on the Lagoon anymore: by the mid-1800s, few enough of them were even coming to Venice itself any longer.[18] An American who ventured out to Burano—a few years after Reunification, but while the area was still sunk in its torpor—wrote of his impression of the place:

[We] soon drew up in a bad-smelling canal, before a dirty, broad street or narrow market-place. It was arranged that we were to go in search of lace; so, leaving one of the party in charge of the boat, we pushed our way through the crowd of begging children to the lace school. We were accompanied all the way by one particularly dirty little girl with her long hair all over her eyes, who begged persistently and dolefully. . . . When we went back to the gondola, we found the landing place swarming with beggars. We were paddled away from Burano, followed along the sides of the canal by the troop of begging children, among whom the frowsy-pated little girl was still prominent and quite unabashed by her utter want of success.[19]

As the nineteenth century wore on, however, there were some signs of revival in this island world. Of particular importance was the flowering of an entirely new form of tourism on the previously deserted dunes and beaches of Lido. During the Serenissima, it had been traditional for Venetians to "resort in great numbers" to the beaches there, especially after a weekend of parties and gambling at home: these excursions were known as Lunì del Lido, "Mondays at Lido." Under Hapsburg rule, however, this twelve-kilometer-long barrier island was no place to visit. Especially after the failed rebellion of 1849, the Austrians discouraged (or explicitly forbade) such pleasure trips to Lido, whose beaches they had rather pointedly converted into gunnery ranges.[20] Nevertheless, Adriatic sea bathing was becoming increasingly fashionable throughout the 1850s, such that, promptly after unification in 1866, entrepreneurs of all sorts were rushing to promote the beaches on this otherwise deserted sandspit.[21] Though the waters there may strike many visitors today as rather tepid and tame, they were much admired in the nineteenth century: "There is a glamour about bathing in the Adriatic which affects one greatly," as one visitor gushed. "There never was such soft, warm water. It feels like warm oil."[22]

The booming enthusiasm for sea bathing soon brought big changes to Lido. When Henry James went there in 1869, he had found only dunes, with a few shacks and bath houses: "a very natural place," he had called it. On his return thirteen years later, he discovered to his dismay that in the meantime Lido had "been made the victim of villainous improvements" and was now crowded with lodging houses, shops, bad restaurants, and a "third-rate boulevard" cutting across the island from the Lagoon to the sea.[23] Initially there was some dispute between the communes of Venice and Malamocco as to which city Lido beaches actually belonged. But the real masters of the new boom town soon emerged as the bathing barons who capitalized it, building the piers, restaurants, hotels, and concert halls that made this once deserted strip of sand into an international tourist haven.[24] By the 1880s they had organized themselves into the Società dei Bagni, which, with the construction of the two great international tourist centers, the Hotel des Bains in 1900 and the Excelsior in 1906, was further consolidated into the Compagnia Italiana Grandi Alberghi, henceforth known as CIGA.[25] Before long, they were virtually masters of Venice itself.

Indeed, from the 1890s until the Second World War, Lido resorts completely shifted the tourist center of gravity in Venice, bringing into life the wholly new and desperately fashionable world that Thomas Mann so adroitly skewered in *Death in Venice*. The hot and humid summer months, once universally dismissed as "extremely unhealthy" in Venice, became trendy and swank, welding together a new Venetian high season that ran for the entire five months between Easter and the fall equinox.[26] Unlike Venice's historic center, which had a constant struggle just to keep its ancient tourist facilities and infrastructure functioning, the new establishments on Lido could offer visitors all the latest comforts and amusements. As the *New York Times* proclaimed to its readers in 1909:

Venice has of late years taken up a new role, and has become a fashionable bathing place. The Lido, which used to be a mere sand bank with a few fishermen's cottages, is now the centre of some of the finest sea baths in the world. Huge hotels have sprung up on this once desolate spot, and far into the Autumn the wide, flat sands are crowded with people in smart costumes or in bathing dresses. For not only are the sea baths fashionable, but also the sun baths, for which the visitors lie about on the sands all day in the scantiest of attire.[27]

CIGA and other Lido boosters were constantly on the lookout for new attractions, suitable for ever more and ever wealthier clientele. Indeed, much of Venice's present-day cultural landscape was established through CIGA initiatives in the years between the two world wars: the golf, tennis, and riding facilities were all in place by 1920, Lido airport was established in 1926, the Venice Film Festival, or Mostra di Cinema, got its start in 1932, and the grand Municipal Casinò, designed to compete with Monte Carlo in fleecing the profligate and wealthy, opened six years later.[28] Between the wars, the hotels and facilities out on Lido functioned in relation to Venice not unlike a modern-day cruise ship, as it anchors at a safe distance off an exotic port of call. Guests could luxuriate in their fully equipped refuge while still connected by steam launch to the art and romance of Venice: the city was undoubtedly more intellectually and aesthetically stimulating than even the most sumptuous beach resort, but it was also crowded, confusing, and not especially comfortable. In CIGA's "modern and elegant hotels, where all the more profitable foreign languages are spoken at the desk," visitors could enjoy the familiar privileges appropriate to their rank, with the restaurants, beaches, and hotels apportioned out into first, second, and third class. Lido was a *cité loisir,* "where no unauthorized person is allowed to trespass," and where prying or chattering locals would be kept safely out of the way.[29]

Blessed with such advantages, Lido reigned supreme over Venetian tourism until well into the 1950s, long after visitors could no longer expect to find its grand hotels "full of Italian princesses and countesses, most of whom were Americans."[30] Travel writers continued to praise the place, though increasingly for its "aura of faded and exclusive elegance." They promised their readers that even middle-class families of the postwar era, driving to town on the new Milan-Venice *autostrada,* could still take advantage of the frequent ferries that set off from the Terraferma at Fusina and from Piazzale Roma, and that would allow the visitor to "drive on broad, well-paved roads to the hotel of his choice, unload the car, send the children off to the beach, and begin to relax and prepare for a real Venetian holiday."[31] Such automotive idylls were not to last, however. As the number of Europeans vacationing in their cars ballooned in recent decades, the ferry system connecting Lido to the mainland has been unable to keep up with demand. Big attractions like the film festival—but also many ordinary summer weekends—now can overload the whole system and leave motorists waiting in line for hours. Nor is it always easy to take an ordinary *vaporetto* from the island into Venice: in the summer, one must often contend with the beach-going day-trippers, who can mob boats to the point where fights break out; in the fall or winter, heavy fogs sometimes shut service down altogether.[32]

Rather than grapple with such inconveniences, which still come with some of the highest prices in the entire Lagoon, many of those who wish to combine Venice and the seaside have begun to take their holidays at the newer beach towns: Cavallino, Jesolo, Caorle, and Bibione to the north, and Sottomarina to the south. All starting up essentially from nothing in the mid-1950s, these resorts together now surpass Lido by fifteen to one in vacationers, who moreover stay nearly three times as long, on the average, as do vacationers at Lido.[33] Venice is still close enough at hand to be reached in an hour or less, either by road, along the landward side of the Lagoon, or on the commodious *motonavi* that ply these waters, connecting the city to Chioggia and Punta Sabbione, the port of Cavallino and Jesolo. This almost ideal proximity of sun, beach (if not exactly surf), and high culture has given these beach towns a significant edge in the highly competitive world of Adriatic resorts. Since 1962 their advantage has been measurably improved, with the opening of the new Marco Polo Airport on the Terraferma at Tessera, a facility that not only superseded the old Lido airport, but is also conveniently situated midway between the twin poles of Venice and the beach towns.[34]

This has all looked good in the tourist brochures, although it has not always been so felicitous in practice, since mixing beach and cultural tourism in fact produces a head-on collision between two very different forms of tourist culture. Attracted by low fares (the *vaporetto* from Punta Sabbione to San Marco costs the same as one from Piazzale Roma), thousands of day-trippers come to the city every summer morning.[35] The boats are especially packed when it is cloudy or rainy at the shore and culture seems to beckon more brightly as a result: "The worse the weather," a tour guide once laughingly told us, "the more of them come to Venice."[36]

By the late 1970s, Venetians were complaining not only about the number of these visitors, but also of their declining quality. The problem was not simply that these were *turisti mordi e fuggi,* "eat-and-run tourists" (though they were that as well). Such day-trippers were after all nothing new and the natives have regularly disparaged them for years: for their ignorance about things Venetian and (above all) for bringing little income to the city's hoteliers or merchants.[37] Rather, this new breed of vacationers rankled because they seemed to be treating Venice as if the city were simply an extension of the seaside. Their transgressive behavior was especially condemned in Piazza San Marco: "San Marco Beach," as the Venetians have derisively taken to calling their most famous square. Here, on a summer's day, thousands can be seen picnicking on the steps of the Procuratie Nuove, tossing Frisbees and playing touch football in front of the Basilica, and sunbathing around the columns on the Piazzetta.[38] In response, many Venetians rallied around the somewhat mercurial Augusto Salvadori, a local lawyer, city councilman, and *assessore al turismo* in the mid-1980s. In 1985, Salvadori issued his manifesto that "Venice is sacred like a temple, and the tourists must respect it as such"; while *assessore,* he managed to push through a number of draconian regulations to curb what he and many others deemed inappropriate tourist behavior for the city.[39]

These *leggi salvadoriane* quickly gained a certain notoriety, though probably more for the embarrassment they caused many Venetians than for their lasting effectiveness. They forbade eating openly or playing ball games in the Piazza; men (and presumably women) were also to be heavily fined (around forty-five dollars, payable immediately) for going shirtless or sunbathing. It was, intoned the *Gazzettino,* "a question of decency and hygiene," which, when it happened "in front of the Basilica of San Marco also becomes a question of respect toward the sacredness of the place."[40] As with many edicts of this sort in Italy, however, the *leggi salvadoriani* were only sporadically enforced: the police were reluctant to issue citations to bewildered young foreigners who generally had no idea what they were doing wrong.[41] After Salvadore's governing coalition had been voted out of office, a more modest proposal was advanced: that the police hand out free "Respect Venice" T-shirts to the shirtless. This, however, had no better success: indeed, it was suspected that many young tourists were purposely stripping down in the hopes of getting a free souvenir of the city.[42]

Many locals have concluded that by now Venice has finally completed its long and sad descent from Queen of the Adriatic down to Belle of the Beach. The spirit of sandy, seaside casualness that already pervades Lido,

FIGURE 22. Casually dressed tourists in a gondola. (Photo by authors.)

Cavallino, and Jesolo has become almost as ubiquitous in the city's historic center, and many Venetians can only concede that trying to force visitors to Venice to dress up for the occasion is doomed by countervailing trends toward casualness that are the essence of both present-day tourism and postmodern living (see figure 22). Indeed, despite strictures by such local institutions as Harry's Bar to keep "beachwear" off their premises, this is the look and attitude that by now prevail throughout the city's public spaces, "where everybody can go around town [dressed] in bathing suits, without anyone finding fault with it." On all the hot summer mornings, one can see such vacationers meandering through the city, dressed in straw hats, sandals, sunglasses; carrying picnic lunches; walking the family dog; and sometimes even toting plastic buckets and shovels: "everybody in tank tops and flip flops . . . too bad their skin is milk white."[43] On top of this, more than a few tourists lately have started motoring over Venice from Jesolo or Punta Sabbione in their own little boats:

So it happens always more often in the Grand Canal [that] one comes across Zodiacs and motorboats of tourists—Germans for the most part, but a few of our own are there too—loaded up with people in their underpants who go on getting a tan while they distractedly check out the palaces that parade along beside them . . . [seemingly] without ever noticing that they are not on the beach but are going along the "main street" of Venice.[44]

Mass tourism also has had its impact on the little island town of Murano (current population just over five thousand), though sometimes in surprising ways. Despite its canals, a few remarkable churches, and the touches of Venetian-style architecture that one encounters here and there, it is unlikely that Murano ever would have found a place for itself in tourist circles if not for its main industry. As one Venetian official put it, "Without glass, Murano would collapse. Without glass, one could say that no tourist would ever set foot on the island."[45] Yet for a while, during the early nineteenth century, the city almost lost this vital lure, as glass making was converted into a industrial process throughout Europe, leaving Murano far behind in terms of both quality and quantity of production. Only toward the later part of the century did this little world begin to make a comeback, due in good part to a drastic reorganization that brought a few of the largest manufactories into positions of overall authority in the industry. Some of this rebound can be ascribed to the modernizing of the process that ensued under the leadership of artist-entrepreneur Antonio Salviati, but a good deal of Murano's success also derived from an increasing middle-class demand for knickknacks.[46] Operating in a sense as ancillaries to the booming vacationing centers on Lido, the Muranese cranked out immense quantities of "touristware"—glass vases, plates, perfume bottles, tumblers, pen holders, and cups, most of them etched or embossed with Lions of Saint Mark, Rialto Bridges, rowing gondoliers, and mugging Pantalones—all destined to grace bourgeois parlors throughout Europe and North America.[47]

Its ability to give visitors to Venice something to take home has made the glass industry dominant in Murano ever since, turning this little island cluster into something of a company town. Presently, more than 60 percent of Murano's workforce, or around eighteen hundred individuals, are involved in some way with glass: in production, supply, sales, or packaging and shipping. About a third of the Muranese work on their own, generally as artisans or shopkeepers heading small concerns of five or fewer employees; the rest labor for one or another of the twenty-seven industrial glassworks there. Altogether, they process some thirteen thousand metric tons of raw glass a year, in about four hundred kilns. A certain portion of their work is what is known as "Muranese art glass": everything from large sculptures to miscellaneous objets d'art, much of it worthy of a gallery, if not a museum. To promote this side of production, the Muranese set up a museum, the Museo del Arte Vetrario, not long after unification. They have also lobbied hard to attract financial support at the state and national level, most recently to fund training facilities in Murano itself, capable of passing on the local traditions and skills in the art.[48]

On the other hand, a goodly portion of the glasswork that does come out of Murano—indeed, probably the largest part, though accurate figures are hard to come by—is simple touristware: tchotchkes, knickknacks, and little collectibles for vacationers to buy and take home as souvenirs or small gifts; clown and harlequin figurines seem to be among the most produced. Thanks to the speed with which even a half-trained glass worker can turn out such trinkets, this end of the business can be hugely profitable, although many Muranese firms have tried to increase profits still further, at least according to common accusations. For some time, in fact, it has become "the rule" that most of this sort of glassware is actually produced in "Taiwan, Bohemia, and the hinterland of Campania [i.e., around Naples]," shipped back to showrooms in Murano, and "sold off as Venetian glass." If, as many accusers would have it, vendors can enjoy a tenfold markup on such sales, this kind of chicanery must simply be too tempting to pass up. Those who have mastered the art of fobbing off foreign-made glass as "genuine Murano" have acquired much of the wealth in the city, at the expense of the artisanal producer-sellers. Meanwhile, the tourists who buy such spurious goods, thinking they are purchasing Muranese glassware at Muranese prices, may then get home to discover not only that these plates, vases, and trinkets were produced somewhere else, but also quite possibly that "they could have bought [the same things] right at home, at decidedly lower prices."[49]

Many of the smaller producers lately have begun raising the specter that their city is "turning into a branch office of Taiwan" whose only function is to resell these foreign-made knickknacks to tourists at a 1,000 percent markup. They have attempted to counter this practice, though without much success so far, by pushing for the requirement that all glassware sold in the city carry its own certificate of origin, in much the same way that Italian wine and cheese are identified.

Despite the profits they can make as clandestine importers, many glass outfits continue to run their furnaces, if only for the central place that glass production plays in the whole packaged tourist experience of Murano. A trip to a Muranese glass factory is a featured part of many tours to Venice and promises tourists a chance not only to shop for glass souvenirs, but also to enjoy the spectacle of glass production itself. Entire busloads of visitors get herded onto the big *lancioni* (often owned by the manufacturers themselves) and taken from Piazzale Roma or Tronchetto onto docks in Murano that lead directly (and only) to one producer's special workshop-theater. Here they can enjoy a demonstration of glass working, which in our experience usually has consisted of an artisan forming a figurine by dabbing on different colored blobs of glass; rarely can one expect to see actual glassblowing. Then the tourists are led out into the vast salesroom for a chance to buy something in glass. In the high season, this process of show-and-shop can move along with something of the same industrial efficiency that the gondoliers employ, as demonstration rooms are filled and emptied of their several score potential customers every half hour or so.

Although the whole process might strike many as rather hucksterish, the efficient twinning of actual glass production with the opportunity to buy the results has proven enduringly popular with tourists. From what we have gathered, visitors tend to rank Murano and its glass as second only to gondoliers as essential to and emblematic of their Venetian experience. Perhaps this explains why so many tourists combine the two metonyms in one: they come back from their Venetian visit with a miniature glass gondola.[50]

Moreover, rich with the profits they reap from this system, the bigger glass companies have managed to extend their influence beyond the modest borders of Murano itself. There is indeed said to be a "cartel" of the largest manufacturers and retailers—said, that is, primarily by those who have felt excluded from the profits and "left only the crumbs." It is an informal but effective organization that has managed not only to monopolize the traffic in foreigners coming to Murano, but also to impose itself on much of Venice's own tourist industry. In 2000, the region's antimafia prosecutors opened an inquiry on assertions made by Roberto Magliocco, president of the Associazione dei Commercianti (Venetian Merchants' Association; ASCOM), that the cartel accomplishes this by maintaining trade-restrictive connections with travel-agency networks in the tourists' countries of origin, allowing it to bid for control of large blocks of visitors, like any other commodity. According to Magliocco,

It's a branching system, which begins right from the foreign travel agencies—for example, the Japanese [agencies]—to whom the tourists apply, and [the system] accompanies them throughout all of their Venetian stay, involving hotel porters, taxi drivers, gondoliers, restaurants, guides, and, naturally, also the glass retailers. . . . Formally, there doesn't seem to be anything illegal here, because this chain begins long before [the tourists] arrive in Venice.[51]

In theory, this has meant that, once a tour group comes to the Lagoon, its entire itinerary (and thus all of its spending) is controlled by elements of this cartel. The process begins with the *intromettitori,* the "meddlers" or agents, who meet the group as it gets off its bus in Piazzale Roma or

Tronchetto and lead the tourists to the taxis or *lancioni,* whose job, in turn, is to ferry these innocents to the appropriate glass shops on Murano. In this fashion, the shopping of tens of thousands of customers, said to spend upward of $25 million on glassware every season, can be controlled, assuring that they never visit or, often, are even aware of competing glass shops, even though these might be right next door. Moreover, the cartel's involvement continues after the tourists leave the showroom, for (again, according to the denunciations) its agents then lead them to a member restaurant, after which they supply the necessary transport and guides for a quick visit to Piazza San Marco, and finally provide the taxis that will take everyone back to the bus—a circuit that can be completed in as little as five or six hours (though overnight lodging can be thrown in, if it is part of the package). Enriched from this largely captive market, which can expand almost without limit, thanks to their "deluge of false glasswork imported from Asia and eastern Europe," the big glass dealers can hold off demands that only Murano-made glass should be sold in Murano through the influence (so it is claimed) that their wealth has bought them among local politicians.[52]

In recent years, the Lagoon's island towns have for the most part followed Venice in its downward path of population decline, the rate of which is somewhat slower than the city center's, but not much.[53] Some of them have long since passed that murky demographic threshold where local residents are too few to provide essential services, beyond the point where such towns can really be said to function any longer as communities: shops close down, schools are practically deserted, houses are left empty.[54] Having lost half their populations since the 1950s, Burano and Sant'Erasmo no longer have the demographic clout to persuade the Azienda del Consorzio Trasporti Veneziano (Venetian Municipal Bus and Vaporetto Service; ACTV) to give them any but the oldest and most beat-up *vaporetti* and *motonavi*.[55] The Buranelli tried unsuccessfully for over a year to get one of their two public telephones repaired; even so, they were better off than the few remaining residents of Torcello, whose only telephone simply disappeared one day, box and all. Lido, home to one of the world's best-known cinema festivals, no longer has any regular cinema of its own: the last one closed years ago, and during the winter months it can be hard even to find an open bar.[56] Sant'Erasmo, home to eight hundred residents, can go for days without mail when its (only) postman falls ill; nearby Vignole (with a population of sixty) no longer has any postal service at all. Neither island has been able to persuade either a pharmacy or a first-aid center to locate there: one can imagine how the residents of

Vignole must have felt when, on top of this, the island's only church became so dilapidated that it had to be closed.[57]

For many of those who live in these beleaguered estuary communities, foreign tourism has long represented a salvation from looming extinction. Though these towns may lack Venice's special glamour and unique place-ness, they still have their own attractions, if only in the sense of offering a kind of anti-Venice: they are island museums conserving the way things used to be in the Lagoon, without the tawdry pandering so linked to Venice itself. Travel writers occasionally pick up on this theme, presenting these backwater communities as alternative destinations for the more discriminating (and leisurely) tourists who might wish to "turn their back on the herd" at San Marco, to enjoy instead the "somnolent, heat-soaked, late afternoon silence" for which they would search in vain back in Venice.[58]

This desire to be near Venice while not actually in it, to sample this extraordinary estuary world before it disappears for good, helps explain the lasting appeal of Burano, which—so far at least—lacks any hotels, though it does feature several good restaurants. Requiring nearly an hour to reach on the *vaporetto,* Burano is beyond the limited schedule of most package tourists. In actual fact, it does not provide very much to do, or that much to look at, beyond its many brashly painted houses and its very few old women who still carry on, in public view, the art of lace making. It is, however, still something of a genuine Venetian Lagoon town, where one can see "real" locals living their lives in a quiet world of canals, *campi,* and boats that seems not so different from the way Venice must once have been: girls are jumping rope and fishermen are bringing in their catch, while women gossip and shop.[59]

This same desire to obtain an idyllic spot, anchored like a secure cruise ship, near attractive, chaotic Venice, also helps explain the recent scramble to buy up or rent and develop the Lagoon's various abandoned islets. Many of them have been deserted for decades: no one wanted anything to do with these lugubrious little anchorages, covered for the most part in tangled mats of vegetation, their mud littorals deep in beached garbage, swarming with rats, "abandoned, ruined, crumbling, disastrous."[60] For much of the last century, it was difficult even to give some of them away, as the local diocese discovered in 1973, when it sought to unload the little islet of San Giorgio in Alga on the very unwilling communal government of Venice.[61]

All this has changed since the late 1990s, however, and the Lagoon's many uninhabited islets have become of great interest to a variety of potential buyers and lessees. Although most of them only amount to be-tween one and five hectares of dry land and none has any natural source of fresh water, these little havens offer the undeniable lure of splendid iso-lation and tranquillity within just a few minutes' boat ride of one of the world's most beautiful, but crowded, cities. "Why go to the Caribbean," asked *La Nuova*, "if two steps from home there is a paradise for sale, within reach?"[62] Many islets have belonged for years to entities that have no use for them and lack the funds to develop or even maintain them: the Church, as already noted, until recently held a number, as the sites of ex-convents; the military had title to various little islands and fortresses of no particu-lar use anymore; likewise, the state health service had inherited a number of island hospitals, some of them dating back to the days of the Serenis-sima. These institutions—along with the city itself, which owns more than a few—have for the most part proved more than willing to unload these rat-infested, heavily vandalized eyesores: if possible for virtuous purposes, but unload them in any case.[63] Fortunately, there has been financial help available, from the city's "Special Fund" and from the European Union, to turn at least some of these islands into parks. These have so far included the fairly large Certosa (lying very near the eastern end of Venice itself), as well as Campalto, Lazzaretto Nuovo, and San Giuliano.[64] Since none of these lies along existing routes for the public *vaporetto*, however, it is unlikely that they will play much of a part in the tourist experience of the Lagoon.

On the other hand, several of the thirty-eight or so available islets have been specifically groomed to serve foreigners. In 2001, the city handed San Giacomo in Palude over to the environmental group Verdi Società Ambiente, which intends to turn it into an environmental-studies center and an archaeological museum of the Lagoon. Anyone who is traveling on the *vaporetto* between Murano and Burano will have a chance to stop off at this little way station for an educational tour of the estuary's ecol-ogy and history.[65] Completely dwarfing this modest educational effort, however, is Venice International University, an entire campus that has been set up on the island of San Servolo, just to the southeast of Venice, by a consortium of European and American universities. Taking over the building complex of an old psychiatric asylum, the VIU, with its con-tingent of students from both sides of the Atlantic, is the first (pre-dominantly) foreign university in decades to set up shop in Venice: its location on San Servolo, midway between Venice and Lido, almost guar-antees an increased foreign interest and involvement in this stretch of the Lagoon.[66]

For the fortunate few, it is (or at least it was, briefly) possible to snap up one of these islets purely for personal use. Tiny Tessera, for example, lying between Murano and the airport, has become a "private cultural center," while a local industrialist bought the islet of Craven, just beyond Burano, for around a million dollars at a state auction; San Giacomo delle Polveri and Santa Cristina have also passed into private hands.[67] Others who want to enjoy the privacy and tranquil isolation of island living without actually buying one of their own will be able to find these attractions in the luxury hotels that are going up on two islands just south of Venice. Having learned from the great success of the exclusive Hotel Cipriani, walled off on the tip of the Giudecca and accessible only by private launches, the financial group Benetton, SpA, has purchased the islands of San Clemente (for around $18 million) and Sacca Sessola (for around $12 million), with the intention of opening elite hotels on each, with nearly a thousand beds between them.[68] Other investment groups interested in bringing tourism out to the Lagoon also have picked up islets: a consortium from Chioggia, for example, took over San Secondo—the Lagoon's original "Rat Island"—on a lease for around six thousand dollars a year, with the intention of developing it "from the environmental and touristic point of view . . . as a meeting place for young people of all ages."[69]

As development projects such as these continue to be realized on the islands, traffic—and especially tourist traffic—will certainly increase throughout the Lagoon. The elite and ecotourism that will be on offer here has produced its share of positive results, as steps have been taken to preserve (or at least survey) the vanishing tidal flats, to protect local wildlife, and (above all) to clean up decades of accumulated trash and pollution.[70] On the other hand, to truly realize the full tourist potential of these islets, some operators have tried to adapt them to accommodate large numbers of visitors, many of whom necessarily have to move through the space fairly quickly. Such practices can readily damage the target islands' fragile ecosystems and what survives of their cultural heritage.

This can nowhere be seen better than on Torcello, in the recent events that many locals referred to as *lo scempio di Torcello,* the ruination of Torcello. Though one of the least populated of the Lagoon's remaining inhabited islands, with a mere twenty residents (the twenty-first died in late 2000), Torcello also boasts one of the estuary's most striking architectural monuments—the thousand-year-old cathedral of Santa Maria Assunta, the oldest and best-preserved example of Byzanto-Venetian architecture in the entire region.[71] For this reason, as well as for its small archaeological museum and five-star restaurant-inn, the Locanda Cipriani, Torcello offers something of an ideal Lagoon getaway after a crowded, jostling day in Venice. It has become increasingly common for tour guides who are looking for something new for their clients to take them there, shaving some vital time off the forty-minute *vaporetto* trip by loading their charges onto the much quicker *lancioni di granturismo*. These big, glassed-in tour boats, coming directly from San Marco, Murano, or Piazzale Roma, have proved capable of shuttling hundreds of visitors an hour in and out of Torcello, leaving this once-tranquil spot "swarming with tourists": upward of around two hundred thousand or so visitors annually, making this the fourth most popular attraction in the area, after the Basilica of San Marco, the Ducal Palace, and the Galleria dell'Accademia.[72] Small as it is, the island, with its narrow canals, miniature *campo*, cathedral complex, and access quay, can seem as crowded as Piazza San Marco in the high season, often presenting a continuous carpet of humanity that stretches from the landing dock to the solidly packed Santa Maria Assunta.

To help accommodate so many tourists, who needed to get to Torcello's church complex from docks on the other side of the island, the Magistrato alle Acque and the Consorzio Venezia Nuova—together responsible for maintaining the Lagoon—decided in 1999 to drastically widen the landing and the quay, known as the Fondamenta dei Borgognoni, essentially doubling both of them in width and rebuilding their foundations with modern materials that were then camouflaged with traditional poles and stones. Luciano Scalari, architect for Italia Nostra, complained that the island was becoming unrecognizable: "Unfortunately these works are deranging the original aspect of the land," he noted, "and their enormous structural foundations do not find, in our opinion, any reasonable justification; it scares us to see that there has been set off a perhaps irreversible process to falsify Torcello." Many Venetians agreed with Scalari, angry that the original wooden polls, tamerisk matting, clay, and stone had been brusquely torn up and replaced by a rigid structure of reinforced concrete. In late May 2000, anonymous posters were plastered up around Venice and the Lagoon towns, lamenting,

Goodbye, Torcello! The Magistrato alle Acque is spending 28 billion [*lire*, or $14 million] to devastate the island with cyclopian works in steel and reinforced concrete. Of the very ancient charm of the island there will not remain a trace. Visit it while you still can, because before too long you won't recognize it anymore.[73]

Such "cyclopian works," many preservationists exclaimed, were "more appropriate to the banks of the Suez Canal" (or the port of Rotterdam) than for an island as "extremely fragile" as Torcello. Many suspected that the main motive behind this massive intervention was not so much to enable tourists to stroll more comfortably from the *vaporetto* landing to the cathedral as to make an all-water access route that would allow taxis and *lancioni* to push right to the center of the island, for those tour groups in too much of a hurry to walk the necessary five hundred meters.[74] The widening and straightening of such access canals had, in fact, been in the cards, and their banks were also to have been reinforced with concrete, in a way that would allow the transit of such "heavy vessels" as the *lancioni*.[75] It did not take much imagination for opponents to visualize what would come of all this: the space right in front of the cathedral, at the heart of what had until recently been called the island paradise of the Lagoon, as jammed up as the Bacino of San Marco with exhaust- and noise-spewing taxis and *lancioni*.

Public mobilization and protests got the work halted, at least temporarily, though it was too late for much of it to be undone. The "Torcello Case," as it would be known, also directed attention toward similar restructuring projects that were designed to expedite tourist flow elsewhere in the Lagoon: it was soon discovered that plans were afoot or already had been carried out for similar "hard" implants in such disparate places as the Giudecca, Fondamenta Nuova, Mazzorbo, and Sant'Erasmo.[76] Some observers simply shrugged at all the furor, saying that something of this sort had to be done to handle and speed up traffic on the increasingly busy Lagoon. Indeed, in 1998 alone it was calculated that over twelve million daytime passengers traveled on those waters, and somewhat more than two million of them were tourists.[77] Traffic problems are compounded by the many bottlenecks the Lagoon presents to the mass movement of people: not only do islets like Torcello, in their natural condition, offer only limited points of access, but the Lagoon itself is for the most part much too shallow for any boat larger than a *sandolo*. Just a very few channels are available for transporting visitors who want to see the sights of the estuary, and these sinuous and often narrow routes have become increasingly jammed up by dozens of *lancioni* and taxis whose pilots are seeking to help them do just that.[78]

Overcoming the Lagoon's stubbornly unhelpful topography has been a primary goal of Venice's rulers (though much less so of the Venetians themselves) since the days when the Austrians first decided to connect the city to the mainland with a railroad bridge. As long ago as 1912, substantial plans were drawn up to build a Venetian subway, an underwater tube whose special advantage would lie in linking the booming resort at Lido more firmly to rest of the city and the mainland.[79] Since that idea collapsed—thanks mostly to the First World War, but also under the sheer weight of its own expense—other schemes have been periodically proposed. These have included an elevated monorail; an underwater railed *sublagunare;* and, more recently, a "people mover" to carry visitors on a sort of moving beltway. The hundreds of millions of dollars most of these engineering fantasies would cost, along with the almost certain opposition they generally would face from environmentalists worldwide and from many Venetians, have generally guaranteed their eventual abandonment, but the very fact that such projects are continually coming up again speaks to the perceived need for getting more people—especially more tourists—around the Lagoon more efficiently.[80]

The problem remains, therefore, of how mass tourism is to be brought into balance with the Venetian Lagoon. It is a problem that will almost assuredly get worse, as more of the estuary's islets are opened up as parks or hotels and as ever more visitors want to see something beyond the increasingly crowded and cliched center of Venice itself. If the Torcello Case is any indicator, the first choice of those who are developing these potentially lucrative attractions will be the massive application of bulldozers and cement, to produce the sorts of island paradises that they see as most adapted to tourist needs. It is true that the use of reinforced concrete is officially forbidden in the Lagoon, but when government magistrates themselves permit it, as happened on Torcello, it is hard to believe that corporate developers will not try something similar in the relative seclusion of their own private islets.[81] In that case, we can expect some massive changes yet to come in the Venetian Lagoon, as one by one these vanished island communities are permanently converted to little havens for the residence and pleasure of foreigners.

CHAPTER 8

Dangerous Waters

We were once lounging on one of the little docks that stick out a short distance into Venice's Grand Canal, watching the traffic go by and enjoying the sun, when we heard a shriek. From a nearby gondola, just emerging from an inner canal out onto the open waters, came the voice of an American mother berating her young son: she had just caught him dragging his hand in the water, as boys will do: "I told you, don't ever do that! I told you this water is filthy! People poop in it! Think of the diseases you could catch! I can't believe you did that; this is so disgusting!" And with that she rummaged around in her knapsack and pulled out a half-dozen premoistened towelettes, ripped them out of their packaging (under the bemused eye of their gondolier), and essentially gave her squirming son a bath all the way up his arm to his neck, and then did his other arm for good measure.

Of course, not every visitor shows such aversion to Venetian waters: on hot summer afternoons it is not unusual to see (mostly younger) tourists dangling their bare feet in canals. Nevertheless, the notion that Venetian waterways are both putrescent and dangerous—"murky canals filled with sewage," as one travel writer put it—is one of the most widely held stereotypes about the city; among foreigners we spoke with, it ranked second only to the belief that "Venice is sinking."[1] This is an old refrain with tourists: well over three hundred years ago, Jean Gailhard was complaining about "there being much filth and corruption cast into the Channels, some of which stink very much."[2] Even in an era when metropolises like London and Paris were renowned for their stench, many British and French travelers still found the Venetian canals particularly revolting and wrote back home of how one's sense of "smell is overwhelmed with the stench, which exhales from the water"; how "the canals are becoming choked up, and the stench from them at low water is often dreadful"; and (rather plaintively) how "the canal in the place where I lodge stinks very much."[3] They gave the city's often disturbing smell as one of the reasons why those on the Grand Tour were less likely to take up residence in Venice for several months, as many did in Florence or Rome: "the stench of the Canalls," noted Nicolas de Fer around 1700, was a principal reason why "travellers care not to reside long in the City, when they have once seen the Curiosities that beautifie it."[4]

This is no longer much of an issue these days, since very few of Venice's visitors stay there for any length of time anyway. Admittedly, this has more to do with the scheduling imperatives of mass tourism than with personal choice, but none of those we spoke with ever said that the canals stank so badly as to make them cut their visit short. Nevertheless, almost all tourists hear about the supposed pollution of Venice's canals before they come, a situation that many find somehow offensive, though in more of a moral than aesthetic sense—an indication of the city's failure to provide them with the right kind of holiday, one with a proper set of romantic memories. This was especially true among those who had booked a gondola ride and then found themselves in a rather more intimate proximity with Venetian waters than they had originally bargained on. Paying all that money to be paddled along through a sewer, as some see it, certainly falls short of a visitor's roseate expectations, however much one may have enjoyed the jaunt itself. Unquestionably, certain tourists seem to be rather gratified to catch Venice coming up short in this fashion: it perhaps helps make up for a gnawing sense of cultural inadequacy that seeing so much historical and aesthetic splendor concentrated in a single place can induce. Others, however, seem to be genuinely saddened as much as repulsed when they encounter floating garbage, human feces, or swimming rats while out on the waters.

Perhaps it is simply a matter of surprise. Nowadays, most of those who can afford Venice come from cities with reasonably effective plumbing, where they have gained blessedly little experience of an antiquated sewage system like Venice's, where the embarrassments produced by human beings are not all carefully concealed and piped away. Since in Venice the same element of infrastructure always has been expected both to (as Thomas Coryat put it) "carryeth away all the garbage and filthinesse" and to "serve the Venetians instead of streetes," it has been inevitable that human movements and human waste often have come into uncomfortably close contact.[5] This situation has been far more likely to disturb the foreigners than the Venetians, however, and even today, on days when the canals really do stink, it is the tourists who make various, sometimes intentionally exaggerated, gestures of revulsion. Venetians, on the other hand, seem to have as part of their cultural heritage a capacity to simply ignore the stench and to keep on using their canals: one often sees them rowing about among the rotting garbage and scooping water from the smelliest of canals to wash off their boats. They like to tell foreigners that the stink comes from the "buildup of naturally occurring algae and silt rather than sewage," which strikes us as a charming, if questionable, piece of dissimulation. No doubt they are simply used to it—another sign that marks the "true Venetian," as Johann Wilhelm von Archenholtz noted over two hundred years ago: "The infectious stink, inseparable from the place, is for some time exceeding loathsome to those who arrive. But by degrees the nerves of smells are hebetated, and this exhalation, which pervades the whole town, becomes supportable."[6]

Yet for centuries, Grand Tourist complaints notwithstanding, Venice was objectively one of Europe's cleanest cities. Like all their contemporaries elsewhere, medieval Venetians dealt with their waste mostly by just flinging it out into the streets; the difference here, of course, was that the streets were made of water, and twice each day the tides conveniently swept all the garbage and sewage out to sea. Visitors as far back as Pero Tafur, in 1438, admired how "the sea rises and falls there . . . and cleans out the filth from the secret places." Without the steady accumulation of trash, animal waste, and sewage that was the norm elsewhere, the landscape of Venice, its *calli* and *campi,* was the marvel of travelers, who enthused about the "dainty smooth neat streets, whereon you may walk most days in silk stocking and satin slippers, without soiling them, nor can the streets of Paris be so foul, as these are so fair."[7]

In our time, however, the flushing action of the tides that once had Venice setting the standards for medieval cleanliness turns out to be holding the city back from bringing its level of hygiene up to present-day norms. Since Venice sits right at sea level, what sewer system it has (and there is one there, almost unimaginably complex and arcane) can only function when the tides are ready to cooperate, which is to say, twice a day, and sometimes not even then. The only real solution would be to install (as the Dutch have done in some towns) an active sewage system, with a pump network powerful enough to actually suck the waste out of the city, after which it would be deposited at some other (undesignated) location. Such a project—for which, not surprisingly, the engineers say "there will be no standard solutions"—is currently estimated to cost at least $250 million; it would also, while work was being done, thoroughly disrupt the city and its tourism for years.[8] Given the price and the upheaval involved in installing such a system, it is no wonder that both Venetians and the Italian state have managed to prevaricate about this for years. An underlying motivation, it would seem, is the long-held conviction that, as more Venetians move from the historic center to mainland suburbs like Mestre (which have modern plumbing), the problem will inevitably diminish on its own.

Unfortunately, this has not exactly been the case. Residents have left the city, of course, but the tourists have flowed in, and, as far as sewage production is concerned, they have more than made up the difference. The city's hotels, with roughly fourteen thousand beds, make just over three million bookings annually, or eighty-two hundred a night on average—the equivalent of a decent-sized town permanently in residence in Venice and a not inconsiderable burden on the local sewage system, such as it is.[9] Even the day-trippers, despite their constant complaints about not being able to find a toilet, still leave a generous share of waste, not only in the public toilets (where they can be seen waiting in long lines on every summer's day) but in whatever restrooms they find in the restaurants, bars, and museums they visit. To deal with this rising problem, some restaurants and hotels have arranged to channel their sewage to outfalls on the outskirts of the historic center; others, in response to recent antipollution laws, have put in septic tanks.

Unfortunately, neither system is really up to dealing with peak tourist flows, which in Venice involves roughly two-thirds of the year. The septic and settling tanks in particular can be overwhelmed in the high season, which is why the city in the summer is increasingly crowded with "an entire flotilla of tanker boats *[bettoline]* . . . that periodically has to take care of these great reservoirs under the floors of restaurants and hotels." When walking along the *fondamente,* one often has to step over the large hoses snaking out of these boats, often stretching for a hundred meters or more to reach the cesspool openings that may lie deep within an adjacent hotel or restaurant. We are used to hearing the big engines of the pumps before we smell what they are doing, but by now the combined smell of diesel exhaust and sewage has become a regular summer feature of Venice. In any event, pedestrians often can count themselves more fortunate than clients of the hotels or restaurants involved: not infrequently the operators of the *bettoline* have to run their tubing right through the dining room or lobby of the establishment.[10]

Unless the authorities keep after them, many restaurants and hotels are either lax or illicit about the way they handle their sewage. Especially in Venice's Tourist Core, where such businesses are constantly changing hands, expanding into neighboring buildings, or adding new facilities, cases often come to light in which "more than one proprietor impudently connects his discharge pipes to those already existing, forcing on them [such] an overload . . . that the sewer network explodes."[11] Others hire cut-rate *bettoline* operators who make their profit by evading the required environmental checks (and charges) on proper disposal of the waste they suck up, simply dumping their loads off in some unsupervised corner of the Lagoon.[12] Septic tanks, moreover, are required only of those hotels with more than one hundred beds, and even establishments of this size might not have them. Smaller hotels and pensions are free to dump the sewage from their guests right into the adjoining canals, though it would seem that some go to the effort (perhaps for the sake of the guests themselves) to pipe this waste a few hundred meters away before dumping it.[13]

Despite the apparently dire state of Venice's sewers and the frequent complaints of tourists about the canals, it does not seem that dumping untreated sewage directly into the water has historically poisoned the Lagoon, even when the city was far more populated than it is today. Jean Gailhard claimed that this was due to "salt water being no friend to corruption," but the Lagoon has in any case always tended to purify itself, with solid residues either settling out or being carried off in fairly short order by the tides.[14] The result was that the deeper waters of the Lagoon were for years considered especially clean and limpid, winning the reputation in the earliest years of the nineteenth-century sea-bathing craze as a safe alternative to the open sea for medicinal and recreational swimming. Indeed, visitors could pursue the sport right inside the city, thanks to any number of bathing establishments that for the most part were set up on the Grand Canal. Some of these were little more than bottomless boathouses, anchored on poles driven into the sediment of the canal and accessible from a gangway connected to a quay or the interior of a hotel. Others were much more substantial, done up in sumptuous Oriental or classical motifs: one of the best known of these in the 1850s was run by the adjoining Hotel Grassi, which occupied the well-known Palazzo Grassi. Such was the draw of these bathing houses that, in 1853, the city administration approved plans for building a mammoth "thermal hotel" on the Riva degli Schiavoni, not far down from Piazza San Marco. Fully six hundred meters long and forty-six meters wide, this behemoth was to offer "three ample pools for beginners and [experienced] swimmers;

150 tubs in proper rooms, more than half of which with attached lavatories; and every opportunity for baths of salt, fresh, sulphurized [water], steam, and mud." To build it would have required substantially widening the Riva degli Schiavoni and rerouting navigation on the Bacino of San Marco. Still, its promoters argued that this was just the attraction to put Venice back on the holiday map: that there would be plenty of "rich bourgeois tourists" wanting to take a bathing break from their cultural vacations and willing to pay for the privilege of doing so in the Lagoon's unique waters.[15]

As it happened, this particular architectural assault on the city was never realized. The Austrian occupiers feared that it might block troop movements along the Riva degli Schiavoni; they also suspected that foreign tourists lodged on the top floor of the hotel might spy into their army's fort on San Giorgio.[16] Thereafter, those who wanted to swim in the Lagoon waters had to make do with floating baths, in particular the one originally constructed by Dr. Tommaso Rima in 1833 and then much amplified and improved over the course of the century. Moored right off the Punta della Dogana, the Grande Stabilimento Galleggiante, Bagni di Rima was itself a good two hundred meters in length and occupied a considerable piece of the Bacino, providing its clientele with all the much-touted benefits of salt-water bathing in a variety of public and private pools, appropriate for both sexes and all abilities.[17]

Not that everyone, foreign or Venetian, felt compelled by modesty to bathe within the protected confines of a bathhouse. Whenever it got hot enough, even respectable visitors (as long as they were male) saw nothing wrong in imitating Lord Byron's legendary swims up and down the Grand Canal: William Dean Howells, American consul in the city between 1861 and 1865, for one, simply used to leap off the landing of his consulate right into the water, to "spend a rather informal half-hour in the midst of my neighbors."[18] In this, they were only imitating the working-class Venetians, especially the children, who routinely used to go swimming in the canals. James Jarvis, visiting the city in the summer of 1880, professed to be captivated by the aquatic capers of the locals:

Venice on a hot Summer's early eve. . . . At this hour Venice becomes doubly amphibious. Her canals are alive with her male population, men and boys, and very many little maidens, too, in cloth fig leaves, sporting in the waters like so many dark-skinned Polynesians. They dive, they gambol, they shout, they splash, they make the old walls and slimey waters merry with their cries and laughter, while their nude, white bodies come out against them in shiny, dripping relief, like so many figures of a far-away primitive world where innocence still rules supreme.

The custom of canal bathing continued well into the twentieth century in Venice, keeping up an old tradition that small children were taught to swim right off their own steps, "with ropes around their waists, securely fastened either to [their mothers] or the door-rings."[19] Many older residents fondly recall such experiences, and some youths still go for a swim today, but unquestionably the old enthusiasm for swimming in and around Venice has waned. One reason is certainly that the practice has been made illegal—as any tourist brash enough to jump in will soon discover.[20] Perhaps also, the local youths, who are traditionally much less worried about the legality of things, have become shy about disporting themselves this way because they would have to do so under the lenses of dozens of tourist cameras. This may be why those who still want to cool off on a hot day usually choose to boat out into the open Lagoon north of the city. If no boat is available, chances are they will take their dip on the edge of town: the "upstream" side of Venice, tidally speaking, on the Grand Canal out past Piazzale Roma or on the Cannaregio beyond San Giobbe. This is where we have seen boys taking a swim as recently as the summer of 2001, enjoying themselves at a comfortable distance from both the center city's effluvia and its prying tourists.

Unquestionably, Venetians also swim there less than they used to because they have finally had to accept what visitors have been saying for years: that the waters around the city really are polluted and dangerous. That this has indeed become the case is somewhat peculiar, however. Even with its permanent tourist presence, Venice's current population is still barely half the 150,000 who typically lived here (and used a much more primitive sewage system) throughout the nineteenth century, when hundreds of tourists swam regularly—in the Lagoon, in the floating baths of the Bacino, and down the Grand Canal—and lived to tell about it.[21] Some of the problem is due to industrial pollution from the great petrochemical production complex at Marghera, on the mainland. For years these factories dumped wastes into the Lagoon almost without supervision, and, although tighter laws now mean that over 80 percent of the effluent of these factories is treated, both the chemical plants and the oil tankers that supply them are still often accused of secretly releasing illegal materials; some of the heavy metals that they dumped in the past will in any case take years to break down.[22]

A more visible problem has been produced by agricultural runoff into the Lagoon, in the form of pesticides and fertilizers from the roughly nineteen hundred square kilometers of surrounding farmland. Thousands of tons of nitrates and hundreds of tons of phosphates flow into the estuary every year, producing a chemical imbalance in the water and some-times triggering eco-crises of the first order. Between 1988 and 1992, in particular, excessive nitrates led to enormous blooms of *Ulva rigida,* a fast-growing, nutrient-loving macroalgae that blocked the sunlight from the Lagoon's shallow waters and quickly consumed all the available oxygen. Soon it was killing off most of the local sea life, and then, as it died and rotted in huge mats, it began to fill the air with throat-clenching concen-trations of hydrogen sulfide. Only a few life forms could prosper in such hostile conditions, and unfortunately one of these was the (fortunately nonbiting) chironomid mosquito, which promptly bred in the billions, filling the evening air throughout the Lagoon with swarms so dense that people found it difficult to breathe. Short of completely restructuring water runoff patterns into the Lagoon, there was little that the authori-ties could do beyond the Sisyphean labor of harvesting and carting off the huge sheets of algae. Fortunately, however, after a few years of repeated and highly disruptive blooms, *Ulva rigida* suddenly subsided back to nor-mal levels, and Venice was given a reprieve from this annual tribulation. Why the algae went into decline has never been fully understood, though some credit must go to more effective controls on agricultural runoff. On the other hand, perhaps the collapse was just a passing population read-justment: indeed, in the fall of 2000, new blooms of *Ulva rigida* were spotted in the central Lagoon.[23]

The problems with water quality that most concern the historic cen-ter of Venice, however, are more likely to be of the city's own making. As with hotel and restaurant sewage, tourism generally plays a part in this degradation, though the connections are not always obvious. It was, for example, partly out of a misguided attempt to cater to foreign sensibili-ties and protect hotel profits that Venetian authorities abandoned the tra-ditional dredging of the city's canals in the postwar years. For centuries, both during the Serenissima and after its fall, Venice's canal network was cyclically cleaned out, with each canal in turn sectioned off, drained, and then shoveled clear of all its accumulated muck.[24] With only shovels and wheelbarrows, the work was slow, but over the space of about twenty years it was possible to do the entire city—at which point it was time to start the whole cycle over again. Laborious as it was, the job had to be done: the more the canals were allowed to clog up, the less efficiently the tides would clean out the city. Moreover, silted-up canals—whether within the city or out in the Lagoon—were a tangible sign that Venice was not doing well in the millennial struggle to keep the entire Lagoon from silting up. When those canals began to stink, according to eighteenth-century wisdom, the Senate would grow nervous that the battle was in danger of being lost.[25]

Despite some lapses, this mucking-out process continued after the fall of the Republic, as first the Austrians and then, after 1866, the Venetian authorities themselves made sure that sewage and sediment never got the upper hand.[26] After the Second World War, however, the whole procedure was abandoned—partly to avoid the expense, but also, we were told, in response to pressure from foreigners. Those non-Venetians who had bought themselves second homes on these canals complained about the stench and inconvenience when their own stretch of waterway had to be cleaned; hotel owners in particular were afraid that a long and smelly canal cleaning near their establishment would drive away their yearly share of customers. As a result, for over thirty years the canals were left to quietly silt up. By the mid-1990s, the whole situation was becoming desperate, and one got the impression that some canals were filled with more mud than water. Finally the communal council voted to start the dredging again, pushed in part by another powerful segment of the tourist industry, the gondoliers, who complained that the foul water was driving off their fares. Insula, the state-private engineering consortium charged with the monumental task, has been progressing steadily, if slowly, through the city—slowly, because so much muck has been allowed to accumulate and because the dredging has to be carried out in tandem with the inspection and restoration of the exposed foundations. By the end of 2000, the job was just over half done, and no less than 180,000 cubic meters of mud and sludge had been dug out and hauled off.[27]

Soon, tides began flowing more effectively through the city, carrying away fresh waste and sewage with their traditional efficiency. At the same time, of course, Insula's work has given rise to another problem: where to unload all this mud? Much of this material stinks and some is toxic, containing dioxins, PCBs, and various anaerobic agents. Plans to use the mud to form new islands in the Lagoon have run into stout opposition from environmentalists, who resist any attempts to alter the topography of the estuary. A suggestion that the dredgings go to enlarge the cemetery island of San Michele, which has been desperately short of space for decades, brought howls of protest from Venetians themselves, who refused to even consider having themselves or their relatives buried in dried sewage and toxic waste. For want of anything better, it was agreed in 2000 to have the muck hauled off by train to Germany, at some expense, where it could be buried in an abandoned salt mine near Leipzig.[28]

Such are the things that float into Venice on the tides, but for the most part these are problems of more concern to tourists than to the world media, which have tended to obsess much more about yet another aspect of the city's dangerous waters, its famous *acque alte*. Such "high waters" are the unusually high tides that periodically fill the Lagoon and, as a result, flood Venice itself to a varying degree. An *acqua alta* is brought about when winds from the southeast—known as the sirocco—coincide with the regular cycle of lunar high tides, an occurrence that can add anywhere from twenty to fifty extra centimeters to the normal tidal crest. The phenomenon has been recorded in Venice for centuries—the first *aqua granda* was mentioned in 782—and, as everyone knows, occasions of flooding are becoming more frequent and higher: in just the twentieth century, the number of "official" *acque alte,* when the tides rose at least sixty centimeters above mean sea level, increased from 385 during the 1920s to 2,464 during the 1990s.[29] Many of these events have little visible effect on Venice and are barely noticed by outsiders: after all, it takes a tide of at least seventy-five to eighty centimeters to begin flooding the lowest-lying parts of the city. Still, with the overall increase in the *acque alte,* the number of more extreme tidal events has also increased apace, and will continue to do so. Unless something is done, the number of tides cresting at over a meter—sufficient to flood nearly half the city—will occur around 15 to 30 times a year by 2020 and 80 to 115 times by 2050; disastrous *acque alte* of over 140 centimeters, which cover the entire city and now occur only once or twice a decade, will become annual events, at the very least.[30]

Despite their disruption and destructiveness, the *acque alte* are also famously popular with foreign visitors to the city, and the papers often make mention—half charmed, half ironic—that the high waters are "the joy of the tourists," who "amuse themselves splashing about"; and of Venetians overhearing the excited chatter of "'That's incredible!'" in English, French, Spanish, or (presumably) Japanese.[31] Many visitors do seem to get positive pleasure out of seeing Venice assaulted in this fashion, and not just because they are glad to see a famously arrogant cultural center get its comeuppance (though that may certainly figure in). As far as we can tell, they are also excited by the dynamic and graphic qualities of an *acqua alta:* several of those we spoke with while watching the waters rise especially stressed their pleasure, saying, "This is *really* something I can tell people back home about!" For most tourists, nothing seems to prove more convincingly their cherished belief that "Venice is sinking" than the sight of waters lapping over the Molo at San Marco or (still more forcefully) the small fountains that begin gushing up out of the drains in mid-Piazza when the higher tides start to move in.

Venetian tourism and the *acqua alta* are closely linked in another way: since Piazza San Marco is one of the lowest places in the city, this most visited part of Venice is also one of the most frequently flooded—even a relatively minor *acqua alta* will suffice to set a sheen of water creeping into the Piazzetta. It takes a tide of around seventy-five to eighty centimeters above mean sea level to begin flooding this area—indeed, the narthex of the Basilica is at a mere sixty-five centimeters—and during the 1990s such levels were typically reached or exceeded about fifty times a year.[32] To deal with pedestrian movement about the city when there is flooding, Venezia Servizi Territoriali Ambientali (VESTA)—the same state agency that collects the garbage—maintains over five kilometers of removable walkways, known as *passerelle*. These wooden planks, usually a meter or so wide and mounted on metal frames that lift them about half a meter above the paving stones, first appear every year around October, stacked in readiness for the season of *acque alte,* traditionally the late fall and early winter. Using the *passerelle,* Venetians and tourists are able to pass along at least some of the major routes around town, although the number of available avenues is of necessity highly limited. For this reason, virtually all Venetians and long-term residents such as students equip themselves with rubber boots, in which they slosh to the many parts of town not served by the walkways.[33]

It is almost inevitable that the system of *passerelle,* such as it is, has greatly served tourist interests, if only because the walkways end up connecting such major tourist nodes as the station, Rialto, San Marco, and the Accademia. It is certainly the case that, without boots or boats of their own, tourists are otherwise almost helpless when the waters start to rise. Yet the system could never be more than barely adequate for any but the lightest traffic, tourist or otherwise. To remain portable, the walkways are limited in size to either a meter or a meter and a half in width, or just barely wide enough for two-way traffic. With just the slightest hitch, the plodding files of pedestrians can become hopelessly snarled: one sees long tie-ups resulting from a few youths toting oversized backpacks, from parents pushing a stroller, from the hesitant steps of an elderly woman. Nothing more irritates Venetians, however, than when they find themselves impeded by tourists who are determined to record this occasion on film and who persist in stopping and photographing their friends just behind them on the *passerelle;* when other tourists stop in turn to keep from ruining their shot, fifty meters of traffic in both directions can come to a dead halt.

To help traffic move along (a little) more smoothly, VESTA decided in early 2000 to set up a double network of *passerelle* in Piazza San Marco.

One route is more or less for tourists, connecting the *vaporetto* and *lancioni* stops along the Molo with the Ducal Palace and the Basilica, passing then to the Calle Canonica. The other is for Venetians who simply wish to cross the Piazza; it has a wider walkway (a meter and a half) and connects the opening of the Piazzetta with the far side of the Piazza, at Calle Larga San Marco.[34] Such makeshift solutions to the city's bottleneck problems have shown some small success, but they do little to alleviate a longer-term difficulty: the season of high waters increasingly is overlapping with that of high tourism. Until the 1970s the "traditional" time of the *acque alte* was November through February, which was also conveniently the dead season for tourism in the city. The *passerelle* were indeed largely developed for the sake of local residents, who had Venice pretty much to themselves during these gray months. In the past few decades, however, the tourist season in Venice has grown considerably.[35] In part this has had to do with the eruption on the scene of the reconstituted Venetian Carnival, which ever since 1979 has turned February from the deadest of months into one of the most lively for the city. At the same time, the core season from spring to fall has also lengthened, reflected in the fact that hotels are now able to charge high-season rates from late March until mid-November. Worse, with the onward progress of global warming, the *acque alte* themselves are occurring ever more frequently outside their "normal" time: in 2000, there were several floods in April, September, and October; once the waters even rose to ninety centimeters— enough to completely cover Piazza San Marco—in July.[36] All of this was handily and lamentably superseded in 2002, however, a year in which the *acque alte* broke many existing records and seemed to indicate a relentless trend for the future. In all, the tides exceeded 80 centimeters above mean (enough to flood Piazza San Marco and ruin the shoes of anyone walking there) no fewer than 111 times—10 percent more than the previous high. There was also one whopping great tide in November of that year— 147 centimeters, the fifth highest ever—and 12 instances of tides over 110 centimeters, the depth that is sufficient to float away the *passerelle* and basically render the city unwalkable. For the first time ever, moreover, one of these extreme tides (121 centimeters) hit in the late spring, on June 6, dead in the middle of the high tourist season.[37]

The result has been that, since 1990 or so, increasing numbers of tourists have collided with the *acque alte*. This has certainly pleased some visitors and made for a good many photo opportunities. It also has created some serious problems, however, especially with large tour groups, which increasingly are coming to Venice in the cooler months and consequently overloading the *passerelle* during the occasions of high water. When there are throngs numbering in the hundreds, even the apparently simple task of moving them between the Ducal Palace and the Basilica can turn into a tour guide's nightmare:

Over a thousand tourists "imprisoned" in the Ducal Palace because of the *acqua alta*. It happened yesterday morning, with an unexceptional tide . . . [when] three cruise ships had unloaded something like two thousand tourists on Piazza San Marco in the course of half an hour. The troubles started at the end of the visit to the Ducal Palace, when the passengers were started off on the only *passerella* that directly connected with the Basilica, second stop on the tour. And immediately everyone ended up jammed, until the time when the tide went down [after several hours].[38]

For visitors to Venice, *acque alte* can bring more than just delays. Serious tides, when the waters crest at over 130 centimeters, cause many of the *passerelle* to come loose from their frames and simply float away, ending movement altogether. The siroccos that usher in the high waters sometimes also blow strongly enough to push unwary tourists right off the walkways. Those who do not know the city and its quays are well advised not to try wading off on their own: even in shallow water it is easy for the inexperienced to slip on the slick paving stones. We also heard stories of unfortunate foreigners who miscalculated the edge of a flooded quay and stepped right off into the adjacent canal. Moreover, when the tides arrive, they do not simply sweep from the Lagoon, but also erupt out of the drains and sewers all over the city. Especially if it has been a while since the last high water, then "all around [there is] the stink, the miasmas, because the water smells of latrines and of sulphur, it washes out the sewers, insinuates itself into the tight spaces and floats out the filth. There also bob along in the *calli* the sacks of garbage."[39]

Several approaches have been proposed to deal the *acque alte,* all sharing one thing in common: they are hideously expensive. In the 1970s, when both Venetians and the world community were still reeling from the shock of the great flood of November 4, 1966 (when tides had reached a stunning 194 centimeters—a height never seen before or since), plans were put forth that mostly drew inspiration from similar pharaonic works underway elsewhere in Europe: the Thames barriers below London and the Dutch Oosterschelde in particular. This was a classic engineer's solution: if too much water was coming into the Lagoon, then block the entrances to the Lagoon and keep it out. Venice being Venice, however, there were other, *aesthetic* issues that came into play. It was not enough for the barriers across the *bocche* of the Lagoon simply to keep the *acque alte* out; they also had to disappear from the public (that is, tourist) gaze when not needed. Those who insisted on this said it was to allow maximum flushing of the Lagoon during normal tides, but they were also clearly concerned with keeping Venice's mythical charms and views uncluttered by masses of modern machinery. In other words, the process of protecting the city from the threats of nature ended up, like almost everything else there, conditioned by the demands and requirements of the tourist industry.

The solution, arrived at after around fifteen years of debate, was known as MOSE (for Modulo Sperimentale Elettromeccanico) and was finally approved at the end of the 1980s. The idea was ingenious in its simplicity: each of the three openings to the Lagoon (which are around five hundred to two thousand meters in width) would be barricaded by twenty or thirty hollow "flap gates"—giant metal caissons hinged along one edge. These normally would fold down flush with the sea bed, where they would be filled with seawater and tucked down in the mud. When an *acqua alta* threatened, however, the water could be pumped out of them and the air that replaced it would make them buoyant enough to swing up until they broke the surface, creating a dam. When the emergency passed, they could then be refilled with water, causing them to sink once again.[40]

As it happened, the MOSE, though it was first tested in prototype form in 1988, fell victim to seemingly infinite wrangling among politicians, environmentalists, fishermen, and the shipping industry, with each group imagining and proclaiming the worst possible results if the project were (or were not) carried out. It was not until mid-2001 that funds were finally "unblocked" in Rome, and only on 14 May 2003 did Prime Minister Silvio Berlusconi actually come to Venice to lay the so-called cornerstone of the new structure. That he chose to do so not at the actual site where the MOSE will be installed, but rather on the Venetian island of Sant'Elena, well protected by a barricade of troops on land, in ships, and in helicopters, was a good indication of the strong opposition the project still aroused, even after so many years, and, indeed, hundreds of Venetians— including fully half the *giunta,* or city council—turned out in small boats (and a *vaporetto* rented for the day) to protest the entire undertaking.[41] One of the main bones of contention remains whether, in blocking out high tides, the dams would also prevent the Lagoon from cleansing itself. Since, by all predictions, both the frequency and height of the *acque alte* are sure to increase over the coming years, environmentalists in particular expressed fears that the MOSE would have to be closed so often that the Lagoon would end up choking to death on its own pollution—a foreboding that the estuary had obligingly and already lived up to with the great algae blooms of 1988–92.

For those in the city's powerful tourist sector, the MOSE has often seemed a Hobson's choice. Without some sort of barrier, their shops and hotels—especially those around San Marco, which are necessarily at ground level and in some of the lowest-lying parts of the city—would be flooded ever more often, even as the tourist flows on which their livings depended were obstructed and diverted away from them by the chance of the tides. On the other hand, no one really knew how much the MOSE might interfere with flushing out the Lagoon, and there was always a possibility that, after spending something like $5 billion for the job, Venice would end up shunned by tourists for having canals that stank so much from trapped pollution that the place was too disgusting to visit.

As the government and the project's various detractors dithered over the MOSE for a decade or more, another defense to the *acque alte* was developed, more or less in parallel. This approach followed the reasoning that the best way to avoid high waters was not to block them out but simply to lift the city itself out of their reach. The concept is not as ludicrous as it may sound (or as many have taken it to be): it was, in fact, the city's primary defense during the Serenissima, as one can readily see by comparing photographs of the Basilica of San Marco or the Ducal Palace today with the way they appear in paintings by Canaletto, when they were clearly much higher relative to the adjoining pavement.[42] As a solution, this approach has demonstrated some obvious limits: since the buildings themselves cannot be raised along with the pavements, their original proportions over time will be destroyed as adjoining streets and *campi* are lifted up. One can clearly see the results in looking at the Ducal Palace, whose supporting columns and overall profile have acquired an increasingly squat appearance as the neighboring Piazzetta has been raised over the centuries.

Still, beginning in 1997, through the organization of Insula, efforts have been underway to raise, if not the entire city, at least as many of the quays and canal banks—the *fondamente* and *rive*—as possible. The idea was to make a new threshold for the city by lifting not entire islets but just their edges, to create effective barriers at least against tides up to 120 centimeters above mean sea level. According to studies in the late 1990s, roughly 84 percent of the city could be protected in this way, at an estimated cost of at least $300 million, with the rest being unfeasible for topographical or architectural reasons.[43] In 2000, after four years of work, Insula reported it had successfully raised around 30 percent, or about twenty-one thousand square meters' worth, of the city's quays, lifting the paving stones an average of twenty to thirty centimeters.[44] Not until the spring of 2003 did Insula finally get started on the sixty-million-euro task of raising the edge of the city's most important islet, where the Basilica and Piazza of San Marco are located, together with the Ducal Palace and the archbishop's residence. As one of the lowest parts of town, the Isola San Marco could not be barricaded to a height of more than 115 centimeters above mean tide, because if the Molo in front of the Piazzetta were built up to the usual 120, the water pressure would simply cause the tide to erupt through the drains—as it already does, making little geysers in the storm drains of the Piazza. Even 115 centimeters should still be enough to spare San Marco from a hundred or more inundations every year, at the current rate of flooding. Just as important, it will also preserve more or less unaltered the panorama of the Bacino from the Molo that is so striking and so beloved of tourists, with San Giorgio Maggiore and the Salute in the background and ranks of gondolas tied up and bobbing close at hand (they will be often bobbing *above* the Piazza itself, however).[45]

Even as various agencies struggle to protect Venice from high waters and pollution, the impact of these dangers is being greatly magnified by yet another of the city's aquatic plagues. In order to service the city, and in particular the millions of foreigners who go there, thousands of boats are constantly plying the canals, churning up waters that were once calm, if not exactly clear. This nautical supply system is also an ancient tradition in Venice, but until the 1960s most of these boats, from the modestly sized *sanpierote* up to even the big transport vessels, known as *topi,* were rowed. In line with the economic imperatives of supplying Venice's hundreds of restaurants and hotels, however, almost all these supply boats have been motorized in the last thirty years—"a little fleet of transport boats for linens and supplies"—and their impact on all levels of the Venetian environment has been considerable.[46] All the major canals, such as the Cannaregio and the Rio di Noale, but above all the Grand Canal (which the papers regularly call an *autostrada,* or motorway), are seriously overloaded with powerboat traffic. Particularly in the morning, when supplies are being brought in, the Grand Canal is an uninterrupted stream of the big powerboats called *moto topi* (because their diesels go *mototopo mototopo mototopo*), carrying everything from food and drink to toilet paper, computer equipment, and construction materials (see figure 23). Jockeying around between these lumbering transports are the public *vaporetti,* which in some stretches pass every two or three minutes,

FIGURE 23. A *moto topo* on the Grand Canal. (Photo by Philip Grabsky.)

along with a vast menagerie of private vessels, taxis, and the oversized, glassed-in tourist boats known as *lancioni di granturismo*. Inevitably, with such a mob, big traffic jams are a frequent result, especially around Piazzale Roma and the Rialto area: at such times all the diesel noise is further enhanced by angry shouts and piercing boat horns as pilots jockey fiercely for right of way.[47]

One wonders how much longer the hotels that front on the Grand Canal will be able to demand their customary (huge) markup for rooms overlooking the water. The constant drumming of motorboats chugging back and forth and the diesel exhaust that often hangs visibly overhead have made a *camera con vista* overlooking the Canal about as alluring as a room adjoining a freeway overpass. The view down from the Rialto Bridge gives much the same impression:

Seven ACTV *vaporetti* in Indian file. Twenty taxis. Fourteen big transport boats with four more tied up under the Rialto Bridge. Plus a few gondolas, government launches, and a couple of speedboats. Try a bit to count the boats that pass during the peak hour under the Rialto Bridge. Yesterday morning at about 11:00, there were around fifty every minute. . . . The uncontrolled growth in the tourist sector and of hotels, the changes of use that have transformed apartments into taverns and little palaces into *pensioni,* the freelance fast-food activity, [all] have produced a predictable, massive increase in [water] traffic. And now the effects are right before the eyes of everyone.[48]

Though some of this traffic supplies the Venetians themselves, probably the majority of the transport vessels and certainly the taxis and *lancioni* are there for the tourists—who, after all, outnumber the locals annually two hundred to one. This powerboat traffic produces more than just diesel-exhaust pollution and racket, however. With their screw propellers, these hundreds of boats also churn up the waters and in the process disturb the canal beds, bringing up rotting sediment and detritus from below. One can see this process in action just by standing alongside the Grand Canal at low tide as the big *moto topi* go by, each leaving a plume of gray-green silt trailing in its wake from the muck its propeller has churned up. The Grand Canal, at least, is four or five meters deep: when these monster boats push their way into some of the smaller waterways, where the bed may lie just a meter or so below the surface, they thrash up so much sludge that the roiled waters they leave behind stay foul and putrid for hours. The problem is compounded by the sheer size of many of the *moto topi*. Frequently the canals their pilots want to enter offer barely a meter of clearance overall: in the course of maneuvering their way in and out, throwing the boats in reverse, and continually revving their often grossly overpowered engines, they roil the waters and the underlying bed like so many giant paint mixers.[49] Even out in the open estuary, powerboats can tear up the Lagoon bed: along many of the small but heavily overtraveled channels that connect Venice with its outlying islands, there are frequent reports of "gurgles of reeking gas bubbles, filthy water, [and] putrescent mud" produced by the constant disturbance of the sediments by so much motor traffic.[50]

From back in the 1950s and 1960s, when these big boats were converted from oar or sails to diesel, up to the 1990s, most *topi* fit well enough even in the smaller canals, their operators respecting the limits imposed by the city, which usually allowed boats with a maximum width of around 220 centimeters (five and a half feet) to travel the internal waterways. Since the 1990s such restrictions have been more violated than observed, however, and it is easy to spot *topi* nearly three meters wide, parked in the canals to unload or struggling backward and forward to work their way around sharp bends or under bridges. Moreover, these *moto topi* are increasingly built of steel instead of the traditional wood, with a huge increase in the damage they do to the *fondamente* and undersides of bridges, as their operators use the marble and brick quays "like billiard cushions in their maneuvers." From what we have seen, these big boats serve overwhelmingly for two primary purposes: construction and renovation (hauling scaffolding, brick, sand, and heavy equipment; taking away debris) or supplying foodstuffs (soft drinks and *acqua minerale* in particular, it seems). Certainly some of these materials go to help Venetians rebuild their own houses, or to markets where Venetians shop, but from our experience, once again, it certainly seems that the ever-increasing number of *moto topi* in Venice—on some internal canals thirty or more may pass in a day—is due to the demands of hotels, restaurants, and builders who are all seeking to either service or attract more tourists to the city. Many are equipped with big cranes amidships, and as their operators work to load or unload their goods, revving their three-hundred-horsepower engines, they typically end up blocking the canal almost completely, leaving just a few meters of open water for the beleaguered gondoliers, who try to slip their craft by and dodge "the waste [cooling] water that is dumped directly on board [and] the cloud of [diesel] smoke that envelops the bewildered tourists."[51]

In a single, average day in Venice, according to a census in April 2000, over twenty-five thousand "boat trips" are being made around the historic center, plus another five thousand out on the Lagoon.[52] Yet, though the pollution and stink all this traffic generates are considerable, they are by far the least of the problem. Much more important for Venice's survival are powerboat wakes, which the locals refer to as the *moto ondoso*. In one sense, waves of this sort have bedeviled Venice ever since the first steam-powered *vaporetto* started running on the Grand Canal back in 1881. At first, however, the economic and aesthetic disruption caused by these powerboats obscured their other potential problems. The city's gondoliers, as we have seen, went on strike primarily out of fear that the *vaporetti* would undercut their fares, though they also claimed the steamboats would soon lead to a "depoetized Lagoon" and in the process would ruin the entire gondola experience that depended on tranquil waters and soothing songs.[53] Foreigners, especially the British, went much further in protesting the aesthetic violation of a world they firmly believed should be kept romantically premodern at all costs. They railed against "the snort of steam-boats" that "tore and puffed up and down [the Grand Canal], churning up the foul water into a deep brown froth" and declared that henceforth there would not be a "tourist whose finer feelings are not lacerated by the smoke-begrimed facade[s]."[54]

In the long term, of course, visitors seem to have adjusted to the increasing level of sound pollution in Venice, perhaps because, as noisy as the city has become, it still strikes many tourists as one of the quietest on their holiday and certainly the most tranquil in Italy. It is also true, however, that it has taken the passing of the years and the steady increase in powerboat traffic around the city to make the real impact of the *moto ondoso* more obvious, since even big boats make fairly small waves, but the effect of these, slapping day after day, is cumulative. Over time, the steady and continual impact of little wavelets can cause cement to crumble, loosen foundation stones, and speed up the deterioration of the *bricole* and *paline,* the wooden poles that mark channels out in the Lagoon and that traditionally brighten the water entrances of palaces along the Grand Canal.[55] The *moto ondoso* undermines the very premise of Venice itself, a city built with delicacy and precision right into water that for centuries was dependably still unless ruffled by the wind or exceptional tides.

Venetian gondoliers are leading the charge against the *moto ondoso* as they did against the original *vaporetti,* and again with mixed motives. Certainly the constant waves hurt them economically: recent studies have shown that the continual buffeting of powerboat wakes has reduced the effective life of the gondolas themselves "from the traditional forty years down to little more than ten." For this reason (or so they claim) some gondoliers have taken to having their hulls painted with a fiberglass resin paint, even though this is strictly forbidden by their own bylaws, as untraditional and polluting.[56] The sheer number of motor vessels now crowding Venetian waters—there are currently estimated to be around five hundred *lancioni* and taxis, along with scores of *moto topi* "like aircraft carriers"—has also forced present-day gondoliers to cope with rowing conditions well beyond the worst nightmares of their ancestors who went on strike back in 1881.[57] The boatmen endlessly complain about having to battle "gale-force eight waters" that the powerboats churn up on the Grand Canal and (especially) on the Bacino, and it is not uncommon for a gondolier to be flipped right off his craft by a motorboat passing too closely. The gondoliers blame their plight mostly on those "eat-and-run" tourists, the *mordi e fuggi,* whose need to see Venice in five hours or less supposedly has them racing about the city in high-powered taxis and launches. Such people, the boatmen say, "should be obliged to move through town by the slow and traditional ways"—meaning, of course, by gondola.[58]

The gondoliers' stance as protectors of genteel tradition in Venice is somewhat undermined by the fact that a fair number of both the taxis (13 percent of those officially licensed) and the *lancioni* are owned and operated by none other than the gondoliers' cooperatives themselves. "Then,

there is the type," *La Nuova* observed, "who, having taken off his gondolier outfit, goes speeding by at the wheel of his taxi, oblivious to the
complaints of his excolleagues."[59] In effect, then, the traditional element
of these watermen is publicly denouncing their own, more modern version. One wonders if some individuals ever denounce themselves.

In any case, we doubt that most day-tripping tourists really want to
race through Venice, on land or by boat, any faster than they have to. The
more obvious culprits are the pilots themselves, "for whom the only valid
equation is: speed equals earnings."[60] Most notorious among these
"Speedy Gonzaleses" (as Venetians like to call them) are the taxi drivers
who shunt tourists from Piazzale Roma, Tronchetto, or the Riva degli
Schiavoni out to Murano. Paid by the glass factories on the basis of how
many bodies they can deliver, these pilots have every reason to go as fast
as they can—"as fast as we used to go for water skiing," one American
told us. When they make the trip, they can turn the Canal of San Cristoforo, connecting Venice with Murano, into a sea so choppy that even the
large passenger ferries can have trouble maneuvering. "The taxi drivers
and the various private speedboat [operators] are just crazy," complained
the president of the ACTV pilots' association. "The Lagoon seems like
the sea in a gale! And in these conditions it is impossible to work safely;
we have to do acrobatics just to tie up [at the dock]."[61]

Attempts to control motorboat traffic in the more chaotic areas have
had only limited success. Since January 1, 1998, nautical speed limits on
the Grand Canal have been set at eleven kilometers per hour for *vaporetti*
and seven kilometers per hour—less than four knots, really just a brisk
walk—for all other craft; on the remaining canals of the city, the limit is
only five kilometers per hour. The police and carabinieri have equipped
themselves with a variety of radar devices for clocking the boat traffic, but
these often break down; moreover, the authorities seem willing to use
them only sporadically, never after dark, and rarely out in the Lagoon. In
any case, boat pilots in Venice react to such attempts to control their self-
perceived right to go fast much like motorists do anywhere in the world:
they use (illegal) radar detectors and cell phones to give each other warn-
ings about speed traps. As a result, the police fail to give out many tick-
ets (only 213 in the entire summer of 2000; an average of barely one every
two days over the rule's first 30 months), and many of those go to pleasure boats belonging to owners with no savvy. "The greater part [of the
drivers] are professionals, who know us and know how we work," observed the commander of the water police for the historic center. "You
take one of them and in a second they pass the word, and then everybody

transforms themselves into snails."[62] Even so, it is not so much the speed as the sheer number of boats that churns up dangerous waters: recent counts indicate that "in the Grand Canal a motor boat [passes by] once every ten seconds." This, more than ever, is linked to the increasing number and requirements of tourists to the city, and some complain that with such massive water traffic it is pointless either to lower the speed limits still further or indeed even to enforce the limits that do exist.[63]

It has proved still more difficult to control the enormous cruise ships that have begun to ply the Lagoon, most of them passing right through the center of the city on their way from the Bocca del Lido to Venice's docks, behind Piazzale Roma. In the first years of the new century, "hundreds" of these monsters regularly come and go from Venice, passing through the Bocca del Lido to dock at the Port of Venice or, if space is lacking, at the Riva dei Sette Martiri, a thousand meters or so to the east of Piazza San Marco. Though the big cruise ships necessarily make their way to their berths at their slowest possible speed, most of them still have such huge draft, displacement, and propeller size that they leave significant signs of their passing.[64] Even at a crawl, they produce a considerable bow wave, and though recent studies have shown this to be only around twenty-six centimeters in height, they carry considerably more energy than those from ordinary boats, as anyone who stands on the Riva degli Schiavoni while one of them chugs by can attest (see figure 24).[65] It is also assumed that their propellers effectively dredge out the parts of the Bacino of San Marco and the Giudecca Canal over which they have to pass; observers express concern that this effect is intensified, along with the risk of impact damage to the quays of Molo and of the island of San Giorgio Maggiore, by the maneuvers that must be made in the Bacino by those liners that have docked at Sette Martiri and have to turn around before departing the Lagoon.[66]

Although the national conservation association, Italia Nostra, has worried about the ecological damage of "the enormous cruise ships that by now we see pass every day in the Bacino and in the Giudecca Canal," these ships' impact on the sediment and the quays has yet to be determined, perhaps in part because the Venetians, who are hugely involved with the burgeoning new cruise business, are in no great hurry to find out.[67] Venice, or rather the nearby industrial town of Porta Marghera (which, though on the mainland, still forms part of the city), has been reborn in recent years as one of the world's great shipbuilding centers. The *Grand Princess,* for example, the first cruise ship over one hundred thousand tonnes, was completed in the Fincantieri shipyards there in 1998, as were Holland

FIGURE 24. The cruise ship *Costa Tropicale* and a gondola on the Bacino of San Marco. (Photo by authors.)

America's sixty-thousand-tonne *Amsterdam* and *Zaandam* two years later. Such heavy industry represents one of the most vital economic elements of the entire region, replacing the Lagoon's fading (and highly polluting) chemical sector and creating thousands of high-paying jobs.[68] When these floating hotels are launched and outfitted, it is also only natural that they should be presented to the world and their shareholders with a grand display against the ever-popular backdrop of Venice itself. For the Venetians concerned, it is equally desirable—for reasons of prestige as much as for the economics involved—that their city be included among these behemoths' ports of call, and the local chamber of commerce has worked hard to add the city to as many cruise itineraries as possible. To insure its appeal for these lines, the Port of Venice was heavily renovated over the course of the 1990s, with millions in Italian and European Union funds. Between 1980 and 2000, the traffic in the port increased 230 percent, making it "among the most productive in Italy, securely the first on the Adriatic"; by the late 1990s, the port was netting over $20 million annually in commercial fees, while also creating hundreds of jobs.[69] Fundamental to all this activity was the cruise trade, which continues to chase its own economies of scale through the commissioning of increasingly gargantuan ships. In the mid-1990s, Venice decided to go after this important sector in a big way, and massive efforts to rebuild and streamline passenger facilities at the port paid off: in 2000, for the first time, the port handled over a million passengers; within the next decade, this figure is expected to double.[70]

Aware of the big attractions Venice offers in terms of ambience, culture, and shopping, many lines schedule an extra day in port for their passengers. When ships like the *Grand Princess,* the *Amsterdam,* or the *Mistral* pull into port, they are able to dump upward of twenty-five hundred passengers—the equivalent of around fifty tour buses—onto the city in the space of just an hour.[71] To help their customers maximize their Venetian time, getting them as quickly as possible to the shops and sights at San Marco, the big ships employ dozens of *lancioni* to shuttle back and forth from their berths, sending them up the Giudecca and "transforming . . . that canal in a kind of autodrome for tourist motorboats, each one uglier than the last." Even when they themselves are tucked away in their slips and their giant engines are stopped, these cruising hotels thus still make their contribution to the *moto ondoso.*[72]

The arrival of truly enormous ships like the *Grand Princess* (and she has already been superseded by several others, still larger) will impact more in Venice than just its quays and waterways: these monsters promise to transform completely the whole relationship of the ancient city to the postmodern world of tourism.[73] Unlike the big CIGA establishments in Lido that once boasted of similar luxuries, these new "slab-sided highrise hotel[s] on a hull" come right into Venice proper, where they dominate seascape and landscape together. As they pass, such ships seem to rip the visual fabric of the city right apart: the *Princess,* for example, at over 950 feet from bow to stern, stretches clear across the Bacino of San Marco as it noses its way through town. For someone standing at the Piazzetta, the passing ship completely cancels out the island of San Giorgio and looms up taller than the Salute as it moves up the Giudecca Canal. On such occasions, a Venetian friend told us, he was reminded of the alien spaceship hovering over the White House in the film *Independence Day.* Such a piece of floating real estate does fit oddly and rather threateningly into Venice. Indeed, according to travel writer Alexander Frater, soon after it was launched, the ship's "senior comedian" used to joke with the passengers that "when our shadow fell across Venice the consequences would be apocalyptic: citizens fleeing in terror, phones ceasing to work, television signals interrupted, gondolas plunging wildly in our wash, minor canals drained by our monstrous gravitational pull. That got a laugh, but also, I sensed, gnawing worries that it might contain a nugget of truth."[74]

It is just this sheer size, however, that liberates those tourists who push their way into Venice while peering down from one of these floating hotels. At 174 feet, or around fifteen stories, the *Princess*'s top deck looms above every *palazzo* and building in the city—only the Campaniles of San Marco and San Giorgio, as she glides by, overtop her.[75] From this angle, Venice presents itself to the cosseted passenger as a panorama rather than a city, without the inconveniences, racket, or smells of an inhabited town. And, indeed, the cruise lines, more than even most holiday agents, lure their guests with images of a Venice that is surface and color, an entirely malleable stage on which tourist dreams can be projected without the slightest interference. The companies seek customers through image-ads that offer up a deck-level view of the glowing, sunset-drenched Piazzetta, the Grand Canal, or the Salute, further enhanced with a distinctly cloying prose that promises a Venice requiring neither effort nor understanding:

AS IF EXPLORING EUROPE WEREN'T LUXURIOUS ENOUGH.

Ahh. The sweet and exotic texture of unexplored lanes, hidden treasures in an ancient marketplace, warm café afternoons. Of time spent traveling, not touristing. But wait, where's my spacious oceanview suite? Where's my 5pm seaweed wrap and my complimentary fine wine with dinner?—Fear not. It's right here where you left it. Aboard your Radisson Seven Seas ship anchored conveniently nearby and ready to whisk you off to yet another day of adventure and discovery.[76]

This, of course, is a Venice of purest fantasy: no one is ever going to discover unexplored lanes or hidden treasures in the most touristed city in the world. It is a fantasy of infantile pleasures, a promised round of hide-and-go-seek that ends up in a candy store. It is, however, a realizable dream for those pampered passengers now able to peer *down* into Venice from their high perches, safely removed from the city itself and free to fantasize any Venice they may desire. Integral to this freedom is keeping intrusive distractions at bay—not just the sort of racket and smoke that kept past tourists from fantasizing this place through the palimpsest of their Shelleys, but also the more modern tourist banes: the milling crowds, the supercilious shopkeepers and waiters, the endless shills, and of course the confusing and enigmatic city of Venice itself.

Interestingly, in order to insure their passengers' fantasies, comfort, and isolation, the cruise lines have had to resort to ever more massive systems of power and technology that, in turn, intrude ever more insistently into the lives and even the homes of the Venetians. It is not just, as residents have increasingly begun to complain, that these floating hotels block the view or intimidate with their presence: they also torment locals at home, even in their own beds. When a cruise ship as large as Royal Caribbean Line's *Brilliance of the Sea* (90,090 metric tons, 962 feet in length, capacity of 2,501 passengers) docks overnight at the Riva dei Sette Martiri—and despite having the facilities of the Port of Venice, the big ships continue to tie up at this more convenient anchorage nearer to San Marco—it can wreak havoc deep inside the city. Thanks to the massive engines that drive such ships' air-conditioning and power systems, residents living several hundred meters from dockside are tormented by a "continuous booming and buzzing [causing] the windows [to] vibrate constantly. It is an annoyance that one notices distractedly during the day but which becomes hammering and unbearable at night." Those who have the misfortune to actually live facing out onto the Riva report having to also put up with "a kind of continuous whistling sound"; one resident also lamented of having the bad luck to find "the part [of the ship] with the elevators . . . right in front of my windows. Try to imagine the noise of those four lifts that were going back and forth all night long." As if all that were not enough, the huge electrical implants of such ships send normal electronic appliances completely haywire, such that there is "the aggravation that, if you can't manage to sleep, you can't even distract yourself with the TV, seeing that the reception becomes terrible."[77]

Its surrounding waters are thus proving dangerous to Venice, probably more dangerous than in any other time in the city's centuries of history. What we have found especially interesting in examining how the Lagoon and its waters impinge on Venice, however, are the ways in which the problems presented by such long-term changes as global warming, rising sea level, and the shifting Lagoon are all made much worse for the city by tourism itself. It is not simply that Venice's millions of casual visitors leave behind sewage and garbage on an Augean scale, polluting the canals beyond their traditional self-cleansing abilities. Or that the need to take care of these foreigners and move them about, in vessels ranging from taxis to monster cruise ships, has turned the city's normally placid waters into the tumultuous *moto ondoso* that acts as a continual battering ram against its crumbling palaces and quays. The real problem is that those who are charged with solving the problems that, to a greater or lesser extent, the tourists themselves have helped inflict on the city are required to do so without significantly changing Venice from the premodern fantasy world that these foreigners come expecting to find. This can be a difficult and often extremely expensive limitation: we have already seen the special complexities and costs that were imposed on the MOSE and on the diking in of Piazza San Marco by the obligation to keep such barriers invisible and unobtrusive to the tourist gaze. It certainly would be far easier to handle such questions as pollution, Lagoon ecology, and the *moto ondoso* if these could be treated primarily as environmental and engineering problems. Instead, in Venice the aesthetic component remains (as it has since at least the late nineteenth century) fundamental, if not paramount. A worthy priority, many would say, and more than a few tourists would recoil from even the hint that the city and its waterways might some day end up as thoroughly engineered—with, say, external piping, wave barriers, and pumping stations—as the Dutch wetlands already are. It may, however, prove to be beyond the skills of even the most cunning of Venetian engineers to produce a city that continues to run on the medieval rhythms of the tides and to depend on the calm of the Lagoon while still managing to absorb the variegated and putrescent leavings of more millions of annual visitors than there ever were subjects in the old empire of the Serenissima.

PART FOUR

Worldscape

CHAPTER 9

Restoration Comedies

For its size and population, Venice probably suffers from more physical problems than any other city: this is one of the few world heritage sites that perpetually runs the risk of vanishing completely. Having been investigated for decades, Venice's travails are well known, even if all their precise causes are still under some debate. One problem is subsidence, which was catastrophic in the 1950s and 1960s, as mainland industries freely drained the local water table for their uses. Even since the practice was banned, the town has continued to sink simply because its buildings are so heavy—this is a city of brick and marble built on mudflats, after all. Subsidence not only causes Venice's monuments to crumble, but also leaves them more vulnerable to the Lagoon waters, which, thanks to the continued rise of world sea levels, encroach ever more frequently. Tides increasingly rise above the impermeable barriers that cap building foundations, and once seawater reaches the overlying brick and porous stone, it wicks up higher, saturating structural walls and leaving behind salt deposits that corrode solid structures into dust. Meanwhile, air pollution from the factories of Marghera and Mestre has stained and eaten away much of Venice's once distinctive bone-white *pietra d'Istria* (Istrian marble) until many buildings have become nearly as black as those in downtown Milan or Turin. Finally, the physical presence of far too many tourists has caused not only the social degradation of the city, but also, as we have seen, its physical erosion—through the water pollution, engine exhaust, and the *moto ondoso* generated by supporting traffic.[1]

All these afflictions continually tear up the fabric of Venice as they cause its buildings, pavements, and quays to shift, crack, and crumble. As a result, the city is constantly under repair. Restoration has become as dominant a theme in Venice as canals and bridges, and hardly a year has passed since the nineteenth century without some major project under way, leaving one or another of the city's principal buildings swathed in scaffolding for lengthy periods. It took a quarter century to restore the Basilica of San Marco in the mid-1800s and another fifteen years after that to take care of the Ducal Palace. The end of 2001 saw the winding down of another large-scale renovation of the Basilica that had taken nearly twenty years; meanwhile, work on the Fenice opera house, gutted by fire in 1996, concluded only in late 2003. These efforts at keeping the city standing, plus the struggle to maintain its disaggregating quays and *campi,* have all reinforced the belief, widespread around the world, that Venice is indeed in a parlous state—barely hanging on until the next big blow or flood. Such convictions even have their own iconography that turns up as one of the more popular souvenir images found at the kiosks and bookstores around Piazza San Marco—fantasy posters of an apocalyptic Venice either submerging beneath gigantic waves or already underwater, populated only by giant, luminescent sea creatures and tourists brought in by glass-hulled submarines.

To most visitors in Venice, restoration is primarily an annoyance: thus, for years, we have been overhearing tourists complain how wooden scaffoldings on the Basilica of San Marco were ruining their photographs. For those more involved with the city, however, questions raised by restoration can be much more fraught, and have over the years regularly fired up acrimonious debates, both about the best ways to save the city's endangered treasures and about what those treasures should become once salvation has restored their beauty, if not their functionality. Most of this goes on completely under the radar of the average tourist, for lovers of Venice's art and its past—including many Venetians and Italians, certainly, but also great numbers of foreigners—speak largely to one another, in terms of a highly nuanced understanding of what Venice should be, or at least what it should look like. Although these sophisticates are relatively few in number, their influence on Venice has been tremendous over the years, and if their debates are less sharp these days than in the past—almost everyone now agrees on at least the major principles involved—they continue to play a fundamental role in a process that we would call the aestheticizing of Venice.

One could safely say that this process began with John Ruskin, who lifted the entire city of Venice to the level of an aesthetic principle. Admittedly, though, Ruskin was not the first to notice Venice's degradation.

Even in the final years of the Serenissima, and especially on into the first decades of the nineteenth century, visitors complained of Venice's neglected state—canals needed cleaning, streets were littered with garbage, the buildings looked scruffy. By mid-century, poverty and ill-usage had so reduced the city that even its Austrian overlords began to worry—not only might monuments like the Basilica of San Marco or the Ducal Palace actually collapse, but Venice could end up losing its tourist appeal if it got any shabbier. In response, architects were commissioned, first by the Austrians and, after unification, by the Italians, to straighten up the mess: among other monuments, the Ca' d'Oro, Fondaco dei Turchi, church of Santi Maria e Donato on Murano, and the Basilica itself were all reworked according to the Neo-Gothic aesthetic of the time. This meant that any design elements from the Renaissance or later had to go, such that walls, windows, frescoes, even whole chapels were stripped away with the aim of returning buildings to what was deemed as their original medieval form, or at least spirit. At the same time, weaknesses and flaws, even if original, were also cheerfully tidied up, as old marble was replaced or ground away, uneven stone mosaics were flattened or switched for glass, and crumbling brickwork was replaced with new.[2]

Such bold (and often highly destructive) interventions set the stage for Ruskin's arrival in the 1830s and inspired his own life's work toward creating a new aesthetic for (and of) Venice. Convinced that the city, at least as far as its monuments went, was "something uniquely precious—a miracle that could not be reworked, a dream that could not be redreamt," and that, indeed, "no second Venice could ever be built," Ruskin blasted the Neo-Goths for having the temerity to attempt this vainglorious goal.[3] He, but even more those who would follow (or rediscover) him, especially crossed swords over Neo-Gothic attempts to produce a sanitized, reworked version of the Basilica, leading in the late 1870s to what might be considered the first full-blown crisis in the aestheticization of Venice.

The argument was not so much over the urgency to do something about the ruinous state of the Basilica—though some purists did insist that the structure had originally been so perfectly designed that it could never fall down. Rather, the fight was about whether the aggressive Neo-Gothic restorations on the building's north and south (Piazzetta) facades should be extended to its main, western flank on the Piazza itself.[4] It was the unveiling in 1875 of the southern facade, its original book-matched marble paneling replaced with pallid substitutions and its weathered decorations reground to a hard-edged newness, that set off the storm of protest, first in Venice and then throughout Italy, Europe, and America.

Though the master himself was by then too far into depression to take an active part, an international cast of Ruskin's adherents joined the battle in his name. Britain's formidable Society for the Preservation of Ancient Buildings, already fundamental in defeating the Neo-Goths at home and in formulating the new conservationist aesthetic, set up its own St. Mark's Committee, the first international group formed to influence aesthetic policy in Venice. Run primarily by Henry Wallis and William Morris, the committee focused its opposition against further restorations on the Basilica by mobilizing what John Pemble has called "all the resources of the Victorian campaigning tradition—letters to the press, public meetings, scientific investigations and reports, and a memorial with 2,000 signatures, some of them very eminent." Also drawn into the fray were supporters in France (led by Charles Yriarte and the baron Adolphe de Rothschild) and in the United States (under Charles Eliot Norton).[5]

These concerned citizens of the world continued to clog the London and New York papers with their protests for much of the 1880s. By all accounts they seem to have accomplished more harm than good, not least because the British in particular displayed a supercilious condescension toward the Italians and their restoration skills—attitudes of cultural colonialism that still persist in the travel writings of more than a few present-day English and Americans. Their outrage against supposed plans to demolish and rebuild the west front of the Basilica was in any case both misdirected and too late: the Italians had already abandoned that approach, insofar as they had ever considered it, more than a year before the protest movement had even started. By the early 1880s the Basilica had a new director of works. Pietro Saccardo, himself a member of the St. Mark's Committee, was committed not only to saving the Basilica structurally, but also to removing the rash and destructive changes (as everyone now agreed) imposed by his predecessor. When Saccardo unveiled his work in 1886, it appeared to the casual observer that the south and west facades, though completely restored structurally, were the same as when work had begun back in 1865. How Saccardo accomplished this has never been clear. Ruskin's credo had been for restorers to leave untouched as much as possible without replacing anything; whatever was too far gone to be saved should be replaced by material that was obviously new—neutral filler, in other words—so that later generations would not be fooled into accepting a modern, mimic cathedral as somehow the original. Obviously Saccardo ignored the master in this regard: secure behind his scaffoldings, he had evidently gone ahead and torn down the west facade just as everyone had feared (though he refused ever to admit this), but he had hidden his work so skillfully that in the end no one was the wiser—or at least not sure enough about what had happened to complain. As such, Saccardo gave Ruskin's followers (if not Ruskin himself) what they wanted: a solidly restored Basilica that still looked appropriately old and idiosyncratically medieval, with the necessary changes hidden behind artfully darkened marble and fake irregularities inserted in the mosaic and stonework. Certainly it was enough for the average tourist at the time, as it has been ever since. As one thoroughly pragmatic American observed in 1880,

> The restored work, though often unworthy, hardly detracts from the general effect. Many lovers of Venice protest against all restoration, and there is no doubt that here to mend has often been to injure. Yet it seems to me that . . . [in] Venice, where many of the principal buildings lean at angles that look most dangerous, and where hardly a church tower stands plumb, continual repairs must be necessary. What wonder if the zeal of restorers has sometimes led them too far?[6]

Much of the existing Basilica is thus not medieval at all but rather a nineteenth-century creation. It was the (dubious?) benefactor of the burgeoning science of "imitation and repair, ever more sophisticated and ever more widely applied," as Pemble has put it—skills that "enabled old Venice to survive . . . but in idea, not in substance." Having seen off both the Neo-Goths and satisfied conservationists, the Italians turned to the Ducal Palace, which they subjected to a restoration as extensive as that of the Basilica, demolishing the building's fundamental southwest corner and replacing over a quarter of its ornate column capitals with copies. When the scaffolding came down in 1889, criticism about these and other drastic repairs was muted, however, since head architect Domenico Rupolo, who had learned his lesson from Saccardo, had artfully stained and weathered his alterations to make them look original. "The uneducated eye at all events cannot distinguish between the old and the new," commented one English observer, and the results seem to have been good enough even for most connoisseurs, perhaps because Rupolo offered something for everybody. To make the palace more "original" and please the tourists, he opened the five arches on the southeast corner that had been bricked up for over three centuries; to satisfy the antiquarians and medievalists, he preserved the capitals and other figurings that he had replaced, putting them on display inside the palace, where they could be admired by those discriminating enough to appreciate the quality of their craftsmanship.[7]

This was by no means the way restorations had been carried out in the days of the Serenissima: when a building had to be repaired or its function reassessed, it might be thoroughly altered even as it was being renovated. As late as 1753, when the Venetians restored the Torre dell'Orologio—certainly a monument if there ever was one—they did not just update the mechanism and repair the roofing, but also added another story to the structure as well.[8] A century later such alterations, no matter how artfully done, would have been unthinkable, not least because foreign lovers of Venice were able to mount intense and coordinated opposition against any changes to the city's monuments.[9] To make their point, these Ruskinites were happy to assert their own economic importance for Venice, claiming that no city so dependent on tourism could afford to ignore foreign tastes when it came to restoring its monuments.[10] Such clout especially derived from the sheer numbers that the town was increasingly attracting: by the late 1800s, five thousand or more package tourists were coming to Venice daily, just by train, during the high season. Henry James may have dismissed them as "trooping barbarians," but these visitors still expected to see Venice's famous monuments, in suitably weathered and romantic condition. The ever larger amounts of pounds, marks, francs, and dollars that they were bringing to Venice in the later 1800s made a persuasive argument that their wishes be satisfied.[11]

This notion—that those who paid for the aesthetic experience of Venice should have some right to determine what that experience should be—was not a new one: as we have seen, Johann Wilhelm von Archenholtz had advanced a similar argument over a century earlier.[12] The Ruskinites mounted a second line of attack during the Basilica controversy that was rather more fraught in its implications when they maintained that Venice's increasingly recognized status as a world treasure gave foreigners like themselves the right to meddle in questions concerning its well-being. To call this city a Treasure of the World was to assert that it belonged to the world—not just to the Venetians who happened to live there (or perhaps even owned bits of it). It was, rather, "the property of all cultured men and women" worldwide.[13] Such a contention reflected an emerging sense of interconnectedness among the aesthetic and cultured classes throughout Europe and America, though it also spoke volumes about the candid participation of those same classes in the dominant imperialist idiom of their day. Indeed, their claims of ownership need not have been based on their living in Italy for years, or perhaps even on having seen the works at all: it was enough simply to have the schooling or national culture to appreciate it. Thus, the *London Times* could pronounce that

St. Mark's is, in a way, the pride and possession of the whole world. Every one that has seen it or pictured it to himself would be pained to hear that it was rudely or ignorantly touched; and to destroy or deface its jeweled facade, to dull the glorious glow and rich phantasies of colour on which generations have feasted, is sacrilege against which there is a universal right to protest.[14]

In the years after the Basilica affair, Venetians would repeatedly find themselves at the wrong end of the hegemonic nose down which cultured Britains and French also viewed their own colonial subjects in Egypt, India, or Africa, likewise treated as unfit guardians of (their own) artistic patrimony. As the Ruskinites saw it, Venetians were saturated from birth in the soft beauty of their own blessed home and were consequently unable to appreciate or even understand the marvels they saw around them every day. For Henry James, living in the constant presence of so much perfection had rendered the Venetians not so much jaded as ahistorical, timeless, and childlike—classically Orientalist figures in need of support and advice from the cultured Westerner:

When, a year ago, people in England were writing to the *Times* about the whole business and holding meetings to protest against [the restoration of the Basilica] the dear children of the lagoon—so far as they heard or heeded the rumour— thought them partly busy-bodies and partly asses. . . . It never occurs to the Venetian mind of to-day that such trouble may be worth taking; the Venetian mind vainly endeavours to conceive of a state of existence in which personal questions are so insipid that people have to look for grievances in the wrongs of brick and marble.[15]

Driven by the assumption that Venetians were a basically feckless lot, this paternalistic, interventionist spirit continued long after the controversies around the Basilica and the Ducal Palace had died down. It had its part to play in the city's next great conservationist crisis, when the Campanile of San Marco abruptly collapsed, on July 14, 1902. The loss of this potent symbol was a clear disaster, both for Venetian morale and for the city's public image, and to reassure both the world at large and Venetians at home, the mayor, Count Filippo Grimani, rapidly and boldly promised that the Campanile would not just be rebuilt as soon as possible, but also that it would be resurrected "as it was and where it was," *com'era e dov'era*. It soon became apparent, however, that Grimani's assertion was rather quixotic: there was little chance of actually copying the original bell tower, since there were no surviving drawings or plans to go by. Nor was there much salvageable material left to work with, the collapse of the old tower having pulverized most of the original bricks. Moreover, an initial survey of the site soon revealed that this particular piece of Venetian terrain was so low-lying and spongy (comparatively speaking, that is) that it would be folly even to think of rebuilding a structure as massive as the old Campanile there.

Count Grimani's brave promise of *com'era e dov'era* aimed partly to reassure Venetians that no one was going to give in to the engineers, the accountants, or the Modernists who had suggesting replacing their beloved Campanile with a smaller tower, one more adapted to the conditions of the soil, one of modern design, or indeed with no tower at all. The count was also speaking to the foreign public, however: those "cultured men and women" of Europe and America who found the image of the bell tower "like an assurance of safety, of our civilisation, of Europe, of our Faith," and for whom "to think of Venice without the Campanile is . . . almost an impossibility."[16] As it happened, *com'era e dov'era* would become both Venice's defining credo and its eventual prison, setting it on the road to becoming the museum city it is today. Except for painstakingly gluing Sansovino's masterful Loggetta back together from thousands of component shards, restorers decided on a facsimile Campanile, essentially rebuilt from scratch. To do so, and still end up with a tower picturesque enough for any tourist's camera, head designer/architect Luca Beltrami had the bricks fired in a special kiln to make them look old and developed a stronger mix for the mortar to hold the whole thing together. Beltrami's determination to produce a replica tower constructed along modern lines led him to make some significant design changes. He hung the structure on an iron framework as if it were a skyscraper, remade the original bell chamber entirely in marble, and replaced the original stepped ramps that had run up the interior walls (so gently, it was said, that one could ride to the top on a horse) with an elevator. Umberto Calcaterra, son of the engineer who installed this device, recently noted how something as commonplace as an elevator could open up the tower as a tourist attraction: "It was an idea that completely changed the way one approached the monument. Until that moment, to get up to the top of the bell tower loggia, you had to climb hundreds of steps on foot. It was an undertaking that was impossible for older people."[17]

Thanks to Beltrami's improvements, the new Campanile weighed a good deal more than the original—perhaps 40 percent more—so it was also necessary both to widen the foundation and make it twice as deep. Completing this modern simulacrum Campanile took nearly ten years and cost over three million lire (around $600,000—perhaps $20 million today). Its perfection as a reproduction was admired by many foreigners as proof of an aggressive new Venetian spirit, though it was an odd sort of spirit in which, as Pemble wryly puts it, "the desire to imitate what had perished was translated as a symptom of revival."[18] So successful was the work that within a few years even those who had been familiar with the original tower were not sure they could spot any differences. Later tourists probably never even suspected that the old one had been replaced, unless they happen to see one of the (heavily retouched) photographs of the collapse, found in many modern guidebooks.[19] Nevertheless, the new Campanile is not the same as what Venetians used to call *El paron de caxa,* the "Master of the House." What had been there—the city's historic watchtower, beacon, alarm clock, and millennial statement of civic pride— disappeared for good on July 14, 1902. What sprouted up in its place was a creation of the industrial age—sleek and functional under its faux exterior, installed primarily for keeping the illusion of Venice going, replacing the necessary pieces so that the tourists would continue to come.

The Campanile crisis—so dramatic and tragic, yet ultimately so reassuring—played an important role in stimulating foreign interest in the enterprise of saving Venice. In the months just after the tower's fall, international papers were filled with stories of other Venetian monuments at peril: Santo Stefano, Santi Giovani e Paolo, San Giobbe, the Frari, and the Procuratie Vecchie were all said to have crumbling roofs, tilting bell towers, or weak foundations. There was even a fresh flurry of jitters that the Basilica of San Marco and the Ducal Palace were in danger of collapsing again.[20] Big disasters like the Campanile's collapse helped broadened foreign interest in Venice beyond the immediate San Marco area, proving marvelously tonic for art clubs and building societies in Britain or America as recruitment and fund-raising occasions of the first order. Within just three months of the bell tower coming down, both New York and London were busy with their own Campanile funds, even though the Italians themselves had initially insisted that they had no need for any outside aid.[21]

Further bruised by the thousand or so bombs the Austrians dropped on it between 1915 and 1918, Venice entered the great era of consumer tourism in the 1920s on the trajectory that it has followed ever since— stimulating foreign concern and attracting ever more visitors, while stumbling from one near-disaster to the next. Ripples of alarm about the city have continued to agitate the world public and press, concerning the imminent collapse of this monument or that; worrying about new or proposed factories, office towers, or apartment blocks ruining the skyline, or about widened canals or streets spoiling the ambience; and (repeatedly) worrying about the gondola disappearing or being motorized or faked up out of fiberglass.[22] Yet it was really only with the great *acqua alta* of November 4, 1966, when the waters hit 194 centimeters above mean tide, that all these disparate fears for Venice coalesced into the permanent, universal conviction that the city actually could be disappearing for good, that indeed *Venice was sinking*.

Although many foreigners criticized it at the time for its dilatory and overpoliticized approach to the save-Venice problem, the Italian state eventually did come to the aid of the city. In 1973 the Parliament passed the so-called Special Law on Venice, meant to fund not only Venice's rescue from the sea, but also the restoration and preservation of the city itself. Unfortunately, much of its generous allotment has remained unspent in the subsequent thirtysome years of endless wrangling over the proper way to deal with such matters as depopulation and flood control, but the Special Law has had some successes. First of these was the big job of converting all Venetian residences from coal to natural-gas heating, which has greatly benefited local air quality and slowed down stone erosion. Subsequent funds have been used to modernize much of the city's often primitive private housing and for new aqueducts that have made it possible to stop pumping water from the local aquifer. The Special Law has also paid for many of the city's most critical restorations, along with most of the day-to-day maintenance of its monuments. Because this sort of work is structural, it is often quite expensive, and costs sometimes have had to be met from other sources as well, such as the regional government, the state lottery, or the semipublic corporations that flourish in Italy.[23]

These sources of funding have by no means taken care of all of Venice's needs, however, and as with earlier restoration crises, the great flood of 1966 soon spawned a host of "save Venice" organizations. These have been charitable groups at least partly prodded into existence by UNESCO, which had set up a sort of command center right on Piazza San Marco. Known as the Association of Private Committees (the APC), they currently number around thirty from eleven nations and have linked funding and foreign donors with the Venetian Curia and the Superintendencies of Art and Architecture.[24] Sometimes they join together to fund a single project or to manage funding from an outside source, but more typically each group runs its own jobs. Since about 1970, the overwhelming focus of APC work has been on the restoration of Venice's rather tattered artistic and architectural heritage. Committees select projects from an annual list, prepared by the Curia and the Superintendencies, of restorations that the Church and the Italian state are either unwilling or unable to fund. These can range from very minor touchups to the complete overhaul of a building—the sort of project that might drag on for a decade or more. On top of this, the committees also actively sponsor studies and colloquia on Venetian art and architecture, while providing grants for their own nationals to study problems of restoration peculiar to the city.

The doyen among the Private Committees is the Venice in Peril Fund, cofounded in 1966 by the British consul to Venice, Sir Ashley Clarke, and Viscount Julian Norwich. VIP, as its chosen initials imply, is a fairly elite organization, and the group's patron list positively bristles with dukes, lords, and sirs—"lords on the board," as the British like to say, much in the fashion of the old Society for the Preservation of Ancient Buildings.[25] The Australian, Dutch, Danish, French, German, Swiss, Austrian, and Swedish committees are more modest in their social profile but are every bit as scholarly and energetic in the work they do. All tend to fall into the shade of the Americans, however, who have no fewer than four organizations in the APC, led—at least when it comes to the serious money—by Save Venice Inc. Formed in 1971 and headquartered in New York City (with regional chapters in Boston and Los Angeles), Save Venice generates far more funds for the projects it chooses than any other Private Committee. In recent years its restoration spending has represented 60 to 70 percent of the total APC outlay: each fall Save Venice takes on at least ten projects from the Superintendencies' list, so that it is running at least two dozen restorations at any given time (thirty-four in 2000–2001), typically disbursing nearly a million dollars annually.[26]

What the current executive director of Save Venice noted about the organization's late founder—that he "adopted Venice after the disastrous flood of November 4, 1966"—could indeed be said for many of those trying to rescue the city. From all over the world, these men and women of the best classes—business, finance, academic, and leisure—have taken up the city, "adopted" it for themselves, as everything from an idle pursuit to a pet project to a driving passion.[27] Americans, more than other nationalities, tend to take up this expensive hobby, perhaps because their contributions, unlike those made to the other Private Committees, are tax deductible. Of course, all American charities offer contributors a tax write-off, so the competition in that country for the tax-exempt dollar can be ferocious. To gain an edge in the charity market, committees like Save Venice remind donors that by contributing they can do good while also enjoying Venice, the object of their charity. Soon after it was founded (and still going by the name of "the Venice Committee"), Save Venice announced that membership (twenty-five dollars at the time) guaranteed a 10 percent discount on goods and services at participating hotels and shops in Venice and promised that "anyone who spends three days in Venice can easily recover this fee in discounts."[28]

Before long, Save Venice was moving on to more elaborate and more frankly elitist means of attracting new contributors. Its techniques of mixing the chic with the charitable have followed the classic New York recipe: "raising funds and rubbing elbows," as the *New York Times* has called it.[29] These have included the regular (and very exclusive) Winter Carnival Ball, held with great pomp at a major Manhattan venue, an ongoing lecture series, and a succession of sponsored luxury cruises. Save Venice's most characteristic event is its biannual Regatta Week Gala in Venice itself, set for the week before the city's own Regatta Storica in early September. For the price of a ticket (in 2001, three thousand dollars for those over forty, and fifteen hundred dollars for "Young Friends," travel and lodging excluded), participants are treated to a number of cultural events: lectures, visits to the committee's recent restorations, day trips to Palladian villas, and chances to dabble in some especially "Venetian" activities— gondola rowing, for example, or wine tasting—all punctuated by a continual string of lunches, dinners, and concerts. Meanwhile, for its more active members, Save Venice kicks off the gala with its signature all-day treasure hunt that sends competitors scurrying all over Venice's famous maze of *calli* and canals.[30]

All these cultural amusements, both back in the States and in Venice during the Gala Week, have been further enhanced by Save Venice's promise that its members and donors will have the chance to "rub elbows" with Venetian and Italian nobility. Luncheons and teas at aristocratic palaces on the Grand Canal, in "spectacular homes the public never sees," feature gracious counts and countesses of antediluvian lineage whose simple presence and supposed insider knowledge of the Venetian social scene offer Save Venice's *arriviste* millionaires from New York or Los Angeles the ultimate backstage experience of true Venetianness. Indeed, the social whirl of the enterprise—mingling with these elegant remnants of the Serenissima, the charity balls, and the frolic about the Venetian *calli* in the Gala Week treasure hunt—has sometimes threatened to overwhelm Save Venice's declared mission of saving Venice's monuments and works of art. The Venetians themselves have protested that all this partying and royal-chasing is turning their city into the frivolous plaything of shallow socialites. Perhaps they are not far wrong: the *New York Times* quoted an "habitual gala-goer in New York" on the local climate of opinion: "'Let's face it: some people don't care whether the party is for Save Venice or for Save Chicago. The parties are fun. Do people care about Venice? It could be Save Fort Lauderdale From Spring Break and they'd go.'" Tensions over this issue eventually grew to the point that in 1998 the organization sundered, with "the social set" pulling out to form a new committee called Venetian Heritage, Inc. As Ralph Guthrie, Jr., the current chairman of Save Venice, put it, "There are two groups. One group thinks we're in the restoration business in order to have parties and have royals around, and the other group—that's ours—is throwing parties to have restorations."[31]

Despite this setback, Save Venice has since reconstituted itself, and its parties glitter as much as ever, complete with a luring new stable of *baronesse* and *contesse*. Still, developing a donor base requires constant cultivation, so lately the organization has tried new programs, such as offering a "Wish List" of projects that need sponsors. Most, although not all, of the jobs are taken from the compendium provided by the Superintendencies: often they are smallish restorations—not the sort that Save Venice would otherwise undertake using its own general fund. Instead, we were told by Melissa Conn, director of the group's Venetian office, Save Venice uses this list to troll for prospective donors by offering, online, attractive little projects for the first-timer, with estimated costs ranging from $130,000 (for rebuilding the *sukkah* tent in the Scola Canton in the Ghetto) down to a very reasonable $2,000 for inspection and upkeep on some lesser-known paintings.[32] The idea, as Conn put it, has been less to get many of these specific projects done, important as they may be, than to attract sponsors and get them involved with the process, in the hope of greater gifts down the road. The directors at Save Venice know that charitable giving, especially for romance-laden Venice, tends to be focused among older, wealthier contributors. To keep the donor base growing, a stream of new, more youthful enthusiasts must therefore be continually cultivated to replace the old, and schemes like the Wish List seem to fit well with younger contributors. For those whose finances might not support an entire project, collective contributions are encouraged as an alternative—especially appropriate for those who give a Wish List item as a collective wedding present, perhaps, or who offer it as a memorial gift for someone who showed particular affection for Venice during his or her life.

Does all this amount to Save Venice or Sell Venice? We were curious about the extent to which donors could have their contributions recognized, not only by the tax authorities but also by posterity, with one of those little bronze or marble plaques posted next to the restored work. Conn assured us that putting up such markers was against the policy of Save Venice, saying that the group "would rather not have plaques all over the city." One crucial exception to this has been the memorial plaque: a major restoration or Wish List item funded in someone's memory may have the deceased commemorated in this way.[33] Although we were not able to clarify the policy of other Private Committees in this regard, nor of the Superintendencies themselves, we would not be surprised if later plaques will eventually feature the donors' names as well: many who give to the APC do so at least in part in the hope that they will be recognized as aesthetics and supporters of the arts. Indeed, for its first quarter century, Save Venice raised its funds almost exclusively through staged events; its donations came almost exclusively from anonymous contributors. Only since 1990 or so has Save Venice named its donors, identifying their gifts with specific projects and sometimes publicizing how much they gave. As restorations become more closely linked with individuals, we would assume that those donors will receive some form of tangible recognition for their contributions: a plaque seems to be the least they might expect.[34]

For others who wish to save Venice while living well, the group also offers various lines of "home furnishing products, including furniture, giftware, lamps, and decorative fabrics . . . inspired by Venice." According to the blurbs, this "Venice Collection" features pieces with "soft and relaxed designs for today's more casual lifestyle . . . decorative accents [that] feature the play of light and color found in Venice." In exchange for the percentage it gets from each sale, Save Venice offers discriminating consumers an alternative to the common touristware available in Venice, by means of an exclusive and reliable mail-order catalogue. In the process, Save Venice hopes that this commodified Venice will further stimulate buyers' interest in the city and, eventually, in Save Venice itself, perhaps enough to join or to make a donation.[35] In addition, the organization has also taken out and dusted off its old discounts-for-members program, first introduced at the time of its foundation. When the concept was brought back in 2002, it was given rather more snob appeal, however: rebaptized the "Serenissima Society," it had become a "newly created program consisting of special benefits in Venice" and was available to donors only "at and above the $1,000 level." As before, the promise was that those who had demonstrated their "loyalty and commitment to Save Venice" should get to enjoy, at a discount, aspects of the city they were dedicated to saving, at some of Venice's most exclusive hotels and restaurants—

including the Hotel and Locanda Cipriani, and Florian's and Quadri's on Piazza San Marco.[36]

Britain's Venice in Peril, probably the second most successful fund-raiser among the Private Committees, managed to bring in around £1.6 million between 1971 and 1996, though using very different techniques than Save Venice's. Unable to offer a tax write-off, and perhaps finding titled Venetians not so impressive to the many milords involved with the organization, the VIP has instead taken an approach that is both more public and more egalitarian than that of Save Venice. Through a licensing agreement, VIP has linked itself with several British companies—most notably Pizza-Express, which offers customers who order its "Pizza Veneziana" (topped with onions, raisins, olives, capers, and pine nuts—supposedly a medieval Venetian recipe) the option of designating twenty-five pence of the purchase price for VIP's restoration fund. This has been a popular pizza: between 1965 and 1995, VIP collected around half a million pounds from PizzaExpress, the proceeds from what must have been two million pies—enough to stretch just about halfway between London and Venice. In a similar fashion, British Airways Holidays agreed to donate a pound from every holiday involving Venice booked through its *Cities* brochure: this brought in some forty-two thousand pounds in 1998–99 alone, a good sign of how popular Venice remains with the British.[37]

However the Private Committees have generated funds, their combined contribution to rebuilding and renovating Venice's cultural treasures has been impressive. Between 1969 and 1999, they supported work on nearly a hundred monuments and around a thousand works of art, projects that took anywhere from a few months to over ten years to complete, at costs ranging from under a thousand to upward of $4 million. In recent years, their collective outlay on restorations has easily exceeded a million dollars annually.[38]

All this money has had its impact on Venice. Buildings that might have collapsed have been restored in many cases to something like their original state, thanks to interventions sponsored by the APC. One could say that Venice's unique artistic patrimony has received a new lease on life and is beginning to flourish in a way that would make other tourist centers envious. Yet was Viscount Norwich justified in asserting that the Private Committees "have been working to save the most beautiful city in the world"? Although some of the committees—Save Venice, Venice in Peril, and the Comité Français pour la Sauvegarde de Venise—have names that imply this, a good deal of the APC's efforts have supported undertakings less than essential for the survival of the city, largely cosmetic projects that involved cleaning stone or the painted surfaces of works already safely indoors or otherwise structurally sound. The Comité Français took on the Portego of Palazzo Querini-Stampalia, for example, because "the accretion of dirt on the walls and ceiling detracted considerably from the aesthetic appearance of the whole"; likewise, the Swiss initiated work on Santa Maria del Giglio with the observation that "a thick black deposit made it impossible to appreciate the decoration and sculpted stonework at ground level."[39] This sort of activity falls under the rubric of "cleaning and consolidation" and accounts for fully half the projects taken on by Save Venice. Other restoration ventures, some fairly substantial, have aimed at even more clearly Ruskinite ends, such as the removal of later, "crude" additions (usually nineteenth century) to bring a piece back to what the experts have judged to be its original Renaissance or medieval state.

The Private Committees thus end up devoting a good deal of their efforts not so much to saving this city as to beautifying it: "preserv[ing] and protect[ing] the art and architecture of Venice" is how Save Venice describes its mission.[40] The two missions are not always mutually exclusive, especially when involving monuments crucial to the city's very identity—the Basilica of San Marco is a good example, as is the Ducal Palace. Nevertheless, many APC projects reflect individual committees' agendas and donor ambitions more than they do any long-term or specific plan to save Venice. Thus, we find that the Comité Français "often sponsors the restoration of Venetian monuments with a French connection" and the Austrian committee, "Venedig Lebt," has an interest in renovations involving the nineteenth-century Hapsburg occupation of Venice. Save Venice's New York donor base includes many Jewish patrons interested in supporting projects with Venetian-Jewish connections, such that, small and peripheral as it is, Venice's Ghetto has become the most restored area of the city, with the exception of the Piazza San Marco area.[41]

Donor-based restoration influences the choice of projects in other ways. Contributors, whether individual or corporate, tend to like high-profile jobs that bring them status and recognition; they prefer projects with a clear-cut end point, so as to produce recognizable results, in a timely fashion if possible. It strikes us that UNESCO was displaying some studied ingenuousness in asserting that the whole field of "preventive preservation" of monuments—that is, monitoring structures that are still in good shape—has been neglected by the APC because "paradoxically, this less striking and less prestigious type of activity has greater difficulty [than straightforward restorations] in finding the necessary funding." Of course it does! That donor money is attracted to prestigious projects rather than routine maintenance work is not paradoxical at all: indeed, fund-raisers have a hard enough time just convincing their patrons to contribute for restoring works by artists of less than the first rank. The use of donor rosters, wish lists, and gala celebrations for completed projects (engraved invitations, white tie, cocktails, guest speakers) seems to confirm that many of their contributors want to be recognized as patrons (and saviors) of the world's artistic canon.

The academic community also has a central role in Venetian restoration. Almost all of the committees use a portion of their funds to support students of art history and restoration from their own countries. Likewise, virtually every committee has a handful of art professionals serving on its board, to advise on which projects to choose and to interact with their opposite numbers on the Superintendencies. For men and women like these, Venice is "one vast laboratory," as Venice in Peril once put it, offering them career-enhancing possibilities to work on some of the world's greatest artistic masterpieces and thorniest problems in conservation. Many are eager to get involved in restructuring work, clambering over scaffolding and probing inside dismantled walls to discover the secrets of each building's inner fabric. Such opportunities are as useful for teaching students as for actually moving the restoration work along, and many sites turn into long-term classrooms, where restoration can take second place to archaeology.[42]

The resulting collaboration in the APC—Italians might call it a *connubio,* a pragmatic bed-sharing—between art professionals and wealthy donors, all eager to associate themselves with Venice's artistic heritage, is reflected in the committees' lists of directors and patrons, a roster of academics, aesthetes, and socialites who could have descended directly from Ruskin, Wallis, and the Society for the Preservation of Ancient Buildings. Fortified by such potent supporters, many committees, especially the smaller ones, have tended to "adopt" a particular building— a church, perhaps—on which they might fund work for years, identifying both their prestige and their scholarship with the ongoing project to the point where both they and the restoration community in general begin to refer to it as "their" church or *palazzo.* This notion of adopting a particular structure goes back to the aftermath of the 1966 flood, when nationwide appeals were made in many countries on behalf of a well-known monument: the French committee got its start in this way, with a nationwide campaign to renovate the church of the Salute.[43] Since then the process has become somewhat more formalized than these one-off programs, and now a steady flow of national funds is funneled through each local Private Committee. Thus, the Dutch Stiching Nederlands Venetië Comité has labored for years on the church of San Zaccaria, Pro Venezia Switzerland spent a good eight years with Santa Maria del Giglio, Venice in Peril concentrated for most of its first decade on Madonna dell'Orto, and so on.

What remains unclear in all of this is the extent to which ordinary Venetians had any input into the process—how a project should be undertaken or whether it should even be done in the first place. Although virtually all of the committees have a few (usually noble) Venetians on their boards, project decisions still tend to be made outside of Venice by elites who, from the viewpoint of those left out of the process (including most Venetians), appear as "a peculiar kind of 'international mafia'": "They are scattered through many countries, operate a score of private committees; they maintain a tight network of telephone and mail communications with 'friends' in the major American and European media. Whenever there is a threat to Venice, great or small, from whatever source, they make it known fast."[44] Years ago, local patriots might write to the papers and complain that "Venetians and Italians in general should be ashamed of having [foreigners] going around begging for funds to save Venice," but nowadays most locals seem content to have their artistic treasures reclaimed, restored, and generally prettied up, regardless of who pays for it. Few Venetians that we spoke with expressed more than mild resignation over having restoration decisions made at the international level, and no one welcomed the well-known alternatives to more neglect: paintings literally falling out of their frames, stonework crumbling, or buildings collapsing into heaps of salt-saturated marble and brick.[45]

Still, by working largely outside of community structures and by confining their restoration activities for the most part to high-prestige, show-piece art and architecture, the Private Committees do not end up making much genuine impact on the daily lives of resident Venetians. The artwork that excites donors and scholars of the APC—paintings by Carpaccio, Titian, or Veronese, for example, or sculpture by Tullio Lombardo or Sansovino—might not have such significance for locals. Restoration work is, unfortunately, most likely to attract neighborhood attention when it disrupts the regular flow of life in the area, as happened in the late 1990s when an Australian-Swedish renovation of a carved medieval arch in an obscure corner of Castello blocked "an extremely popular shortcut" for several years in the late 1990s.[46] When it has come to ruffling local sensibilities, however, no one has yet outdone the Comité Français, which recently followed its francophile enthusiasms a bit too forcefully by buying (at Sotheby's) and restoring a neoclassical statue of Napoleon by nineteenth-century sculptor Domenico Banti, with the intention of donating the work to the Museo Correr. Not surprisingly, when word got out, there was a storm of protest among Venetians, who were outraged that the man who had once boasted that he would be "Attila the Hun to Venice" should be so honored by the city whose thousand years of independence he had personally extinguished. Letters and e-mails poured into the newspapers, and conferences were held, denouncing Napoleon as a "highwayman," a "barbarian," and a "tyrant." "It wouldn't come to anyone's mind," growled one commentator, "to present a statue of Hitler to the Ghetto," and some of the *antibonapartisti* darkly warned that "they can mobilize the army to bring [the statue] to Venice, but they can't garrison it night and day: we will smash it down with sledgehammers."[47]

Despite years of having their artistic heritage largely in the care of outsiders, some Venetians obviously are still able to reassert authority over their own birthright. In a similar, if less aggressive, fashion, some locals continue to tend to their parish churches, concerning themselves with the spiritual intent of the artwork in their care, as opposed to its purely aesthetic value. Ordinary Venetians have, for example, continued to tend their city's five hundred or so *capitelli,* or street-side shrines, "that adorn the intersection of a great many *calli* and canals, that illuminate dark *sottoportici,* or that reign on remote channel markers in the Lagoon."[48] These little altars are part of a cultural tradition in Venice that dates back a millennium or more, though as works of art they have yet to appear on the aesthetic radar screens of UNESCO or the APC. Still, modest as they are, these shrines require regular upkeep, generally from parishioners, who not only keep them in flowers and candles, but also clean, paint, and repair them when necessary. How well this work is done depends to a great extent on the "sensitivity" of the volunteers—we never heard of anyone getting any special training in this sort of thing—but then fixing up of such "popular" artwork is not usually subject to the same standards as those demanded for art in the recognized canon. However well the restoration is done, neighborhood people still appreciate the improvement, and they still come to make use of their local *capitello*—for personal prayers, to leave *ex voti,* or to join together in the somewhat more formal business of chanting orations or reciting the rosary in the evenings.[49]

Such organic linkages between a restoration and subsequent local use are less sure when it comes to projects undertaken by foreigners primarily intent on reclaiming Venice's more famous artwork. It sometimes happens that restored works cannot be seen by anyone, local or tourist, because the Curia cannot afford to keep the building open. "After having restored the churches with enormous financial efforts," as don Aldo Marangoni, spokesman of the patriarchate, has warned, "it will be necessary to close them once more," unless the Curia comes up with more operating funds than what it currently generates from its dwindling number of faithful and its other limited sources of income: there are simply too many churches in Venice to hire the necessary guards and to pay for the heating and lighting for all of them, regardless of the art treasures they may contain.[50]

Even when the Curia manages to keep a restored church open, the structure's new status as a world-heritage site can problematize its relationship to its surrounding neighborhood. The little church of Santa Maria dei Miracoli, near the border of Castello and Cannaregio, offers a good example of what can happen. Massively restored by Save Venice between 1987 and 1997 at a cost approaching four million dollars, the Miracoli was a big enough undertaking to make it something of a defining experience even for the robust American organization: to tackle something so big, the group had to mature "from a volunteer organization, haphazardly making $40,000 a year, to a professional one with an office and a full-time staff." Perhaps it was also a defining experience for the Miracoli, since, in the eyes of many in the restoration community, this has subsequently become "Save Venice's church."

The job itself required assembling a "team of highly skilled craftsmen [to] literally take the building apart and put it back together again." The Miracoli's extreme deterioration was due partly to the fact that one wall of its nave was anchored in the adjoining canal: as tidal levels have risen over the centuries, the structure's porous brick and marble walls were invaded by salt, wicked up from the seawater. The result was that in many places along the interior of this wall, as we ourselves discovered in the 1980s, the carved stonework crumbled at a touch. Such advanced decay gave Save Venice's team a welcome opportunity to use the latest in high-tech devices—harmonic hammers, special cleansing solutions, computer studies, and the like. For this part of the church, the technicians had to disassemble the wall and rebuild it, inserting a higher barrier against the *acque alte* and redoing many bungled nineteenth-century repairs. In such places, and along the exterior where the deterioration was advanced, the restorers replaced rotted figurations with facsimiles, following the tradition already established with the Ducal Palace capitals and installing replicas that were "recreated exactly as they had been in every aspect with the same materials worked in the same ways."[51]

All this effort (and expense) produced results that, as Save Venice has justifiably boasted in its literature, were nothing short of miraculous. Now renewed and rebuilt within and without, "the church gleams in its little campo." Its interior has been called "a treasure chest" of refined stonework, and the space one encounters there is "free of the usual encrustations of mosaic, the jutting tombs, the side-chapels and undistinguished annunciations."[52] This, however, is not all to the good: one soon notices that the Miracoli, although still consecrated space, is free of almost anything that would indicate that it is an actual church. Mass is held in this antiseptic space only once a week, and the newly glistening interior walls are innocent of all those posters and announcements that typically adorn (or clutter, depending on one's point of view) churches that are still in active use. The number of pews is kept to a minimum, presumably to avoid obscuring the dazzling inlaid-marble floor.[53]

Having gone through its ten-year salvation, the Miracoli has emerged on much the same terms that the Basilica of San Marco did over a century ago: as a facsimile church. Certainly this has not stopped it from becoming, as Save Venice asserts, "a favorite church of Venetian brides"; thanks to recent changes in matrimonial laws, it is now possible for foreigners to wed there too, and in the spring and fall all the available slots are booked, as couples from all over the world vie to get married in one of the most romantic spots in the Most Romantic City in the World.[54] Through much of the summer, we have noticed, the *campiello* in front of the church is perpetually strewn with rice.

Indeed, the Miracoli appeals to more than just brides-to-be: it has emerged from its restoration as a new reference point on the Venetian tourist grid. Just twenty or thirty years ago, foreigners generally ignored this area, and travel writers had to cajole their readers into going there. Now, by contrast, as Save Venice notes with pride, the Miracoli has become "a popular tourist attraction," and "gondoliers make a point of taking their tourists through the tiny canal to observe its beauty from the water."[55] Furthermore, the Miracoli has been joined with twelve other churches to make up part of the CHORUS tour: one ticket will do for all thirteen, and visitors to this church have increased accordingly. Thanks to the so-called Marco Polo house and the recently renovated Malibran theater, which are both nearby, this area on the border of Castello and Cannaregio has become a new node of attraction. The adjacent little *campo* of Santa Maria Nova, once a quiet, neighborly place, now echoes with tourist groups, while *carovane* of gondolas continually crawl past the *fondamenta* that faces the church.

The Miracoli owes part of its success with foreigners to the marvelous acoustics created by its bare marble walls, making it an excellent venue for concerts; chamber orchestras and choirs go there from all over to play. In just a single month (June 2001), one could go there to enjoy the New York Catholic Chorale, the Athens Drive Jaguar Singers (both from the USA), and the Anglia Chamber Choir (from Britain). These musical possibilities have not been lost on others in the tourist sector: Norwegian Cruise Lines, for one, has arranged performances at the Miracoli and boasts that its passengers can disembark to attend "an exclusive private concert held by professional musicians, in this charming and relaxing atmosphere . . . [where] you will be treated to works by Vivaldi, Pachelbel & Bach."[56]

Not only has the Miracoli become a facsimile church, then, but it also has begun to establish itself as both a secular temple of art (the "earliest masterpiece of the new Renaissance style in Venice," according to the guidebooks) and a venue for public events. Some transfigurations have been eased along by the unusual nature of the Miracoli itself. Constructed as a convent rather than a parish church, it was not intended to have a prominent role in neighborhood life; its highly finished marble interior and exterior have made it uniquely suited as both a temple of art tourism and a concert hall. Nevertheless, the way that restoration has put the Miracoli on the tourist map, and in turn opened up a new part of town to tourism, can be seen as a model for the process by which Venice has been aestheticized, in particular through its churches. Restoration work in such diverse locations as Madonna dell'Orto (in Cannaregio), San Nicolò (Dorsoduro), Santi Maria e Donato (Murano), Santa Maria Assunta (Torcello), and San Giovanni in Bragora and San Pietro (both in Castello) not only has reclaimed endangered structures, but also has turned them into tourist attractions. Their once remote and rarely seen local neighborhoods are now regularly visited by tourists, whose presence has stimulated the usual flourishing of mask shops, gift stores, and fast-food outlets.

Many of these churches are part of Venice's CHORUS package, which has allowed the Curia to charge admission to its churches—something that was virtually unheard of in Italy even ten years ago, but which is becoming increasingly common. This may seem sensible, since it lures tourists out of the San Marco–Rialto rut while getting them to subsidize the artistic heritage that, in a sense, they claim as their own. There are dangers here, however. We have already seen that spreading the tourists more evenly around town is not always satisfactory for those Venetians who lose the privacy of their neighborhoods as a result. Moreover, charging admission to churches promotes the aestheticizing of

Venice, further turning the place into a museum city. The more tourists experience these churches as minimuseums, the less they will be inclined to treat them as places of worship, or *luoghi di culto,* as the Italians say. We once overheard an American in the church of San Zaccaria refer to the artworks there as "exhibits," and this tendency to ignore the sacred character of these places has become almost universal. Many tourists consequently behave in ways that would be more appropriate for a museum than a church: they cluster around important artworks, while their guide lectures them; they take photographs or shoot videos; they sit and rest in the pews while reading their guidebooks, chatting, and sometimes eating their lunches. In addition, charging admission to enter these churches has radically changed their function and meaning for many of those who may need them for their original purpose of providing a haven. A petition drive in 2002 to have the admission charges revoked collected over seven hundred signatures in just eight days: over 90 percent of those approached—"Venetians, [and] tourists, but above all the many immigrants who work in the city but live elsewhere"—supported a return to free churches, opposing a process that some claimed had "turned the house of charity into a church-business." As Nadia De Lazzari, leader of the petition drive, explained to the press, "The pew of a church, like a park bench, [is] a corner of liberty, repose, and meditation for many people, among whom are, above all, the hundreds of caregivers, maids, and baby-sitters who during the day have only a few free hours in which they can take advantage of a moment for reflection and prayer in church." A good many such service workers, in Venice as elsewhere in Italy, are women of Philippine or South American descent, but unlike those in other Italian cities, if they wish to visit their neighborhood church in Venice outside of the official time for mass, they may well have to pay two euros for the privilege.[57]

As more churches in Venice began to have newly restored, top-quality artworks, the demand among tourists to get in and see these attractions becomes stronger and more diffused across the city. Conflicts inevitably break out between outsiders who feel that it is their right to see sites they regard as world treasures and Venetian locals who insist that their churches are *luoghi di culto.* Such a clash made its way into the papers in the winter of 2000, when a woman from Turin wrote to complain that while she and her husband were on a trip to Venice, the local beadle refused to let them into the Frari, not only claiming that there was a wedding in progress, but also adding sarcastically, "Unfortunately, even here in Venice we get married too." Big as it is, the woman asserted, the church of the Frari ought not be closed to art lovers like herself, even temporarily, just so (as she put it) "a chosen few" could have the use of it.

In a written response, the parish priest of the Frari picked up the argument, shooting back that in his parish those "chosen few" did indeed amount to just three thousand parishioners, compared to the three hundred thousand or more tourists who want to visit the church every year. To accommodate the demand of so many non-Venetians pressing to see the works of Titian, Vivarini, and Bellini, he noted, the parishioners have to squeeze all the religious functions of their church down to just two and a half out of the nine hours that the building is open daily (a bit more on Sundays). Still, the *parocco* noted, this is ultimately a church, not a museum; should his parishioners wish to have the place to themselves for few extra hours to celebrate special events like weddings or funerals, "We claim that as our right," he said. "It doesn't seem to us at all a matter of acting arrogantly or being insensitive."[58]

Perhaps we should consider sites like the Frari that are blessed (or cursed, if one prefers) with world-class art by Bellini, Carpaccio, Titian, Tintoretto, or Veronese as simply museums that happen to be located inside a church. One thing is certain, though: the restoration projects in Venetian churches over the past thirty years, whether involving single works or entire buildings, have helped to make the sites themselves and the entire city more attractive to visitors, playing a central role in boosting the overall tourist appeal of Venice. It is all the more ironic, then, that UNESCO, working through the APC to promote this process of the aestheticization of Venice, should then lament these inevitable consequences of its own efforts: "[The] milling crowds, the reduced availability of services for normal [that is, resident] customers and inflated prices . . . We now have reached the breaking-point for the preservation of a proper network of public and private services. Venice is becoming a museum-city, and is no longer a residential one."[59]

Some of the Private Committees do seem aware that saving Venice must mean more than simply saving Venetian art. Venice in Peril, in particular, boasts a charter (called its terms of reference) that charges the organization, as a registered charity, to "meet the needs of the people of Venice, as well as of its monuments." Though VIP has had to admit, after thirty years in business, that "we have so far concentrated our energies and resources on restoration," some of the group's more recent projects still have managed to tackle situations in which restoration could lead to a certain amount of social utility. Thus, after spending most of the 1970s, and upward of £125,000, on reroofing and restoring the church of San Nicolò dei Mendicoli, in Dorsoduro, the group was receptive to requests from parishioners a decade later to restore the adjoining oratory of San Filippo Neri. The upper floor of the finished oratory now serves the parish as a meeting room, useful both for local gatherings and for the income it brings the parish when rented out—in particular, to the University of Venice. More recently, VIP has gotten involved in an ambitious project to rebuild and restore working-class housing in Venice in such a way that some six-teenth- or seventeenth-century apartment blocks might be occupied again by tenants while still preserving much of their original character and floor plans.[60]

Recognizing that they are, after all, foreign charities, APC members generally assert that they can do only limited work of this nature, re-maining "naturally careful not to interfere in matters which are properly the concern of the Italian authorities." Nevertheless, VIP has turned its attentions to "both the restoration of the city's ancient buildings and ar-tifacts, and the investigation into those environmental factors that are pro-gressively damaging the fabric of Venice itself," meaning in particular the *acque alte*.[61] Save Venice Inc. has also shown interest in problems con-cerning water damage, by sponsoring the design of a new, low-wave pas-senger boat to combat the *moto ondoso*. Called by its designer the Mangia Onda, or Wave Eater, the craft comes in taxi, tourboat, and waterbus sizes, for 10, 25, and 150 passengers, respectively. The latest versions are also elec-tric-powered, meaning that the Mangia Onda may have the potential not only for solving Venice's *moto ondoso* problems, but also for easing its difficulties with air, water, and noise pollution.[62]

Such efforts by Venice in Peril and Save Venice (which has also put the wife of the Mangia Onda's creator on its board of directors) are hearten-ing.[63] There is, however, a great deal more that APC members could do, still staying strictly within the realm of the "bricks and mortar" work that most of them prefer. While their experts worry about the proper shad-ings and the exact nuances of composition of various masterpieces, a good deal of the Venice that Venetians continue to live in is falling apart. This is hardly surprising: the same afflictions of subsidence, seepage, and pol-lution that plague Venice's churches and public buildings have also taken their toll on the city's private housing—indeed, often more so, since the great monuments here, as elsewhere, were generally built with more care and out of better, more expensive materials. The papers are increasingly full of stories of buildings that have had to be hastily evacuated, their oc-cupants moved to emergency housing in a hotel in Mestre, because the structures were judged in serious danger of falling down. The fire department in Venice estimates that it gets around ten calls a month about this sort of thing: big cracks appearing in walls, massively leaking roofs, subsiding floors, or even big chunks of stonework—gutters or carvings—simply falling off buildings into the adjoining street or canal.[64] When the city finally started draining and cleaning out the city's canals, fully a third of the buildings along these waterways turned out to have foundations in a state of either significant or serious degradation, for the most part caused by the continuous impact of the *moto ondoso,* which has been generated in its turn by the flotilla of power boats serving Venetian tourism.[65]

Although this is not a prescriptive book, we would suggest that this is the Venice that increasingly needs to be saved and that should attract the eye and the interest of the Private Committees. Residential buildings, many of quite respectable age, even without architectural fame, are falling down all over the city. The result is that still more Venetians—many of them elderly tenants, for years accustomed to living on controlled rent—are being forced to leave the city for the mainland. Their permanent departure—not many can afford to return—assures more than ever that the fate of Venice's historic center will turn out to be exactly what both UNESCO and the APC claim to dread: a museum city, abandoned by its residents and turned over to tourists and to a few wealthy outsiders who have the funds to tackle these expensive restoration jobs and who wish to set themselves up in a comfortable second (or third) home in Venice for four weeks out of the year. In this sense, Saving Venice would mean Saving Venetians.

CHAPTER 10

Ships and Fools

For centuries Venice has been called the Most Beautiful City in the World, a title that can make one forget that not all the attractions it offers are strictly physical. Generations of visitors have gone there to bask in the reflected glow of the place's palaces, churches, *campi,* and canals, but they have also sought out (or had thrust upon them) other allures, including elaborate, sacro-civic processions; music, theater, and opera; the Carnival; boat races, bull-baitings, and boxing matches; and a host of other attractions, both high and low. Though all these cultural productions can be found elsewhere in the world, they take on a special form in Venice, and not just because of the city's unique topography. The particular genius of the Venetians themselves has also played a role, in their intuitive blending of land and water, of elite and popular mentalities, of piety and profanity. In their festivities, the men and women of the city made a point of welcoming outsiders, if only to stand in the wings and admire, with some slight comprehension, the many celebratory expressions of what locals still refer to as Venetianness *(venezianità).* After the fall of the Serenissima, visitors may have come primarily to admire the paintings and architecture, or to visit Lido baths in the summer, but they also continued to have a look at the festivals and were spectators at the various regattas, which Baedeker praised as "characteristic and interesting." Yet throughout the nineteenth and early twentieth centuries visitors tended to keep apart from the Venetians—off in their seaside hotels, their hired gondolas and tour groups, their separate sections of the Piazza during the evening *listòn.* As long as Venice continued as "a city teaming with life," Venetian culture also remained lively: this was the product and glue that kept these people together—and effectively kept most tourists at bay.[1]

This all has changed since about 1960. With the thinning and aging of the local population, Venetian culture has become like Venetian space, available for appropriation by foreigners. Outsiders have encroached, as we have seen, into such uniquely Venetian experiences as the gondola ride to the extent that their seemingly endless demand has not only priced this simple yet exquisite form of transportation out of the reach of most Venetians, but has also so altered the whole gondola experience that most locals now find it alienating and somehow embarrassing. This is not the only example of a Venetian cultural product submerged in a tide of foreign enthusiasm, suffocated by embrace of the outside world: such cultural expropriation is everywhere in this city, as the withered elements of local culture—the glue that once held society together there—continue functioning as surreal imitations of themselves, right before the eyes and often much to the annoyance of the Venetians.

Here we shall explore three recent examples of this process: the Vogalonga, the Festa del Redentore, and the Carnival of Venice. By doing so, we hope not only to lay out the dynamics at work in such cultural expropriation, but also to detail some of the ferment and enthusiasm that originally lay behind these events before they were converted into cultural commodities. We should note that all three celebrations were conceived and set in motion within just a few years of one another in the later 1970s, representing concrete realizations of a spirit of idealistic communitarianism that was especially diffuse in Italy in those years. It was also the spirit that expressed itself in the *cantautore* movement of Italian folk-popular music, representing a (primarily youthful) refusal to give in to the seemingly unstoppable waves of extreme right- and left-wing terrorist violence, weak and corruptible government, and the definitive end of the postwar boom.

The first of these events, the Vogalonga, or "Long Row," was initiated in 1975 by local aficionados of Venetian rowing.[2] Conceived at a time when many feared that this traditional skill was dying out, the Vogalonga was also presented as an ecological manifesto, championing the rower's art as it protested the ever more numerous power boats and their damaging *moto ondoso*. It was set out along a course that would demand endurance as well as adroitness: from the Bacino, participants headed out across the northern Lagoon as far as Burano and then came back into the city past Murano and down the Cannaregio and then the Grand Canal. Venetians who had been involved with the event at its outset told us that they had envisioned the Vogalonga as a friendly match among locals, in which winning was less important than simply managing to finish the grueling thirty-kilometer course. The pace could be competitive or leisurely, as one chose. The point was to showcase and share an experience they considered uniquely Venetian: the art of rowing while standing and the many sorts of boats to which it could be applied. While single individuals rowed many of the boats, a great number of those who enrolled did so as clubs, many associated (as is often the case in Italy) with their place of work; many, though not all of these, rowed in multioared craft.[3]

The Vogalonga enjoyed immediate success among Venetians, and over a thousand of them participated in the first year, cheered on by relatives and friends lining the quays of Cannaregio and Burano or packing boats moored along the route. Before too many years had passed, however, the event began to shift in nature, in a classic example of foreign expropriation of a Venetian cultural product. Such was the appeal of the Vogalonga among visitors—the event was held in mid-May, on the Day of the Ascension, during the peak tourist season—that foreigners clamored for a chance to learn the art of Venetian rowing and to take part. The organizers were happy to let these outsiders participate, and before long "Venetian rowing clubs" had sprung up in a number of European cities. These established links with analogous clubs in Venice and either brought their own gondolas and *sandoli* to the Vogalonga or borrowed extra ones from the Venetian clubs themselves.

Unfortunately for the Venetianness of the event, foreign participants also meant foreign spectators, who before long were filling the *fondamente* along the Cannaregio Canal to cheer on their favorites, dampening somewhat the enthusiasm of the Venetian onlookers.[4] Other outsiders won permission to take part, in other kinds of boats: we assume that the organizers encouraged this because they wished to maintain a big turnout despite having only a relatively limited number of gondolas and *sandoli* available for the many who wished to row. The result has been a Vogalonga afflicted with gigantism: some years there have been over five thousand participants in a chaotic bacchanal of a thousand or more gondolas and *gondoloni; sandoli, puparini,* and *caorline;* kayaks, canoes, dragon boats, racing sculls, and rowboats; in 2003 a "Polynesian princess" showed up in a long outrigger canoe.[5] At the same time, the involvement of Venetians has steadily declined, in both relative and absolute terms. In part, this has only reflected the continued aging, loss of population, and the decline of a traditional rowing culture in Venice and the Lagoon islands: as recently as the 1960s even most large transport boats—the *topi, caorli,* and *peote*—were still rowed; by last count, in 1998, barely 3 percent of the city's aquatic traffic was powered by oars. Yet what the Vogalonga has become also testifies to the overwhelming, distorting impact that the outside world can have on those Venetian cultural productions it embraces.

The Vogalonga is now not only dominated by foreigners, but also operating under somewhat different premises. The event has been caught up with a sense of competition that it was not originally intended to have: squads and individuals have to "train for months in advance" if they are to have any hopes of victory, and guidebooks make no bones about telling their readers that the Vogalonga is a "race."[6] There has been talk of separating out the "serious" contenders, rather like in major urban marathons; cash prizes and commercial sponsorship cannot be far off. This being Italy, however, the event has also been politicized, to the extent that in the 2000 and 2001 editions the Venetian participants delayed their start for fifteen minutes in protest over the *moto ondoso*. At the same time, the city government, in a rather heavy-handed attempt to right the balance between foreign and local participation, announced at the last minute in 2000 that the boats promised to a number of foreign clubs would be given only to Venetians, something the disgruntled head of the Viennese Erster Wiener Gondelverein called "not in the least sense corresponding to a European spirit."[7] Many Venetians see things differently, however, maintaining that

the Vogalonga's original message has weakened over time, becoming in the perception of the many foreigners who come to row in the Lagoon simply a rowing marathon, if not precisely some sort of long-distance race. . . . The Venetians are bit by bit abandoning the Vogalonga, [which is] always less a demonstration against the *moto ondoso,* always more an international (and touristic) rowing festival.[8]

After the Vogalonga of 2001, many Venetians, both organizers and participants, continued their complaints about the direction things were going. They worried about an event that supposedly celebrated *venezianità,* yet in which "the majority of participants are foreigners, and the boats rowed *alla veneta* are in the minority." They lamented that "this is one of the few moments that remain to us to demonstrate that we are still—despite everything, although always fewer in number—yes, still alive." Yet even as they have increasingly fretted that the Vogalonga "could be hijacked to be handed over to the tourists as a folkloric event in which we Venetians participate only as actors," locals had to admit that "in Venice everything that we invent, as soon as it enjoys a bit of success, ends up by becoming a tourist attraction."[9] Three years after the inception of the

Vogalonga, perhaps inspired by the event's initial success, the city government decided to invest in another festival as a reflection and stimulus
of Venetianness. The Festa del Redentore, or Festival of Christ the Re-
deemer, had its origins as a sacro-civic celebration intended to commemorate the passing of the especially devastating plague of 1575–77.[10]
Held on the third weekend of every July, the Redentore belonged to a
select group of six annual celebrations that, during the Serenissima, had
been known as *feste di ponte,* or "bridge festivals." On such occasions, a
special pontoon bridge was thrown up as a votive highway to connect
the church or shrine on which the festival was focused with the more central parts of the city. Along these bridges would proceed the doge, the patriarch, the procurators, and the rest of Venice's patrician hierarchy, to
signal the opening of the celebration; the bridges also provided an important, if fleeting, psychological alteration of the city's topography, presenting a brief, liminal avenue through which Venetians could enter and
leave the appropriately defined sacral space. One pious observer offered
an exegesis of the events: "The bridge is Christ, connecting God to
mankind and humans among themselves; it is Christ who permits men
to go to God and to their brothers. To undertake the traditional crossing
recalls the reality of mankind as wayfarers who do not have a fixed home
in the terrestrial life."[11]

Of the six bridge festivals celebrated under the Serenissima, the Redentore was probably the most important; it is also one of the two that
has survived. The Redentore traditionally has had not one but two temporary bridges: one that crosses the Grand Canal at Santa Maria Zobenigo, and a second—at 331 meters easily the longest of all the bridges—
that stretches across the mouth of the Giudecca Canal from the Zattere,
in Dorsoduro, to the Palladian church of the Redentore on the Giudecca.[12]
More than just connecting spans, the Redentore bridges also define the
special sacro-civic space in which the festival is meant to develop: not only
on the bridges themselves or on the quays, but also and especially on the
body of water they help enclose. This they do by temporarily cutting off
normal boat traffic along the Grand and Giudecca Canals, creating a protected pond of about forty hectares—essentially the area known as the
Bacino di San Marco, bounded by the bridges, the two quays, the Punta
della Dogana, San Giorgio Maggiore, and San Marco.

Historically, the Festa del Redentore has followed a strongly religious
text, appropriate both as a commemoration of the dead and as a thanks-
giving for relief from this particular disaster. No fewer than eleven memorial
masses for the occasion are held at the church of the Redentore, the first

beginning within a half hour of the opening of the votive bridge at 7 P.M. on the Saturday night and the last, on Sunday evening, led by the patriarch himself; a sacred concert follows to close the festival. But many other secular and civic activities also have attached themselves to the Redentore over the centuries. All around the church, charity stands are set up, offering sweets, games, and the ever-popular *pesca di beneficenza*—a lottery in which every number wins some sort of prize—which Italians seem to find so fascinating. On the Sunday afternoon, a regatta is held, bringing together rowers from all the traditional communities of the Lagoon—Venice, Burano, Lido, Sant'Erasmo, Malamocco—in the customary competition.

At the same time, the Redentore traditionally has been a popular festival with strong do-it-yourself overtones, in which Venetians of all classes, all over the city, hold parties and feasts. Choral groups serenade in various *campi* and *campielli* of the city during the Saturday afternoon, and in the early 1980s there were even ad hoc concerts by singing gondoliers. By sundown the quays of the Zattere and Giudecca are strung with colored lanterns (*baloni,* the Venetians call them) and set up with long banqueting tables near the water for Saturday-night dinners that can last for hours. Traditional Venetian dishes—the sort that tourists seldom request in restaurants—are served: sardines in onion sauce, stuffed roast duck, beans and onions, iced melon, all washed down by the "rivers of wine" that are also traditional for this evening.[13]

Most important in this regard, however, has been the custom among Venetians of taking over the Bacino with their boats, which they prepare and decorate especially for the festival. Many of them erect a trellis amidships, which they string with vines and colorful Chinese lanterns; *baloni* are also mounted at the prow and stern. The efforts of individuals are reflected and focused by one especially grand barge—really a raft—known as the *galleggiante:* "twenty-three meters long, illuminated by two thousand lanterns of multicolored glass that form an authentic pagoda of light." Loaded with Venetian high society and guests, the *galleggiante* is towed around the Bacino over the course of the evening, as its occupants salute their fellow citizens relaxing in boats and on shore. Altogether, the celebration turns the Bacino into a vivid carpet of lights, reflected upon the black of the water, their "craft by the hundreds, big and small, almost all of them adorned," a pageant of color and sociability and a reaffirmation of *venezianità*.[14] Following another ancient tradition, boaters might then head on out to Lido to dance in the cool night air. For those who feel up to it, at dawn it is also customary to row out to the open sea, at the Bocca del Lido, and greet the rising sun with a swim.

This more or less describes the Festa del Redentore as it has been celebrated in Venice since the time of Reunification in 1866. Venetians like to call the event on the water La Notte Famossissima, the "Most Famous Night." Judging by what friends and the local papers have told us, their special affection for the occasion derives from the great chance it has given them—even if only for a night—to be the people of the water that Venetians are supposed to be, socializing, courting, and harmonizing together in all senses of the word. Allied to this was the special pleasure that here, at least, was one Venetian celebration that the tourists could not invade, subvert, and take over. By the very nature of the affair, foreigners were kept at the fringes, both behind the dining tables lining the quays and off the Bacino filled (for a change) with just the boats of Venetians. As the journalist Giorgio Nardo noted with satisfaction in 1983, "The tourists admire the festival, the Venetians live it: as in tradition; above all, in boats."[15]

Unfortunately for this revanchist spirit, and for reasons that have never been made too clear to us, in 1978 the city council decided to improve on the Redentore festival by capping off the evening on the Bacino with a midnight musical fireworks display. This was not exactly a new idea: a tinted etching in the Museo Correr shows just such a spectacle underway sometime in the 1860s. But it also seems fairly clear that the *foghi,* as Venetians have come to call them, were not a regular part of the Redentore before 1978: Baedeker, who termed the festival "pretty," never mentions them, and several fairly detailed foreign and local newspaper accounts of earlier Redentore say nothing about fireworks.[16] What is certain, however, is that right from the beginning the *foghi,* accompanied by the music of Vivaldi or (of course) Handel, were a big success with the Venetians and soon became a key, even defining moment of the Redentore festival. As with the Vogalonga, the *foghi* also rapidly revealed a tendency toward gigantism, with each year's organizers trying to outdo those of the year before. By the end of the 1990s, something like thirty-five hundred kilos of sky rockets and cascades were being shot off, in a forty-five- or fifty-minute show that culminated just after midnight.

Some Venetians were not happy with the impact of such pyrotechnics on their fragile city, complaining that loud explosions could make the mosaic tiles fall off the walls of the Basilica and that the acrid smoke could corrode the ubiquitous Istrian marble.[17] A greater problem, however, has been the ability of the *foghi* to simplify and refocus the Redentore celebration, turning a socially complex, communal, and votive observance into a straightforward spectacle: restructured this way, the Redentore was ready to be launched as a new tourist attraction for the city. It took a few years for the word to get around, but by the early 1980s outsiders (mostly other Italians at first) were beginning to show up in large numbers—not, of course, for the religious observances, dining, boating, or sociability, but for the midnight spectacle. Unable to find a place, either on the Bacino itself or at the adjoining quays, these visitors have had to make do with standing room in the Piazzetta and along the Riva degli Schiavoni, where they have started their own parties and dances. The process typically was repeated (and intensified) after the fireworks, when outsiders joined Venetians flocking to Lido, often crowding into the beach-side discos for a night of revelry that often degenerates into "troubles . . . and some violence." "About the rest," the *Gazzettino* observed sourly, "[there are] the usual footnotes: bivouacs in Piazza, a tide of garbage . . . a great many young people in sleeping bags in the woods at Sant'Elena; festivities a bit intemperate at Lido."[18]

Already by the mid-1980s, there were many such signs that the Redentore was slipping out of Venetian control. Continual attempts by the city council, the hoteliers' association, and the tourist board to pile ever more activities into the afternoon and evening leading up to the *foghi* only seemed to increase the crowds and the mess while diminishing the Venetianness of the event. Unable to see where all this was headed, the local government went fatally over the line in 1989, when it voted to allow a full-blown rock concert by the British band Pink Floyd, which was set to play from a floating platform in the Bacino, in a performance timed to climax with the beginning of the fireworks. This was a long way from the music of Handel and Vivaldi, and it should have surprised no one (but apparently did) that the spectacle attracted something like two hundred thousand youthful fans, whose unexpected presence carried the city up to the point of collapse and, many said, beyond it. Once the enormous crowd had filled the Piazzetta and Riva to overflow, many onlookers began scrambling onto the roofs of adjoining buildings, which at the time (and seemingly always) were covered with easy-to-climb scaffolding for restoration. These fans alone caused an estimated thirty thousand dollars in damage to the roof tiles. About as much had to be spent to clean up the city and haul off over a thousand tons of garbage that all the visitors left behind.[19]

The Pink Floyd debacle at least had the positive result of bringing an abrupt, if temporary, halt to rampant boosterism in Venice: most of the city council promptly resigned in disgrace and a more conservationist element took over for a while. It was not so easy to squelch the increasing enthusiasm worldwide for the Redentore festival, however. Promoters at the beach resorts of Jesolo and Cavallino found that the *foghi* represented an attractive and lucrative evening diversion for their tourists and soon began organizing excursions from Punta Sabbione to Venice in the big *lancioni di granturismo.* In these long, glassed-in craft, tourists no longer had to stay ashore for the spectacle and indeed could now get right in among the decorated Venetian boats, as close as possible to the fireworks barges. The pilots of these big boats seem to have no compunction whatsoever about placing their craft between the Venetians and the fireworks, and before long Venetians were complaining that their own colored Chinese lanterns were being overwhelmed by the constant sparkle of tourist flashbulbs.[20]

For the Venetian boat owners, who had felt like an essential element of the Redentore, it has been "a discordant note, these great motor boats, crammed full of tourists," nudging up to the fore to consume the festival rather than make it. This is not always obvious to the casual observer who is standing on shore, since the Bacino still seems "as full as an egg . . . a carpet of boats." Yet as *La Nuova* pointed out in 2001, "observing more closely, one notices that there are always fewer [than in past years] little boats, with families and friends (those [that are] rowed have completely disappeared) and always more of those that host a hundred, hundred and fifty people, often having come from outside [Venice: that is, from Caorle or Jesolo]."[21] As a result, and in a manner not unlike the Vogalonga, locals have begun withdrawing from La Notte Famossissima. The papers have increasingly complained about how "there are fewer [decorated] boats in the Bacino than in past years"; instances are reported of Venetians throwing fruit at *lancioni* that get in their way. But it is not just the Venetian boaters who are unhappy with the Redentore: those on shore, who have been accustomed to banqueting on the quays adjoining the Bacino, also have found their festival expropriated by foreigners, and sometimes in very graphic ways.

Some of these Venetians have complained to us that simply dining at the quayside has begun to make them nervous, sandwiched as they are between the water and the jostling masses of tourists behind them. More vexing, however, has been the experience of those who have set up their decorated tables in the afternoon only to find that some rich yacht owner—one of the *nababbi,* or "nabobs," the Italians like to call them—has managed to moor his boat right in front of them, blocking their view of the Bacino and the fireworks completely. They lament to the press: "If things continue like this, and ever-fewer people have the desire to prepare banquets, then the Redentore will turn into a bogus spectacle, [put on] only to the advantage of tourists, the big hotels, and the VIPs."[22] Such was very much the case in 1998, when the port authorities allowed the *Grand Princess,* then the largest cruise ship in the world, to tie up alongside the Riva dei Sette Martiri for the evening of the Redentore.[23] This meant that the residents of eastern Castello—one of the few remaining "popular" districts in the city—found that their community's traditional spot for dining and viewing the *foghi* had been completely blocked by this "'naval wall'" of a party crasher, 289 meters long and 60 meters high. They also realized that, high above them, the paying passengers of the *Princess* were enjoying the fireworks from a perfect vantage point, "without ever having realized that with their floating city they had blocked out part of the houses of Castello and set off not a few polemics with their cumbersome presence."[24] Not surprisingly, the locals were furious, announcing:

We will not stand here with our hands in our pockets. We have started collecting signatures against this decision by the Capitaneria di Porto. The city is ripping us off . . . they are kicking us off the quay, privileging luxury tourism. . . . [T]he Redentore is the festival of the Venetians, the only one where they [should be] the protagonists and not the spectators, and instead we only serve for paying taxes.[25]

Complaints and petitions may prevent such truly grotesque foreign intrusions into festivals like the Redentore, but they do little to diminish the distorting effects that the weight and focus of tourist enthusiasm can have on the overall structure of such an event. The elements of spectacle—in particular the *foghi,* but also the all-night disco parties at Lido—continue to grow in centrality for the Redentore, at the expense of its participatory and communitarian aspects. Moreover, from the point of view of many outsiders, the Redentore's fireworks on the Bacino are fungible and available for private use, as an American billionaire staying at the Cipriani demonstrated in June 2000, when he staged his own version of the *foghi* to honor himself on his birthday.[26] In response to such affronts, Venetians have begun treat the Redentore, their oldest continuous surviving festival, more and more warily. They try to stress its religious and communal aspects, even as they also remember the important role they once played in its festivities. When, in 2000, it looked for a while like the Redentore would be called off for lack of a bridge, many seemed willing to shift their affections to the smaller, late-November Festa della Salute, hailing it as "the last festival that is completely Venetian" and pointedly noting that, though dreadfully off-season, at least the Salute features "no fireworks and few foreigners."[27]

The culturally rich and inventive 1970s concluded in Venice with the reintroduction of the Carnival, in 1979. This ancient celebration, with its traditional potpourri of mass fantasy, social inversion, personal liberation, and life affirmation, was once observed throughout much of Catholic Europe but by the twentieth century had been driven out of the Continent's cities to survive only in what had once been Europe's creole fringes: most famously, New Orleans, Trinidad, and the megacities of Brazil. Despite Venice's historic association with Carnival, it was far from foregone that the festival should be reborn in that city as opposed to, say, Rome. Indeed, there already had been attempts to bring the Venetian Carnival back in the decades just after Italian Reunification, and the results had been none too encouraging. Despite—or, more likely, because of—the best efforts of the city's cultural elite to finance and stage a celebration that drew heavily on the fabled festivities of the Serenissima, Venetians in general were apparently not too impressed: for the most part they turned out as spectators rather than participants, and the first big production, in 1886, was judged "cold and lifeless, even if, as a show, it was a good success."[28] A decade later, Baedeker was dismissing the whole endeavor as having "of late entirely lost its significance." For most of the next century, Carnival in Venice, as in most of Italy, was reduced to some private masquerades among the wealthy and "children in fancy dress throwing confetti at one another."[29]

In the context of its timing, Venice's decision to bring back its Carnival might at first sight seem strictly commercial and touristic: throughout the 1970s it had become fashionable for Italian resort towns—most famously Viareggio and Ivrea on the Tyrrhenian coast, but also Venice's own island neighbors of Burano and Malamocco—to revive their own moribund carnivals in attempts to fill out a segment of the tourist year that was otherwise shuttered and dead.[30] In Venice, however, the inspiration seems to have been more genuinely communitarian, with those who planned the festival's rebirth claiming aspirations that ranged from stimulating the city's arts and traditional winter theater season to providing some culturally worthwhile diversions for local teenagers. These latter, apparently, had of late been getting a bit too rowdy in the days before Lent, and it was hoped the spirit of Carnival might channel their energies away from assaulting passersby with bags of flour or chalk and toward more constructive expressions and fantasies.[31]

The first reborn Carnival of Venice was programmed for just four days in February 1979. It offered only a modest schedule of events and amusements but was deemed by both the organizers and most Venetians as a great success—especially after the city's teenagers overcame their initial shyness (or fear of the law) and turned out en masse and with considerable enthusiasm for the final, open-air costume ball of Fat Tuesday in Piazza San Marco.[32] Buoyed by a sense of triumph, the organizers—a somewhat informal union of delegates from the city council, the local tourist board (the Azienda di Promozione Turistica, or APT), and a specially formed group called the Scuola Grande di San Marco—immediately began planning for 1980.[33] Everyone agreed there should be plenty of theater. This had been the centerpiece of the Carnival during the Serenissima, and to fill Venice with comedies, opera, and music, all six of the city's playhouses were cleaned up—some had been closed for decades—and productions were readied. Local cinemas also joined in, running continuous and thematic cycles of seventy old films; the Museo Correr likewise put on a special show, and a number of private costume balls were planned. The scheduled activities were expanded to cover seven days—having as peak moments the opening Sunday ceremonies, Fat Thursday, the following Saturday and Sunday, and the concluding Fat Tuesday—and spread out to more parts of town. The organizers were aware, however, that there was always the danger of turning the Carnival from a festival of expression and community into a staged spectacle before a passive audience. They took as their watchword "Ma varda che poco che basta" ("But look at how little suffices") and invited all Venetians to join the celebration in whatever guise of personal and group expression caught their individual and collective fancies. Beyond this, planners aimed to schedule only a few formal events, for the most part drawn from or inspired by old Carnival traditions. These included a "children's carnival" in San Marco on Fat Thursday; a costumed, sixteenth-century-style soccer game in Campo Santo Stefano on Friday; the *vogada,* a procession of Venetian boats rowed by maskers the length of the Grand Canal on Fat Saturday (also known as *sabo grasso* in Venetian); the *volo della colombina,* a papier-mâché dove that spewed confetti as it "flew" on a wire between the Campanile of San Marco and the Ducal Palace at noon on Sunday; and two general bacchanalian balls in Piazza San Marco on Sunday and Fat Tuesday evenings.[34] As to the rest, it was expected that the comic genius of the Venetians would fill in the cracks, causing "a thousand spectacles to pop off here and there, numerous and unpredictable like mushrooms."[35]

The result—perhaps just because organization was kept to a minimum—was a Carnival that many Venetians still think of as the best ever held (indeed, some told us they remembered it as the first, rather than the second).[36] Good weather guaranteed a massive turnout for all the scheduled events, and thousands of northern Italians, speaking "every dialect of Padania," poured into town. On the Sunday evening, Piazza San Marco fairly boiled over with "crowds, with vivacity, with serenity and with rediscovered human ties"; on Fat Tuesday an estimated fifty thousand or more showed up, producing what the *Gazzettino* called "the biggest ruckus ever seen in Piazza."[37] Amazed journalists wrote about the "frenetic dances amid deafening music and lights as bright as day" and "the chaotic and joyous ferment of improvisations," calling this "monster Carnival *[Carnevalone]* . . . chaotic and joyous," "a phantasmagoric spectacle," "the biggest *happening* ever seen in Venice," which was "incredibly superior to every sane prediction."[38]

When it was over, many Venetians believed their city could never be the same again. "We have come out of [our] lethargy," exclaimed Fabio Marangoni in the *Gazzettino,* proclaiming that the city had rediscovered "its winter vivacity." It was reported that upward of twenty-five thousand Venetians took part in the festivities, turning Piazza San Marco back into "the meeting point of a city, a testimony of its desire for vitality." Some claimed that "the Carnival had come back in a moment [that was] biologically necessary": it provided the "great shake, the whiplash" needed to jolt Venetians out of their "winter sluggishness": "in the streets one feels a new flow of people; faces are circulating that are not our usual gray winter faces." "Is this the festival of a city that has rediscovered its way of being, its reference point?" Leopardo Pietragnoli asked. Many Venetians evidently thought so, adding that their city had something new to offer the world: "Venice," Paolo Rizzi announced, "has confirmed itself as a city of play—a *città ludica*—by vocation . . . a free city, a place of permissiveness," where all the world could come and rediscover itself.[39]

As it happened, the rest of the world was more than ready to take up the offer. With breathtaking speed foreigners of every sort and from every direction rushed to take part. Any Venetians who had been concerned in 1980 that too many of the city's restaurants and hotels were closed down during the festival or that word about the celebration was not getting to the outside world need not have worried.[40] By 1982, there were already over a hundred restaurants in town that not only were staying open, but also were adding special fixed-price menus for visitors—soon they were doing more business than during the height of the summer season. Hotels too were booked solid, and their operators suddenly realized that there had "opened a new 'winter front'" in the city.[41] For many, tourists and Venetians alike, Venice and the Carnival appeared to be a perfect fit: here, in a pedestrian city with low crime and high romance, was the ideal stage for the buffoonery, spontaneity, and self-display that seemed the essence of the Carnival spirit. Special trains and organized tours were soon bringing maskers from all over Italy, Europe, the Americas, and eventually the Far East; the crowds ballooned accordingly—from 60,000 to 70,000 visitors on peak days (primarily the weekend before Fat Tuesday) in 1982 and 1983 to well over 120,000 at such times in 1984 and 1985.[42]

The Venetians, in these early years, energetically dreamed up new activities, both to amuse themselves and the foreigners, and (as much as possible) to lure everyone out of the Piazza. As a take-off on the Vogalonga, for example, the Ombralonga, or "Long Drink," was introduced in 1984 and enjoyed an immediate success. Participants in this loosely organized pub crawl followed a map of participating Venetian wine bars, known as *osterie* or *bacari,* where they would find endless *ombre* (that is, "shadows," as the Venetians refer to glasses of wine) awaiting them. In 1985 no fewer than five thousand of these "racing drunks" *(maricatore-cioncatori)* signed up, received their chest numbers, and set out to cover the city's dozens of traditional pubs, while also sampling various *cicchetti,* the snacks that, like Spanish *tapas,* are customarily consumed with wine in Venice. That year, only two thousand managed to finish the event eight hours later, the rest having evidently "gotten lost along the way, stuck to the spigot of some demijohn."[43]

Such agreeable diversions have helped spread awareness of Venice and its traditional culture—not only to foreigners, but also to some of those locals who had long since left the historic center for Mestre or the Terraferma. Planners have availed themselves of other traditions of the Serenissima for this purpose, embellishing and adjusting them as necessary to fit modern appetites. Beginning in 1986, each edition of the Carnival has been dedicated to an overriding theme that was intended to pay homage to some aspect of the city's history: "Venice and the Orient" was chosen for the first year; others have saluted Venice's multiethnic nature, its *commedia dell'arte* traditions, and (of course) Casanova.[44] Variations on the traditional Venetian bull fight (the *caccia di toro*) have also been staged over the years, while most recently the Festival of the Marys has been brought back. One of medieval Venice's favorite religious celebrations, the Marys Festival had originally been held for the Purification of the Virgin (February 2). Brought back to life in 1999 and adroitly converted into a combined fancy-dress beauty contest and musical procession, the Marys Festival has now been taken up as a new way to open the Carnival, two days before the Flight of the Colombina. In just a few years, the festival has begun to show the usual symptoms of gigantism: initially, the Marys featured a queen and her court of six, chosen from applicants from each district of the city plus Mestre; by the 2001 edition, there had been added two "maids of honor" and an additional twenty "marriagable virgins," to produce a contingent of twenty-nine, all carried in sedan chairs by young men and accompanied by musicians and at least five thousand spectators.[45]

Other programming elements have had less to do with Venetian culture than with the imperatives of spectacle in the electronic age. Eager to spread the word about their reborn festival, Carnival organizers assiduously courted the media, only in part (and at the beginning) in an effort to attract more visitors: more central was a desire to vindicate Venice before the world, to burnish its tarnished image as "a rip-off city, accustomed to squeeze the tourist for some months and then stretch out with a full belly in a comfortable sleep [for the winter]."[46] By the mid-1980s, three or four hundred accredited journalists were showing up to report the scene, providing direct and packaged television coverage worldwide.[47] Inevitably a media presence so powerful, representing and catering to so vast an audience, has done much to reshape the Carnival itself. In 1994, for example, the organizers began naming a "godmother" *(madrina)* for each edition of the Carnival, choosing some Italian diva or movie or television star who would be willing to grace the climatic last few days of the festivities with her presence and, not coincidentally, act as a willing focus for the television cameras.[48] Likewise, after more than twenty years as the symbol both of every Carnival's opening and of the event's peaceful intent, the *colombina* was "put in the attic" in 2001, replaced by a high-wire artist intended to be both more of a crowd-pleaser and more telegenic than a papier-mâché mechanical dove.[49]

The world media have preferred and promoted a Carnival that plays to television's strengths: more spectacle than happening, more focused than city-wide—visual and emotional rather than verbal and intellectual. In 2001, the growing dominance of all these elements was incarnated by the planting of a giant TV screen on the Campanile of San Marco, further reinforcing the process already underway whereby participants become spectators of themselves.[50] Media needs can run counter to the intentions of the Carnival organizers, however. In 1988, for example, plans to decentralize the festivities to ease some of the human congestion in the San Marco area had to be abandoned to satisfy ABC. The American network had come to Venice in its full tele-invasion mode, expecting to offer its viewers a Carnival of Venice conforming to popular expectations, which is to say, one set in Piazza San Marco. To satisfy ABC, the Venetians hastily threw up a stage at one end of the Piazza and programmed some appropriately telegenic spectacles.[51]

In line with their postmodern role of making the news they cover, media interests have been able to turn the Venetian Carnival into a fairly shameless vehicle for their own promotion. Thus, in 1995, Eutestat ("the most important European satellite operator") ran a competition called Statmania for maskers whose costumes had some tie-in to television and telecommunications. For prizes, the ten winners were given both Eutestat broadcast receivers and a chance to mug at some length on Italian state television's (RAI) fifteen-minute special on the Carnival that evening.[52] On a more modest scale, the *Gazzettino,* which cosponsors the Festival of the Marys, also never fails to note its own role in the event, even as it reports it.[53]

Through their Carnival, Venetians have had ambiguous relations not only with the world media but with all business interests. The city can barely meet even the costs of maintaining order and the infrastructure during Carnival (just the overtime for police and garbage collectors costs an extra hundred thousand dollars) and so has from the early 1980s looked for outside financial help. This was never easy, however, as long as some Venetians insisted that donating businesses would promise to keep an advertising profile so low as to be almost invisible.[54] As costs for the festival ballooned from around half a million dollars in the mid-1980s to over a million ten years later, the search for such amenable sponsors grew ever more desperate, until finally, in 1995, the city threw in the towel and agreed to allow corporate sponsors to advertise openly.[55]

Ever since, the festival has become significantly different, and Venetians now call it rather disparagingly the "Privatized Carnival."[56] The whole festival has taken on the air of a giant trade fair, with corporate advertising banners strung up along the Grand Canal and anywhere else that might bring names and logos before the eyes of tourists and (above all) the cameras of the world media. The costumed *vogada,* for example, now serves as an advertising vehicle for the Italian sparkling wine producer Mionetto, which has managed to "project its logo with a laser beam in various points in the city and the Grand Canal and . . . send a[n advertising] boat around, defined by the organizers as 'Byzantine' in style." In just six years, Coca-Cola has also gained an "[advertising] presence that is by now traditional," while Volkswagen, which likewise jumped in at the beginning, has virtually made the Venetian Carnival an advertising subsidiary. With the donation of only a relatively modest $150,000 in 1996, the German car maker was able "to practically carpet-bomb the city with

SHIPS AND FOOLS   253

its Polo Harlequin: helium balloons in the form of cars; real cars and full-sized reproductions [placed] in the most picturesque and touristic spots in the city, even on barges on the canals. Not to mention the brochures of the Carnival Consortium, which feature the 'VW' along with the other sponsors."[57]

As the "Privatized Carnival" became the norm in Venice, there were complaints among locals that the city government had "sold out" the festival—some said "at a discount" as well, considering the fairly small sums that sponsors had to come up with. This thought was apparently not as disconcerting to many Venetians as was the realization that they themselves were becoming increasingly irrelevant to their own Carnival. As with the Vogalonga and the Redentore, the sheer numbers of outsiders showing up had already made it obvious by 1982 or 1983 that this event was no longer really about Venetians—not in the way that the Carnival of Rio is about the residents and culture of Rio de Janeiro, or that Mardi Gras, despite the masses of tourists that go there, is still about New Orleans. This disjuncture between what had originally been the Carnival *of* Venice and what became the Carnival *in* Venice was made still more clear in 1988, the first of several years in which the city council, unwilling to keep on taxing citizens to pay for the pleasures of foreign visitors, failed to either come up with a Carnival program on its own or allocate funds for an organizing committee.[58] Although the influx of outsiders was low during the first days of that year—apparently reflecting an initial uncertainty on the part of potential tourists and travel agencies as to what was going on—by the final weekend the city was as mobbed as usual, giving the *Gazzettino* the chance to run the headline: "Great Crowd around Nothing."[59] In fact, it had already been obvious for some time that foreigners were quite happy to mob Venice during Carnival whether the Venetians wanted to take part or not. World tourism had so taken over the festival by the late 1980s that many Venetians were not even sure that either they or their government still had the power to stop the celebration should they have wanted to. And, indeed, attempts by the city council to shut down the festivities altogether during the Gulf War in 1991 (there was talk of fining maskers) failed to stop some thousands of costumed foreigners from showing up anyway.[60]

Since 1994, the Carnival of Venice has come roaring back bigger than ever, helped along by eager outside sponsors. Somewhere around 450,000 visitors showed up for the festivities that year, over 600,000 in 1995, around 650,000 annually between 1996 and 1999, over 700,000 in 2000, and nearly 800,000 in 2002.[61] All these visitors normally leave ample signs of their stay in the city. The garbage-collecting division of VESTA, the city's maintenance agency, reported having to haul off an average of several hundred extra cubic meters of trash per day during the final ten days of the Carnival of 2002, with a peak of five hundred cubic meters on the final Sunday. A quick pocket-calculator estimate indicates that, on that single day, Carnival tourists thus left behind enough rubbish in the city to fill Piazza San Marco and the Piazzetta—and the Molo, for good measure—to a depth of around fifteen meters: enough to swamp the bronze horses that adorn the facade of the Basilica and nearly cover the Torre dell'Orologio as well.[62] The celebration has been spread out to twelve days (starting two Fridays before Fat Tuesday), though peak moments remain the same as always: on each of the final weekend days the number of visitors now regularly surpasses 100,000—in 2000, over 270,000 came in these forty-eight hours alone—while Fat Thursday and Fat Tuesday each typically see about 50,000 arrivals. Likewise, Piazza San Marco remains, as it has from the beginning, the place to go, despite repeated attempts to lure visitors elsewhere with other attractions. Although the police do their best to direct the pedestrian flow, there arise endless and exhausting human traffic jams, especially on the canal bridges near San Marco and the narrow alleys of the Mercerie, where there regularly break out "dozens of arguments, altercations, and shoving matches." Such tie-ups are also common around the train station, which really only has one exit leading into the city; when a host of special trains are dumping ("vomiting," say the papers) upward of ten thousand people an hour into this limited space, enormous human snarls can result.[63] Sometimes a "stunning jam-up" of new arrivals can also block the nearby Ponte degli Scalzi over the Grand Canal, leaving the more adventurous maskers to scamper across on the parapets.[64]

Unquestionably such mobs take their toll on the fragile Venetian infrastructure. Though the ACTV routinely "puts out every unit of its fleet that can float," there are never enough *vaporetti* to handle the demand; anyway, it is physically impossible for the vessels to come and go fast enough.[65] The crowds trying to push their way onto the already jam-packed boats provoke lengthy delays in the delicately balanced schedule: as successive *vaporetti* fall still further behind, the increasing and impatient throngs often get unruly; fights regularly break out.[66] Meanwhile, the thousands of (mostly younger) visitors who have come to Venice with no place to stay typically end up dossing down in corners around the Piazza, under the arcades of the Procuratie or around the Basilica. If the police are not constantly on patrol around San Marco, someone late at night inevitably starts a bonfire, using trash or breaking up the wooden crowd-control barriers protecting the Basilica. These fires have sometimes badly blackened the Piazza's Istrian-marble steps and the adjacent walls.[67] After midnight, when the loudspeakers are turned off at San Marco, maskers improvise concerts at such secondary *campi* as Santo Stefano or Santa Margherita, holding parties that go on until dawn, to the annoyance of nearby residents. On top of all this are the acts of vandalism and souvenir hunting: doorbells and door knockers are stolen, doors, walls, and windows defaced, pools of vomit and urine splashed on the *calli* and *campielli* all over the city.[68]

Before too long many Venetians started changing their minds about what the Carnival could do for them and their city. A *Gazzettino* readership poll of 1987 revealed that nearly seven-eighths of the respondents were unhappy with how the festival had developed. Some complained that it was too much "to be obliged to amuse oneself on command," even if it were for only ten or eleven days a year. The great majority, however, blamed their discontent on the highly inconvenient presence of so many foreigners in their city—"[Carnival] for the Venetians!" as one of them wrote. "Out with the Barbarians!"[69] Older residents, who make up such a large proportion of the population and who may have gone to the Piazza with their grandchildren during the first few editions of the Carnival, were soon marginalized by the tens of thousands of foreigners ("Don't these people have jobs?") and the roaring disco music that filled San Marco to the bursting point. Many now treat the peak weekend before Fat Tuesday as something to be avoided at all costs, and those who can afford to will leave the city for Carnival week—skiing in the Dolomites is a popular alternative; as for those who have to stay behind: "These have been two truly exhausting days: many have decided to not go out, choosing to imprison themselves in their own houses. Those who went out had to get in line just to reach the store next door and to arm themselves with patience to wait for a *vaporetto* that they could board."[70]

For those Venetians who have persisted in continuing Carnival even in the face of this foreign invasion, the response mostly has been to put energy into much smaller, neighborhood festivals. As early as 1983, having already seen which way the tides of costumed tourists were flowing, Leopoldo Pietragnoli advised *Gazzettino* readers that "perhaps for the next few days it would be well to leave the Piazza to the foreigners and spread ourselves around the *calli* and *campielli,* whether those right at home, with friends from forever, or [those] on the other side of the city, discovering new friends and new tastes."[71]

For most of those that have followed this route, this has meant putting on little carnivals that draw on the participation and enthusiasm of family and immediate neighbors, not unlike those that are staged every year in Burano or Malamocco.[72] The communal government has made a point of helping such efforts financially, in recent years handing out about a third of its available funding to neighborhood-level festivities.[73] The results have been positive: Castello and Cannaregio, the two most populous districts remaining in Venice, each have by now something of a tradition of staging their own "people's Carnival." Castello's—which calls itself *el vecio Carneval,* the old Carnival—is funded by the shopkeepers along the via Garibaldi, where most of the events are staged; that of Cannaregio often takes advantage of the ideal and evocative space available in the *campo* of the Ghetto Nuovo. Still smaller neighborhood affairs are staged in places like San Francesco della Vigna (by the Arsenal), San Giacomo dall'Orio, and the Giudecca.[74]

Like most Venetians, many who put on these miniature celebrations seem to have an ambiguous or contradictory attitude toward foreigners and the risk that even their modest efforts might end up attracting hordes of outsiders. Everyone seems to willing to hire mimes, jugglers, and foreign dance troupes to perform in their parish, but many locals are only irritated when such shows have then escalated into, for example, the full-scale rock concerts that are sometimes piggy-backed onto their festivities.[75] Some organizers, like those in San Francesco della Vigna, have pointedly asked outsiders to stay away; many other groups program overwhelmingly for local children, as if hoping to turn the clock back to the days when the whole affair was much calmer and aimed only at the very young.[76] On the other hand, funded as it is by local merchants, *el vecio Carneval* has always pitched itself and Castello at least partly to the tourists, perfectly willing to attract those who would otherwise crowd Piazza San Marco with such eye-catching stunts as hot-air balloon rides and parade routes along the main street, via Garibaldi.[77]

Such efforts, though they may bring comfort to many Venetians, have not slowed the grand commercial juggernaut that the Carnival of Venice has become. Regardless of what ordinary Venetians may wish to do about it—in any case their numbers have dropped by a third since Carnival was reborn—their city and its commercial sector have simply become too attached to all the cash the festival generates to ever give it up. By the end of the millennium, the profits had become staggering: with announced expenses of around $650,000 for the various productions it had staged, the city estimated that "just the historic center alone" had seen a (declared) cash inflow of something over $80 million, to enrich all those who live off the tourist economy—the hoteliers, restauranteurs, taxi-pilots, gondoliers, mimes, and pickpockets involved have all have found Carnival to be what the Italians call *affari d'oro:* "golden business."[78]

Obviously the Carnival is profitable to Venice and some Venetians, but what has made it so persistently popular with tourists worldwide? The city in February is, after all, often far from agreeable. When they still took part in Carnival, Venetians spoke with chagrin about its seemingly malignant god, "Giove pluvio" or "Giove guastafesta"—Rainy Jupiter or Jupiter the Party Pooper—who was constantly dampening events. More than a few public balls in San Marco have turned into sodden affairs of bedraggled clowns and maskers grimly trying to dance under umbrellas. Often, the rain, snow, or sleet are further exacerbated by the *acque alte,* whose effects are especially felt on the Carnival's focal point at San Marco, and by the strikes that police, *vaporetto* workers, and garbage collectors regularly stage at this time of year.[79] On such occasions, the tourist turnout—especially from nearby Italian cities—can drop considerably. Yet, even in the teeth of so many *guastafeste,* tens of thousands of visitors persist in coming to the gloomy and uncomfortable city, obviously preferring the winter sun, but quite as ready to take whatever they get.

Nor is Venice really the best of places for staging a "real" Carnival: that is, for putting on the type of pre-Lenten festivities that so typify New Orleans or (in particular) such Brazilian cities as Rio, São Paolo, or Salvador di Bahia. In those places, Carnival continues to center on inversion, transgression, sensual overindulgence, and sheer fantasy, much as it did in premodern Europe, and in particular in the Venice of the Serenissima. Equally important, however, are the elements of competition and territoriality through which both the carnivals of the past and the Latin/ Creole ones that flourish today have found their transgressive energies. Behind all the display and mummery that delight the tourist crowds in New Orleans or Rio there lies the often ferocious rivalry between the competing wards or *favelas* that put on their public shows at least as much to shame their neighbors as to win the applause of wealthy foreigners.[80]

No such thing as rivalries between neighborhoods (much less *favelas*) are to be found in present-day Venice, although they clearly had an important role to play in the Carnivals of the Serenissima, when youths from different quarters competed in such semiritualized contests as bull-baiting and the forming of human pyramids, the so-called *forze d'Ercole.*[81] Nor can Venice, by the very nature of its topography, really play host to grand Carnival processions, though these have been the primary and customary avenue of such competition in the Latin/Creole Carnival. In Brazil, New Orleans, Trinidad, and even London's own Notting Hill Carnival, jazz bands, drummers, mummers, samba dancers, floats, or some elements of all of these provide the competitive framework and often the climax of the entire festival. Participants who have trained or decorated for months—often since the end of the previous Carnival—have a chance to parade their creative energies before admiring, if uncomprehending, tourists, their knowledgeable fellow citizens, and a review panel of local notables; through the nature of that year's skit, the music they play, or the costumes they wear, they also aim to put a finger in the eye of rival groups. In Venice, the topography simply does not allow for the grand parades of the sort that march down Bourbon Street or in Rio's Sambó-dromo. The city's one ceremonial route is the Grand Canal, and it has been years since there were enough boats in Venice to make a worthy processional show on that avenue: the sixty-five decorated craft that were mustered for the first Carnival *vogada*, in 1979, left even the Venetian spectators asking, "Cussì pochi?" ("So few?").[82]

Not that most Venetians seem to have ever really wanted to involve themselves in the explosive excesses and grand processionals that have made New Orleans and Rio so notorious and such perennial tourist draws. Having come up with the idea in the first place as a means of channeling youthful impulses that they saw as antisocial into more wholesome and communal forms, many Venetians in the early 1980s continued to cling to the patently paradoxical fantasy that they could produce a Carnival "truly gone wild, but in a form (almost always) well-behaved and civil."[83] What they wanted were not so much carnivalesque transgressions—these they have tended to associate with the other excesses of the Latin/Creole Carnival: the "knifings, muggings, thefts, drunkenness, annoyances"—but rather *happenings* (the Italians have adopted the English word), in a '60s, Woodstock fashion that might promise peaceful and communal rebirth and rediscovery.[84]

Few of them have been willing to admit it, but it soon became obvious that the Venetians are no longer the carnivalesque people of the legendary Serenissima, if indeed they ever really were. These days, they are as bourgeois as Europeans anywhere—probably more so, if one takes into account their greater average age. They did not generate the new Carnival out of local rivalries, class conflict, or racial antagonism: what they staged they did to celebrate their own community and their *venezianità*. For many of them, in the beginning, the outside world was almost an afterthought. They succeeded, briefly, at what they had set out to do, but in the process they opened their city up to another, unexpected kind of carnival, of a sort not previously seen anywhere else in the world.

Some inklings of this genuinely new Carnival could be detected by the mid-1980s. Venetians began to notice that not only were ever more foreigners coming to their city, but also that many of these were taking unexpected advantage of the enclosed pedestrian venue of Piazza San Marco to dress themselves in increasingly complex and expensive costumes. This was not traditional carnivalesque masking, in which one disguised his or her identity the better to transgress or act the buffoon. Rather, identity and recognition, whether the face was exposed or covered, had become everything, and woe to any passing celebrant who dared to dull the luster of a costume that cost "an eye out of the head" with a puckish handful of confetti.[85] As the Carnival of Venice evolved over the '80s, such maskers increasingly flocked to San Marco solely to show themselves off, "with their elegant, stately walk and their languid poses," before one another and (above all) before a steadily growing mob of amateur and professional photographers. By the '90s, this tight, symbiotic relationship between maskers and photographers had developed into one of the defining dynamics of Carnival, as it was largely expropriated and reworked by foreigners.[86]

This annual consummation between poseurs and picture takers in San Marco also helps explain why Carnival organizers have found it so difficult to spread the festivities around the city. For tens of thousands of visitors every year, the Carnival of Venice is fundamentally this exercise in narcissism that they themselves stage in the Piazza, many with the special hope that their picture might show up on world television or in some glossy publication. Venetians tend to sneer at these outsiders who come to their city only to *fare comparsate* and *pavoneggiare,* or strut like a peacock, as they call it: this was not the Carnival they had imagined back in 1980. Many of these foreign maskers come as individuals or couples, but they also flood into the city in much larger groups, often all in matching costumes, from communities elsewhere. These companies can be quite substantial: even if the 1983 plan of some gay Americans to come twenty thousand strong and in drag never panned out, other thematic groups have come, often as musicians or street actors, and made quite a splash. For example, 130 West Germans came in 1983, saying they wanted "to see and be seen"; two years later a contingent showed up from Taragona, in Catalonia, boasting an expertise in making human pyramids, like the old Venetian *forze d'Ercole.* Indeed, every year between forty and sixty bands and mummer groups from outside Venice register as Carnival entertainers; many receive some financial support, either from the Venetian organizers or from their hometowns. A number of (mostly) northern Italian towns have their own "Friends of the Carnival" societies that practice and work throughout the year in anticipation of their eventual appearance in Venice.[87]

Foreign expropriation has thus transformed the Carnival of Venice into a Carnival of the World, with the city itself made to function as "a theater of affectation, like a scenery backdrop, like a 'container,' good for all contents and every size."[88] In this World Carnival, costumes and masques cease to serve as props for transgression or buffoonery and speak of a United Nations, in their own language of rococo high fashion. It is not for nothing the papers refer to the processions of these ornate mannequins not as *processioni,* but rather as *sfilate:* parades or fashion shows. And the Piazza has become their catwalk. In all this, we have glimpsed a return of that competitive urge that has given life to Carnival through the centuries and still fires up rivalries at the Sambódromo and on Bourbon Street. Not, of course, the rivalries of Venetian neighborhoods, or even of Venice at all, but of groups and communities from throughout the world, having traveled, many of them, thousands of miles to challenge and confound one another with their imaginative flights, transgressive fantasies, and visual bravura.

Insofar as one accepts the notion of Carnival as having (or having had) a functional, even fundamental, role in European, Catholic culture—as a vital human response to the weight of obligations of faith, family, and society—the story of the reborn Venetian Carnival has something of a postmodern tragedy about it. For a few years Venetians thought that the festival, this *happening* that they had reinvented, could actually be the magic spell that would arouse them and their city from the deepening lethargy that had been gathering around them for years. In a way, their experience elaborated and incarnated the inchoate efforts of many an embattled society that has struggled to hold on to its ebbing vitality in the face of an omnivorous globalization. Yet the Venetians, despite their almost legendary savvy in dealing with tourism, soon lost their Carnival— even more quickly than they lost the Vogalonga and the Redentore—to a cultural hijacking on a scale and of a completeness that can only be described as breathtaking. They have been marginalized and left redundant, with little choice other than "to hide themselves, abandoning their city, even if indeed only after a heroic and glorious resistance . . . sacrificing themselves before the cameras of the catwalk to the invaders dressed as Coca-Cola cans."[89]

CHAPTER 11

Taking It All Home

At the end of the day—usually quite literally—Venice's millions of tourists have to leave. Knowing that they have to go, most of them seek out something to take home from the city—a reaction seemingly universal among tourists and deeply rooted in their relationship to this or any other tourist site. Venice itself cannot be taken home, but there are aspects of the Venice experience that can be carried off, and in this impulse two orientations are at work: one toward the present, which tries to preserve the immediacy of the gaze, the actual experience of the place; the other toward the future, to complete a record for later use, in which tourist experiences are converted into tourist *memories*. The tourist experience of Venice cannot be understood without keeping this homeward orientation in mind, just as Dean MacCannell has observed in a more general sense: "Sightseers buy and take home an 'advertisement' (marker or memory) for a 'commodity' (sight-experience) which they leave behind for reuse by other tourists. 'The souvenir market' depends on the perpetuation of authentic attractions which themselves are not for sale."[1]

Tourists are not limited only to forging this relationship through what they buy, of course: they can also create it on their own, through photography. Cameras and camcorders are fundamental to tourism in Venice, as indeed they are to tourism everywhere. "It seems positively unnatural to travel for pleasure without taking a camera along," Susan Sontag once commented. "Most tourists feel compelled to put a camera between themselves and whatever is remarkable they encounter."[2] This is indeed very much the way in which the amateur photography industry proudly saw (and promoted) itself at the dawn of mass tourism:

[From 1904] Vacation days are Kodak Days. The Kodaker has all the vacation delights that others have—and has the pictures besides.

[From 1909] Vacation without a Kodak is a vacation without memories . . . [and] is a vacation wasted. A Kodak doubles the value of every journey and adds to the pleasure, present and future, of every outing. Take a Kodak with you![3]

The visual is at the heart of the tourist experience and the linchpin of tourist memory. It is a particular way of seeing, however, constructed through signs: John Urry argues that tourism itself amounts to searching for and collecting such signs. Tourists sites are indeed only recognized as sights because of the signs that have framed and thereby created them. Familiarity with the signs of places—the Eiffel Tower, Big Ben, the Winter Palace, the Leaning Tower—is what draws tourists to Paris, London, St. Petersburg, and Pisa, to find what is original, authentic, and supposedly existing behind the sign: sign, site, and sight are inextricably linked.[4] When MacCannell writes of "sight sacralization," he points to the importance of images—models, advertisements, pictures, brochures, books, postcards, and souvenirs—of a given site/sight, and how such reproductions both mark out and, in their way, authenticate the original.[5]

Venice has its own, very long imagistic history. Centuries before Paris was semiotically linked to the Eiffel Tower or London to Big Ben, Venice was awash in such ineluctable signs as the Ducal Palace, the Campanile, the Basilica of San Marco, the Rialto Bridge, the Grand Canal, and the gondolier. These images of the city circulated throughout early modern Europe in paintings, etchings, woodcuts, and stage sets: "All the views and prospects have been engraved so often that anyone who likes prints can easily get a vivid idea of them," Goethe wrote of the place, and in visual terms it was then probably the best-known city on the Continent.[6] Such fame could cause some problems for a public—even an educated public—still inexperienced in routinely discounting the disparities, the inevitable gaps, that open between the sign, reproduced in its perfection, and the actual place. Venice proved something of a letdown for first-time visitors in this regard, and the ever-splenetic George Ayscough's reaction, in 1778, seems to have differed only in his degree of bluntness from that of many of his contemporaries: "A City built on the ocean, is surely one of the wonders of the world! Such is Venice. I had even longed to see it beyond all others, and had raised so very high my expectations, that I own, when I saw it, I was greatly disappointed. The truth is, it makes a much finer appearance in prints and pictures, than it does in reality."[7]

Despite the letdown they could cause, idealized images of Venice did at least prepare and condition potential visitors for what they would find

FIGURE 25. Group photograph before San Giorgio Maggiore. (Photo by authors.)

in the World's Most Romantic City, and for how they should react once they got there. By the late 1800s, with the combined arrivals of mass tourism and mass communications, Venice had become familiar all over the world, its unique topography and architecture part of the common currency of late Victorian culture, both high- and middle-brow. As one American visitor put it: "Paintings, photographs, and word pictures have reproduced its every nook and corner, until it may be doubted if any 10-year-old school-boy or school-girl, transported thither by magic, would fail to recognize its salient features."[8] Ever since the early 1900s, when tourists began arriving armed with their own Kodaks, they have sought out these "salient features"—the signs they would hunt and collect on film. Indeed, they continue to do so to the present day, although the images of Venice they bring with them in their minds now derive from a wider array of sources—books, magazines, movies, television documentaries, and (perhaps above all) advertisements—and their means of hunting down these features are more sophisticated, including video and digital cameras and film cameras ranging from the most highly engineered down to the simplest throw-away (see figure 25).

Beyond this, however, the visual fame of the place has continued to structure how most tourists will see Venice, determining their reaction to it and actions within it even before they get there. "Much tourist photography is a quotation," as Peter Osborne has commented, "a reprising of the contents of the brochures or the reproduction of a view that as likely as not came into existence as a consequence of photography."[9] The process, according to Albers and James, traces out a "hermeneutic circle" that "begins with the photographic appearances that advertise and anticipate a trip, moves on to a search for these pictures in the experience of travel itself, and ends up with travelers certifying and sealing the very same images in their own photographic productions. In this process, tourists reaffirm the privileged position of photography as a source of their own awareness."[10]

"Tourist photography is more a process of confirmation than of discovery," observes Osborne, since when actually taking pictures, "we are not so much looking as looking at images or looking for images."[11] Yet even as tourists in, say, Piazza San Marco pursue its requisite images, they are also working to personalize the images they produce. This they do by whatever way they frame and aim their shots, but also (and more significantly) by making sure to place themselves or their friends and family within the framed image. Either way, they are validating a presence: they have not simply been there, but they come home with proof of having been there. These two forms of personalization are, in fact, opposites, and while the first is the direct one, with the photograph reproducing the tourist gaze, the second is that gaze reversed, with the person or people within the frame now "properly" placed in the photograph, no longer looking at the sight but rather back at the camera. In this latter sort—certainly the archetype among tourist snapshots—both photographer and subject(s) are present at the site, but they are also in the process of converting that presence into the privacy of the future. As Osborne puts it, "The photographic framing extracts the building from its own history and context and transforms it into a backdrop for a drama of the tourists' personal histories."[12]

Tourists pursue this second sort of photographic personalization in particular ways in such perpetually crowded parts of Venice as Piazza San Marco or the Rialto. We noticed that few try to capture friends or family as part of the crowd, immersed in the scene and unaware of the camera. Instead, regardless of nationality or background, visitors compose themselves pictorially with a knowingness about photographs, entering into an engagement—through their nods, grins, waves, and postures, or simply by their gaze—with the photographer: their attention is not directed toward the Piazza where they find themselves, but rather toward that later time and place where the photograph will gain its full significance. Those photographed must turn their back on the sight and disengage with it as they face the camera: the location is converted to a backdrop that confirms their presence. The sight (or site) behind the subjects in a photo confirms them as travelers, witnessing the fact that they were there, even as they, with their presence, validate the sight itself. Sights need tourists as much as tourists need sights, after all, giving those in the photographs certain proprietary rights over the attractions they appear with: as Osborne neatly puts it, both "tourists and *their* sights exist in order to be photographed, indeed are photographed in order to attain their existence."[13]

Photography is an essential element of the larger ritual process of sight-seeing, itself an activity that Graburn terms a "sacred journey."[14] There is an appropriate way for a tourist to respond when in the presence of the sight: to gaze and to photograph. This includes deference to the sight, but the genuflection we see is not, as in other responses to the sacred, a bending of the knee, but rather a raising of the arm, as tourists bring the camera to their eye. One can observe this response at the Bocca del Piazza, on the far side of the Piazza from the Basilica, where the Calle Larga di San Marco spills out the continuous tide of tourists who have come to San Marco from the Accademia or by *vaporetto*. Even before they have really gotten their bearings in this abrupt and vast space, we see many of them putting their viewfinders to their eyes, reminding us that the very term *snapshot,* according to the *Oxford English Dictionary,* originally meant "a quick or hurried [rifle] shot taken without deliberate aim, esp. one at a rising bird or quickly moving animal": a very appropriate expression for this sort of tourist photography. Such visitors demonstrate an immediate engagement with the subject, recognizing that this is something important, a key sign of Venice (and the view, looking the length of the Piazza toward the Basilica, *is* impressive), even if they often do not seem to know what exactly is in front of them (*after* the photo, we see them looking at their guidebooks).[15]

Eudora Welty once observed that "a photograph keeps the moment from running away," but in Venice tourists have to run after their photographs, as if the opportunities were fleeing prey. Up in the bow seats of the Linea 1 *vaporetto* one can see recent arrivals who are clearly taking their first (and probably only) trip on this slow boat from the train station down the Grand Canal, armed with cameras and so excited by the views offered from their special vantage point that they take pictures of everything in sight: every passing gondola, the fairly undistinguished buildings at the beginning of their voyage, and all the rest of Venice's daily mundanity: taxis, garbage boats, even the tourists on other *vaporetti*. By the time their boat has gotten a few hundred meters past the Scalzi bridge, some visitors have already shot all their film, leaving them with nothing for the Rialto Bridge or the much more striking palaces that await them further down toward San Marco.

This scatter-gun, snapshot approach to photographing Venice has found an apotheosis of sorts in the video camera, which has, like everything electronic, become more portable and affordable in recent years. For those who own one, a camcorder can conjoin the tourist gaze with the active production of memory, as tourists can now observe the passing scene through the viewfinder, even while filming it. Here we have the "meandering gaze"—a gaze without a focus or perhaps in search of a focus, where the tourist wanders along seeing the world through the mini-TV screen on the back of the camera. With video cameras (but also with digital still cameras) it is no longer necessary to gaze directly at the sights at all anymore: instead, the view immediately appears as a reproduction, already framed and displayed in a familiar and reassuring televised format. Whereas film photography is primarily directed toward future consumption, producing a disjuncture between time and place, with video the view is consumed in the immediacy of its creation: we have noticed how much video operators direct their attention not at the sights around them but toward a mediated view, even though the original is in front of them. In an age of the consumption of images, this makes perfect postmodern tourist sense: it is somehow not real enough simply to be in the presence of the "real," but rather, such sights, to be properly seen (or indeed seen at all?), must be transformed into a framed image. The only danger in pursuing Venice on the small screen—as we have seen more than once—is that the camera operators often trip on unseen things underfoot.

Rituals of tourist photography also involve claiming space, however temporarily. Photographers generally try to clear the image plain just around their subject (be it people, sight, or both) of extraneous distractions—in particular of other tourists. At the same time, the space between photographer and subject—the pyramid whose point begins at the camera lens and whose base encompasses the subject plain—also needs to be cleared of interloping passersby. Taking control of this space is not especially difficult at "natural" sites like the Grand Canyon or Niagara Falls, where plenty of viewing areas are set aside to give photographers unobstructed access to sight and subject; even in crowded urban areas, the presence of locals within the photographer's space need not be obtrusive—

indigenous people can even form an essential part of the sight, in places like a Cairo souk or Moscow's Red Square.

In Venice, however, such space is at a premium and can be highly contested. We have estimated (without any particular methodological precision) that, at a minimum, a hundred million photographs are taken annually in Venice's historic center, and for many of these shots photographer and subject try to establish fleeting control over the space where the shot is taken.[16] This is not so difficult: although some passersby—both Venetian and foreign—will just walk in front of the incipient photographer, most will pause for the necessary seconds to allow the photo to be consummated; some good-hearted souls might offer to take up the camera and shoot the whole group. In Venice's tight landscape, such generosity has its consequences, however: bridges, in particular, can get seriously blocked by those looking to have a canal view for a backdrop; we have already noted the troubles that result when tourists try to photograph each other on the *passerelle* during an *acqua alta*. Even out on the wide expanse of Piazza San Marco, one has to run a continual slalom to avoid stepping into someone else's shot.

There is also the difficulty that Venice's most heavily touristed spots— San Marco and the Rialto, above all—present most visitors with a different Venice than the one they had come expecting to see. The daily mundanity of these areas is neither the pristine emptiness generally presented in cinema and advertising nor the lively crowds of locals one might expect to find in the squares of Paris, London, or New York. Instead, when visitors arrive in Venice they find spaces totally dominated by other foreigners: the problem is not that these areas are full of people, but rather that they are full of the wrong people. When tourists come to capture these sights, they find that they are forced to include the density of their own presences in their photos. Those who want to keep fellow tourists out of their photos have to work hard in Venice, getting to the Piazza very early to avoid the heaving, jostling crowds that block the best views of the Bridge of Sighs or the Basilica for the rest of the day. One can leave the center, of course, and penetrate Venice's back alleys, searching for empty and mysterious alleys, small courtyards, or shady canals, but such intrepid photographers must be speedy in filming such scenes, since before long a tour group will inevitably arrive, or perhaps another individual traveler, also seeking out Venice's empty spaces.

It may be their difficulty in capturing on film the idealized images they have brought from home that makes tourist photographers behave so aggressively in Venice. We have regularly witnessed would-be picture takers on the Ponte della Paglia or Ponte dell'Accademia lose their temper and even start shouting over the impossibility of clearing other tourists out of the range of their viewfinder. Minor violence can erupt especially when these photographers scramble to capture an actual event, a particularly "Venetian" incident that promises a picture with future narrative import to enhance the ordinary tourist claim of "I was there" with "I was there during [some event]." During a Venetian wedding reception held at Piazza San Marco, for example, we once saw two Gore-Tex-clad American matrons literally shove the wedding photographer out of their way so they could get a shot of the married couple—complete strangers to them, of course, but by their presence turning the Piazza into a more romantic, and thus more photographable, site. The papers reported a similar sort of scuffle during the Carnival of 2000, at the beginning of the procession of the Seven Marys: "The atmosphere was typical of a popular festival, to which there rushed up a hundred or so Japanese on via Garibaldi. . . . 'Armed' with their cameras and digital videos, they entered into shoving matches with the Venetians, in contention over a place in the front lines."[17]

In the end, the overwhelming majority of tourist photographs taken in Venice (or anywhere else)—whether of remarkable monuments, true-life events, or smiling friends and relations—are just junk: poorly framed, blurry, overexposed, or bungled in any of a dozen other ways that camera manufacturers have struggled for years to abolish. Even with the most foolproof of cameras or camcorders it is difficult if not impossible in Venice to match the work of a skilled professional in capturing this most artfully crafted and cliched of cities. With the now familiar one-hour developing places in the city, tourists to Venice can find out the bad news about their efforts even before they leave town (something that never discourages some from trying again). Those, on the other hand, who seek to take away images of Venice more in keeping with the idyllic city they had always imagined probably do better to content themselves with postcards.

The postcard hovers somewhere between a photograph and a souvenir. Some tourists collect them as a pictorial record of their visit, probably realizing that they could never produce such polished-looking images with their own cameras. In this they are following an old tradition established over a century ago—buying and bringing home premade or prepackaged photo collections of a site's key scenes. Such bundles of photo-postcards can still be found at most tourist kiosks in Venice, but by now the primary purpose of the postcard is that it be sent to someone else (see figure 26). In this sense postcards are gifts, even if they are probably the least of all the gifts a tourist expects to purchase. Still, they guarantee an immediate

FIGURE 26. Postcard and guidebook kiosk near Rialto Bridge.
(Photo by Philip Grabsky.)

connection with the particular place in which they were purchased; and
though very likely manufactured somewhere else, postcards generally still
must be bought in the location they illustrate, and in this sense they are
expressions of the quintessence of placeness.

All sorts of tourists habitually send postcards back to friends, family, and colleagues. This is a key aspect of "tourist performance," that is, of that behavioral totality, voluntary or obligatory, that goes into making a human a tourist. It is, moreover, a two-sided affair, for just as many tourists feel compelled to send a card, many of those at home expect to receive one. This tends to turn postcard writing—as many tourists can testify—into something of an obligation, and one often sees tourists in bars and cafés dutifully working through piles of postcards and lists of names and addresses. Indeed, although travel writers have long encouraged readers to pay the exorbitant prices at Florian's or Quadri in Piazza San Marco, for the great opportunity one supposedly has there for "people watching," what we actually have noticed is that most tourists seated there are not watching anything but are instead assiduously writing the day's postcards.

This "postcard duty" has been made easier by modern technology: some tourists come equipped with computer printouts of address labels to attach to their cards. More than ever, then, the postcard is a convenient vehicle for linking here with there, the other with the familiar, providing one of the most efficient ways possible for tourists to demonstrate that they are, in fact, away from home and in the realm of the foreign. One could, of course, make a telephone call or send an e-mail, fax, or letter, but these media would fail to convey the particularity that a postcard offers. Though cheap and mundane, postcards still provide an apparently genuine sense of the place being visited, a reproduction of what the tourist has seen or (even better) is seeing as he sits at Florian's, for example, gazing at twin Basilicas—one before his eyes and one on the card he is filling out. As Susan Stewart has pointed out, the postcard initiates a significant aspect of tourist performance by transforming the public into the private. The tourist visits sites that are available to everyone; with the postcard, however, such experiences become personalized and then communicated to others, through "inscribing the handwriting of the personal beneath the more uniform caption of the social."[18]

But if postcards convert public image into private communication, they also provide the sender a chance to complexify this communication by offering a variety of manipulated visions of the sight they present. Few modern postcards even try to provide an unmediated image. At the very least, they must, by necessity, impose a frame around the subject they present, but most of them go well beyond simple framing to outright and (these days, increasingly) blatant distortion of what they frame: multiple shots, overlapping, or highly colorized views, all forms of eye-catching graphics. The card manufacturers may simply wish to attract potential buyers by making their cards jump out from the many on the rack, yet by visually intensifying and often emotionalizing their subject matter in this fashion, such postcards can end up offering a tacit or explicit commentary on the scene they present. When tourists come to select from the cards on offer, they (however fleetingly) have a chance to select those that best represent—one might say, editorialize—their own reactions to what they have seen. The available selection of postcards changes noticeably—if not in variety, then in approach—from city to city: offerings range from sexy and extravagant (in beach towns) to the stolid and monumental (in cultural capitals). In this regard, postcards for sale in Venice tend toward the deliberately "serious": there are fewer of those openly humorous, frivolous, kitschy, or ironic postcards that one finds on sale in many other tourist locations—sexual antics on Florida beaches, for example, or Dada shots of "London at night" (all black). Even in Florence one can come across campy renditions of Michelangelo's *David* wearing briefs or some other transgressive outfit. Postcard subtexts available in Venice are rather less flippant, largely limited to variations on a Romantic Venice theme, presumably because these sell more successfully than cards that stress Venice as a beach or shopping town, though a good many modern visitors clearly experience the place in those terms.[19]

As cultural notions of "romance" have shifted over the years, so have the visual references employed by postcard manufacturers to capture the concept in its Venetian context. In the mid-nineteenth century, Romantic Venice meant Venice of the Romantics, the place where artists and poets came to contemplate the remnants of a collapsed culture. Postcards of the 1880s and 1890s indicate that this paradigm was giving way to a much more middle-class, domesticated notion of Venice for Sweethearts, idealized by staged shots of young lovers/honeymooners feeding the pigeons in the Piazza or riding in a gondola, adrift in the moonlight under the watchful care of their gondolier. With the rise of the great Lido hotels after 1900, the postcard version of Romantic Venice shifted again, conflating romance with glamour, prefiguring the images of high-society-at-play that would dominate in these Venetian miniatures during the 1920s and 1930s.[20]

By the 1960s these earlier means for encoding Romantic Venice in a four-by-six format had become dated and untenable, so new approaches were explored. Through the use of soft-focus photography, postproduction techniques, and image manipulation, postcard makers have begun producing romantic visions of Venice based on exaggerated atmospheric colors—blazing sunsets, above all, but also shimmery moonlight and limpid blue skies as needed.[21] More significantly, they have presented a city that is essentially empty: few postcards indeed offer views of Venetian daily life or of the tourist mobs in and around San Marco or the Rialto. While this aesthetic convention is a mainstay of the genre—postcards of the Piazza della Signoria in Florence and the Trevi Fountain in Rome are also strangely devoid of humanity—with Venice the practice seems to have been pushed further than elsewhere, as a means of underscoring the city's dual nature: romantic and scenic. Especially common are cards that feature just an empty canal, a solitary gondola and gondolier, or perhaps a line of unoccupied gondolas or their *ferri* in the sunset (sunrise?) on the Grand Canal or Bacino. An analogous card from Florence or Rome would show just an empty street; such cards exist, but they are rare, and no wonder: even the most medieval of street scenes fails to convey that same sense of melancholy tranquillity one finds in a Venetian canal shot—the "dreamy watery realm," so often evoked in travel blurbs about Venice, as the essence of the city's romantic allure.[22] Such a figuring of the city rejects the more social images of the 1880s and 1920s of couples or groups indulging in Venice's romantic whirl. Instead we return to the romanticism of Byron and Shelley—solitary and somewhat narcotized. Never mind that the realities of mass tourism in Venice are very different from the Venice of the Romantic poets: most canals are jammed with motorized supply boats and gondola rides are now jostling *carovane* under the burning midday sun. In postcard-speak, at least, the viewer (sender or receiver) is invited into a miniature world rich in visual silence, slightly melancholy solitude, and floating comfort: a little scenario of fantasy made all the more romantic just for being so unlikely.

Since the 1980s, a new Romantic Venice has also been imaged in the form of the revived Carnival. Postcard racks about the city are now jammed with shots of flamboyantly costumed mimes and harlequins, often alone and languidly posing in bursts of tulle and color by the Basilica, on a bridge, or in a gondola. Since most tourists come to Venice outside Carnival season, those who want such images generally have to content themselves with postcards, but even those who do visit the city in February would be hard pressed to cut through the chaos of those days to produce photographs as artfully and ironically posed as those taken by professionals. For mannequins such as these, postcard manufacturers use techniques more appropriate to a fashion shoot, and in their arch display the models respond in kind, aggressively mugging for the camera and offering a new kind of glamorous romance, at once infantile and insouciant.

Textually rich though they are as photographs (and their semiotics in this regard have yet to be thoroughly explored), postcards also function as souvenirs, though among the most minor examples of that huge genre.[23] The two realms do have much in common: souvenirs, like photographs, need a narrative, for in themselves they tend not to be self-explanatory. Rather, they need to be related to another (past) time, place, and set of experiences, as a story told to others (or as a form of self-communing) by the person who was there and had the experiences. The souvenir, like the tourist snapshot, is a vehicle for nostalgia, taken by its owner into the future so that it might, in its turn, take him or her on a journey to the past. This interplay between future and past generates a centrifugal movement, as Stewart notes: "The souvenir speaks to a context of origin through a language of longing, for it is not an object arising out of need or use value; it is an object arising out of the necessarily insatiable demands of nostalgia. The souvenir generates a narrative which reaches only 'behind,' spiraling in a continually inward movement rather than outward toward the future."[24]

MacCannell has argued that any object, once imbued with personal meaning, can be cherished by someone as a souvenir.[25] Such "blank slate" souvenirs, items of neither intrinsic nor apparent value, will have their nostalgic worth imposed upon them by their owner: a ticket stub from a special event, a book of matches, a wine bottle from a memorable meal, a lock of hair, even a bottle cap. Such objects are completely idiosyncratic as souvenirs, encoding mnemonic texts that can be deciphered only by those individuals who can link them to a specific experience, or by the select and intimate group that has shared the experience secondhand.

There is also, of course, the premanufactured souvenir, produced by a memory industry dedicated to making objects specifically designed to trigger nostalgia. As many writers have suggested, tourism involves both commoditization and consumption, an equation that gives souvenirs a vital role: all tourist sights are sites for consumption, first in visual terms, as tourists gaze on the sight itself, but then also in commercial terms, as tourists purchase products that will recreate the sight in memory.[26] Without such objects to bolster sight, the tourist experience remains truncated and incomplete, since, as MacCannell points out, "the authentic attraction cannot be purchased" but inevitably must be left behind. Souvenirs, fortunately, can be bought by anyone, allowing the experience to be preserved and prolonged through time; tourists acquire them for themselves, as place markers, and also as gifts for friends who could not be there to share the experience.[27]

High-density tourist markets like Venice are awash in what might be called generic souvenirs—objects that have been imprinted or in some other way marked with a sign of Venice, but which just as easily could carry the mark of another place. T-shirts perhaps reign supreme in this category, but many such otherwise ordinary objects—sweatshirts, caps, glasses and mugs, ashtrays, plates, lamps, pens and pencils, aprons and tea towels—have also been pressed into service by souvenir manufacturers for carrying a place-specific indicator. Open to the imprint of any place in the world, these objects are almost totally interchangeable and can be found in the Plaka in Athens, Fisherman's Wharf in San Francisco, or the Ginza in Tokyo as readily as in the tourist kiosks around Piazzale Roma and the Riva degli Schiavoni. Enough for these Venetian variants to have one or two primary indicators of place: the city name itself ("Venice" or "Venezia," though not "Venise" or "Venedig") and some approximation of the winged lion of the Serenissima. Without such marks, these are not souvenirs of Venice, even if they were purchased there: such objects cannot be generic, but *must* refer to particular places, and out of that context they have no significance. A plain sweatshirt or an ordinary glass ashtray thus would not constitute a souvenir of anywhere. There is, after all, no "natural" or culturally specific connection between a sweatshirt or an ashtray and Venice (unless the ashtray has been stolen from a Venetian restaurant): rather, such neutral objects must have images and words added to them; they must be *made* to act as markers.

With souvenirs of this sort, it is the site of purchase rather than site of production that counts. Few care that a shirt emblazoned with the winged lion of San Marco is actually made in the Philippines instead of Venice, that the ashtray featuring the Campanile is from Poland, or that the plastic gondola is from Taiwan. Although such items may travel thousands of miles to get to Venice, their journey of significance—of becoming souvenirs—only begins in that city. Even sitting on tourist-shop shelves, they are still only *potential* souvenirs, and so they will remain until purchased and personally connected with a tourist: only when possessed and especially when taken home do these things begin to take on the character and function of a genuine souvenir.

By contrast, Venetian tourist shops are also cluttered with items of the sort we call touristware—that is, objects that are specifically designed, produced, and purchased to act as markers of place, defined as souvenirs long before any tourist actually buys them. Much touristware has multiple links to the specific purchase site, and, in the minds of those many tourists who see travel as one long opportunity for shopping, such objects can turn out to be vital in defining and giving sense to the place itself. One of us, working in the past as a tour guide for student groups, was continually asked by his charges, when approaching a new country or town, "What should we buy here?" More than simple naivete or youthful helplessness, the question implied both that these students felt an obligation to buy *something* and that they were concerned to buy something appropriate to that specific place. They knew that, in tourist terms, it would be sense-less to buy castanets in Germany, cuckoo clocks in Portugal, or model gon-dolas in Sweden. Rather, they needed to buy those sorts of things that would properly represent or connect with the place they were about to visit: "characteristic" foodstuffs and drinks, for example, local artifacts, or special items of clothing specifically identified with the site. Some of these objects are functional in nature and are therefore regularly bought by locals—one thinks of loden coats in Tyrol, grappa in the Veneto, or truffles in Umbria; tourists may buy such things to use or consume them on the spot, though more typically they take them home, still in the pack-age, as gifts or souvenirs.

Merchandisers of touristware in any given locale direct considerable effort toward promoting those objects they think a visitor ought to pur-chase if he or she is to take away something authentic. In doing so, how-ever, they have to answer a tourist concern that does not arise with the more generic souvenirs, like printed T-shirts or snow globes, for the au-thenticity of such "traditional" touristware resides as much in where it is produced as in where it is purchased. Typically, a visitor wants to buy these goods at the place with which they are most associated, to give them an additional symbolic and cultural power; the special value of such objects derives not only from their ability to represent a place in later memory, as generic souvenirs do, but also because, in their particularity, they are imbued with an authenticity that speaks to the culture of their origin, even after the visitor has taken them home.

Appropriately for one of Europe's oldest tourist centers, Venice has long produced appropriately Venetian souvenirs to offer its visitors. Three centuries ago, this might have been an oil painting of the Grand Canal or the Bacino: Canaletto was only the best among a number of *ve-dutisti,* or cityscape painters, all of whom cranked out dozens of nearly identical paintings of Venice's most renowned views. For less wealthy vis-itors there were other knickknacks available: as a child in the 1750s, Goethe played with a "pretty model gondola that [his father] had brought back"—the distant progenitor of millions of later souvenir gondolas, made locally out of iron or plaster during the nineteenth century and more re-cently mostly cast in plastic (with lights inside) in the Far East.[28]

The souvenir tourists most associate with Venice, however—like steel with Toledo or tulips with Holland—has always been glass. Specifically, this has meant so-called Murano glass, made in the little island town just north of Venice and characterized by bright colors, blown forms, and its decoration with enameling or colored glass overlays. Glass working is one of Venice's oldest industries, dating as both a manufacturing sector and a tourist attraction back to 1291, when, as legend has it, the glass makers and their furnaces were moved to Murano from Venice proper to protect the more built-up city from fire. By the fifteenth century, the glass furnaces of Murano had become at least as interesting to tourists as the holy relics and patrician gardens of that town, and many pilgrim-travelers toured the workshops, "in which glass vessels of divers forms are wrought with the most exquisite art." For its sheer spectacle, glassblowing seems to have appealed to onlookers as much five hundred years ago as it does today, though the value of the finished goods was evidently much higher in the fifteenth century, something that Pietro Casola found quite intimidating: "I stood to watch the work at the various furnaces, and I saw, above everything else a glass chalice, the price of which was ten ducats. It was noble and very subtly worked, but I would not touch it, fearing it might fall out of my hand."[29]

Interestingly, though the glass masters of Murano were running genuine factories, capable of producing large orders of goblets, "rummers," beads, and (especially) mirrors for clients around the world, they were also aware that their manufactories could provide visitors with souvenirs appropriate to Venice. Presumably this was why they had their furnaces and blowing operations open to the public, and by 1600 many glass makers had also started enhancing their display areas with elaborate glass follies to demonstrate their skills and attract the curious and wealthy. One showroom offered "a whole castle of crystal, with ordnance on the bulwarks and bastions, as also towers of defence, which is to be sold for twelve-hundred crowns"; in another, one could see a fully rigged glass ship, while in yet another there was an "Organ in Glass three Cubits high, so justly contrived, that by blowing into it, and touching the stops, it sounded musically."[30] Tourists who were sufficiently beguiled by such curios to want to buy something, perhaps to add to their own "Collection of divers Curiosities & Glasses," would be readily accommodated. Already by the seventeenth century the Muranese were practiced in satisfying tourist whims, willing to both make pieces to order and then to package them up and send them to the visitor's home—"by long sea," if necessary.[31]

Originally little more than a sideline, the production of glass curios and knickknacks gradually took over from Murano's primary business of mirrors, glassware vessels, and mosaic tiles. Goethe complained about this shift as early as 1786: "The art which gave the ancients their floors and the Christians their vaulted church ceilings is now wasted on boxes and bracelets. These times are worse than one thinks."[32] Much of Murano's recovery from the near collapse of its glass industry after 1797 was, as we have already seen, due to a few of the larger glass houses going into tourist-ware in a serious way around 1850.

"It's all such gawd-awful schlock," an American academic friend once protested to us as we walked past the glass shops clustered behind San Marco. "Why would anybody even want to buy this crap, much less take it home?" She was, of course, talking of the little glass harlequins, clowns, ballerinas, ashtrays, horses, and occasional copulating nudes that crowd these store windows, not the finer pieces of glass sculpture or dinnerware that one can still find in a few shops around Venice or in Murano. These latter continue to find buyers, as they always have, among those who seek objets d'art in the particularly Muranese style and who have the money to pay for such labor- and skill-intensive work. As to the rest, we already know that the factories crank out these tchotchkes (or import them from the Far East) because they are so cheap to produce and permit enormous markups, but why do tourists buy them? We have concluded that such trinkets are not desired so much for their inherent quality as for their role as markers, giving those who buy them—whether for a gift or to keep— an appropriate and authentic object from Venice, rather than a piece of Muranese artwork. Seen in this light, tourists are not so much showing bad taste (though one would never want to underestimate that) as good economic sense, picking up the cheapest marker of the city that they can find. For most who come here, Venice is just one of many cities on their route, and they may want to buy several object-markers from each place for themselves and their friends back home. Some can afford expensive works of Murano art glass, but the proportion of tourists who are con-noisseurs in this field is minuscule indeed. For the rest, it is enough to purchase the most mundane of items, the most ridiculous of figurines, since these, simply by being dislocated from their original context, be-come "exotic objects." For such items, "use value is separate from display value," but display value moreover resides less in the quality (skill of crafts-manship, value of materials, fineness of imagery) of the object itself than in the relationship between tourist and souvenir, and the latter's place of production and/or purchase.[33] Its auratic quality, then, derives more from its past, its narrative power, than from any present consumption of it.

Like many others who sell place-linked touristware—those who work inlaid steel in Toledo come to mind—the Muranese realize that the narrative power (and thus the attraction) of their goods can be enhanced if tourists are allowed to enter into the production process. In such situations, what visitors are offered and what they see make up a theater of authenticity for the products they are expected to purchase. In the bigger glass factories a skilled worker is usually taken out of the normal production area to make one or two items for the tourist groups. This is not highly demanding work: most of what they do amounts to little more than slapping together a figurine (clowns, usually) by dabbing on different colored blobs of glass. Rarely can one see any actual glassblowing or other skilled tasks, and we wonder if the workers themselves see this assignment as a normal job or more as a punishment.

These glass-working demonstrations do not take place in the main factory but in a special area set apart for the purpose. Tourists are not, therefore, taken to see the main production process as they might in a tour of a winery, a car plant, or (if they are lucky) a chocolate factory. As such, they get no experience of the social interactions or work practices of a functioning workforce, nor do they see the various stages in which different products are created. Rather, they are exposed to a representation, an idealization, in which the whole glass-making process is reduced to the miniature and to the individual. Tourists might hope that, in being treated to a glimpse of the artisanal efforts that go into the souvenirs they buy, they are entering into the backstage of glass production, enjoying an insider view that will allow them to transcend the normal consumer-retailer nexus. As Dean MacCannell has famously noted, however, such performances are really just "staged authenticity," acts put on especially for them as tourists. These demonstrations serve to reinforce, rather than undermine or supplant, the normal buyer-seller relationship by authenticating not just the object tourists witness being produced but, by extension, all Venetian glass products. The fictive display of craftsmanship suggests that what is going on here is a living, working tradition: one man making a single glass figurine represents all those, seen or (mostly) unseen, who are involved in the process. His act thus guarantees the validity of all products that await the tourists in the vast salesrooms maintained by each of the factories, despite the fact (fairly well known around Venice) that fully half the touristware on offer in Murano is imported from factories in Taiwan, Poland, or Bohemia.[34]

And, indeed, purchase and consumption are the key elements here. Unlike, say, a winery tour, in which the interest of the vintner is typically to spread the name of his label by educating consumers about wine making, these short shows—little skits, really—are put on fundamentally as a stimulus to shopping. Each tourist group—typically brought, as we have seen, to the factory in the manufacturers' own boats—is guided smoothly out of the work area into the huge showroom. The tourists arrive already primed to buy, the little artisanal drama put on for them having strongly suggested that a craftsman exactly like the one they just saw made any piece they might wish to buy, and in much the same way. Tourists do not need to see their own particular purchase being made by a Muranese artisan to "know" that it was crafted in this fashion.

An interesting, if ultimately failed, parallel to Murano's touristware industry can be found in Burano lacework. As with glass, the lace of this outer island once enjoyed a Europe-wide reputation, only to fall on hard times after the collapse of the Serenissima. As with glass, a few visionary individuals managed to bring lace making back to life in the later nineteenth century, also largely for the tourist market. In saving Burano lace, the credit goes mostly to the Countess Marcello, who in 1872 opened the Reale Scuola di Merletto (Royal Lace-Making School) on the island to collect and teach the secrets of the art. In short order, upward of five thousand lace makers (virtually all women) were at work, mainly on Burano, and their output was both the height of European fashion and much in demand for tourist souvenirs. Completely ignored in Baedeker's first editions on Venice (an American who went to the lace school in 1880 found that "there was nothing completed for sale"), Burano made its first appearance in the 1895 volume, specifically for its "interesting lace factories"; by 1913, the guide was devoting a full paragraph to the island and its by now best-known activity.[35]

Despite its much increased profile and the regular mention it still receives in present-day guidebooks, Burano lace never really managed to achieve the same status as glass, as a touristware that was place-specific for Venice.[36] Efforts have been made, ever since the founding of the lace-making school in 1872, to provide tourists with the sort of performance-authenticating show that the glass factories stage, but unfortunately lace-making-in-action cannot offer the kind of dynamic and colorful spectacle implicit in even the simplest of glass daubing. Moreover, the skill and labor required for even a small scrap of quality hand-knotted lace price this product outside the touristware category: at the moment barely a handful of elderly Buranelle are still willing to put on the necessary performance, and even their most modest handiwork is so expensive that only the wealthiest and most discriminating tourists can afford their efforts.

The rest of the lace one sees—and it is to be found in shops and kiosks all over Venice and Burano—is nowadays either machine made or hand crafted in Asia. Tourists buy it because of its association with Venice, but without the visible demonstration of its production, lace is necessarily a weaker marker of this place.

Even as fine Venetian lace—the *punta in aria* technique—begins in the tourist mind to lose its connection with the city, glasswork, from the finest candelabra to the cheapest daubed tchotchkes, also risks losing its place as the ultimate Venetian marker. Since the resurrection of the Carnival of Venice in 1979, the city has been virtually awash in shops selling masks, which are now vying with, if not surpassing, glass as the characteristic must-buy thing in Venice. Lightweight and for the most part more durable than glass, masks also are available in a wide range of prices, from just a few dollars up to several hundred. They also come in all sizes, from little white or speckled pierrot faces that can fit in the palm of the hand to massive, multicolored extravagances—moons, fantastical beasts, or huge, elaborate demiurges that could barely fit over a mantelpiece, much less on someone's face. Flexible at meeting the requirements of Venice's tourists, implicitly whimsical and cheap to make, masks are an ideal souvenir, even though most visitors come to the city outside the Carnival season and probably never expect to (or even can) wear the things.

Like glass and lace sellers, Venice's mask makers recognize the value of offering tourists a chance to see their product "actually being made." The work requires little more than elementary-school level skills of papier-mâché, paint, white glue, and sparkle manipulation—nothing as complex as with either glass or lace making—but, for tourists, seeing a craftsperson painting and decorating mask blanks seems to produce much the same effect as the glass-working demonstration. Though every mask offered for sale may not be made in the shop, or even in Venice, the display of such skills suggests that it is, and the craft worker seen touching one thus touches all. Thanks to the low technology of their craft, mask makers, unlike glassblowers, also can set up shop in very small enclaves almost anywhere in the city, enjoying a great advantage in Venice's essentially shopping-mall ambience, since they can sit right in their own shop windows. Tourists do not have to be hauled in groups to see such productions, as they do in Murano, but rather can just drift by on their own, along the *calli* of the city, able to gaze on these authenticators who exhibit themselves through the separating screen of plate glass.[37]

Interestingly, though the mask makers' presence as fabricators appears to render their product authentic, and thus appropriate as a marker of

Venice, there is very little authentic about the masks themselves. Unlike glass or lace, there was virtually no artistic tradition of mask craftsmanship in the city: during the glory days of the seventeenth- and eighteenth-century Carnival, masks were mass-produced in just two or three rather plain, unembellished types and soon discarded. Since the desire in those days was more to achieve anonymity than to stand out, both costumes and masks—especially the very Venetian combination of the *bauta* and *tabarro*—aimed at making everyone look more or less like everyone else; there was, in consequence, little call for the elaborate, eye-catching outfits that are the mask maker's pride today.[38] Modern mask craftsmanship is thus a tradition that has been cut from whole cloth, invented, and then presented to the tourist public as an authentically Venetian craft; instead it is a genuine representation of postmodern consumerism, freighted with overtones of infantile self-absorption and promises of ready-made fantasy. In that sense, at least, we could say that masks really are the souvenirs most representative of the Venice of today.

Glass, lace, masks, and a few other Venetian specialty productions, such as marbled paper, do not need the sort of additional sign markers that are required by T-shirts or ashtrays. Although there has been some effort, especially among glass manufacturers, to label or certify their product to assure its local production, all of these objects, whether made in Venice or not, are still appropriate as tourist souvenirs simply because Venice is "known" for their production. In the world-public mind, they are seen as essential representations of local tradition, which is what gives them their semiotic authority. Since tourists, by definition, cannot remain in Venice or make this space part of their everyday (home) life, their consumption of the city is necessarily fleeting; since they must return home, the souvenir speaks directly to this inevitable separation by connecting the present, the future, and the past. Photographs are taken and souvenirs are purchased in the present to be used in the future to conjure up the past. Souvenirs do not relate to ongoing experiences but only to those that have passed. The souvenir must be removed from its original context in order to be a marker of it, taken home and away from the experiences it will later encapsulate. Susan Stewart writes of how the

capacity of objects to serve as traces of authentic experience is, in fact, exemplified in the souvenir. The souvenir distinguishes experiences. We do not need or desire souvenirs of events that are repeatable. Rather we need and desire souvenirs of events that are reportable, events whose materiality has escaped us, events that thereby exist only through the invention of narrative. Through narrative the souvenir substitutes a context of perpetual consumption for its context of origin.[39]

Insofar as they are commodified nostalgia, souvenirs naturally fall into a hierarchy of cost, and thus of prestige. Whereas most tourists to Venice can (and must) satisfy their souvenir needs with postcards, cheap glassware, or masks, a few are wealthy and discriminating enough to immure memory in more costly containers—elaborate masks, art glass, or (for the very rich indeed) a genuine view of the Grand Canal by Canaletto or his contemporaries. There is also the option of simply buying a piece of the city, something that wealthy and even not-so-wealthy foreigners began doing soon after the Serenissima collapsed, snapping up cheap, run-down *palazzi* on the Grand Canal throughout much of the nineteenth century. In more recent years movie stars and millionaires have bought themselves nice pieces of the city: the Benetton brothers, among others, have acquired a whole string of palaces, shops, office buildings, and hotels since 1990 or so.[40] But whoever would buy into Venice also has to put up with the city's many problems: the pollution, the failing services, and—certainly most of all—the horrendous and increasing flood of tourists. As the long-time chairman of one of the Private Committees for restoration once admitted to us, having bought and restored for himself a fine palace on the Grand Canal, what he really wanted now was to have everyone—Venetians and tourists alike—kicked out of the city. "That way my friends and I could enjoy in peace all the wonderful work we've accomplished here."

Some plutocrats, as we have already seen, have gotten around these inconvenient disruptions of their Venetian fantasies by buying and restoring one of the Lagoon's outlying islands, but these islets are obviously few in number, and even fewer are up for sale anymore. A completely different approach has been taken over the years by a number of entrepreneurs and entertainment specialists, men (for the most part) who have fallen in love with the city and decided to quite literally bring Venice home—though not in the manner of William Randolph Hearst's pastiche castle at San Simeon, or of Robert McCulloch's London Bridge, shipped piecemeal from London in 1962, for reconstruction in Lake Havasu City, Arizona. Rather, some have tried to build a completely new and, if possible, improved Venice, making themselves in the process owners, if not of *the* Venice, at least of *a* Venice, placed not in the middle of an unsanitary and inconvenient lagoon in a remote corner of an incomprehensible country, but comfortably close to home. Working in the name of convenience, sentiment, and profit, these entrepreneurs have no difficulty with a Venice swamped with millions of tourists—on the contrary, they want to control and profit from those tourists' spending themselves, rather than leave something so vital up to the Venetians.

This is not a new idea. As long ago as 1843, a group of British architects signaled the direction such ambitions would take when they laboriously crafted a 1:540 scale model of Venice, which they then put on public display at the Egyptian Hall, in Piccadilly. Measuring something like five by seven meters, the model

contain[ed] the exact miniature; 110 churches, which in the original [city] exist in their full dimensions; the 122 towers, the 340 bridges, the 135 large palaces, and the 927 smaller ones, and lastly, the 471 canals; and besides all this, the more minute houses and places of abode, at least to the number of 18,000 and upwards, all perfectly represented in *shape, colour, and appearance*. The Adriatic and part of the Italian continent are described, and the whole is made a perfect representation of the place itself.[41]

A similar venture, though on a much grander (and more consumeristic) scale, was undertaken—again in London, at the Olympia Exhibition Hall—a half century later by the brothers Imre and Bolossy Kiralfy, professional organizers of late Victorian spectacles. "Venice, Bride of the Sea," as it was called, could pack in as many as thirty thousand paying visitors on a single summer's day—about as many as typically come to modern-day Venice during the high season. Once through the turnstiles, the "curious and often bewildered visitors" found themselves in a five-acre recreation of the lagoon city, filled with cafés, souvenir and snack vendors, a hundred or more gondolas and gondoliers, and apparently a good deal of singing: "The air never ceased to reverberate with the refrain of the ever popular 'Venetian Song.' There were gondolier musicians trolling the melody on the canals; there were 'Venetian' ladies singing it from balconies; pianos, mandolins, and guitars were all busy with it, and the crowd themselves unconsciously hummed it."[42]

The Kiralfys' show was quite a success, running for well over a year. After it finally closed, the Grand Canal was converted into the Bosporus and the whole spectacle brought back to life as "Constantinople, or Revels of the East." Other, somewhat more enduring, monuments to Venice were realized during the twentieth century. Disney's Epcot Center is one of several theme parks that features a replica Venice, offering its visitors a scaled-down, bare-bones (but scrupulously clean) rendition of the city's most touristic features as part of a "world tour" of famous urban sites. Other examples of the Campanile of San Marco, the Ducal Palace, and the Rialto Bridge, ranging from half to one-tenth the original size, crop up in places as diverse as Brussels, Tochigi (Japan), and Shenzhen (China).[43]

These all fall in the shade of one of the most recent and quixotic taking-Venice-home projects—The Venetian Hotel and Casino in Las Vegas. Rather in the spirit of the entrepreneur Victor Kiam, who was fond of saying (and advertising) that he had liked Remington shavers so much that he bought the company, the American marketing billionaire Sheldon Adelson realized in the 1990s that he liked Venice so much that he too had to buy one. Since even his wealth could not stretch to buying the whole city, Adelson decided instead to build his own Venice, choosing to set this new Queen of the Adriatic out in the Nevada desert. Having formed the necessary corporate apparatus, Adelson acquired the Sands Hotel in Las Vegas and, in 1996, had the whole structure blown up. In its place, in barely two years, he got this new Venice built, a new hotel and casino complex called (naturally enough) The Venetian. The new landmark, which opened in the early summer of 1999, featured essentially full-scale renditions of the Ducal Palace and Piazzetta, the Campanile and Torre dell'Orologio, the Rialto Bridge, the Ca' d'Oro, and the Bridge of Sighs, along with truncated segments of Piazza San Marco and of the Grand Canal (365 meters long).[44]

The whole thing has cost around $1.5 billion—so far. Work is also under way on multiple additions and enlargements that supposedly will double the complex's original size. The figures as they now exist are far from negligible, however: three thousand luxury suites, 161,000 square meters of convention space, 47,000 square meters of "themed, indoor retail mall," and (of course) 11,400 square meters of "gaming space."[45] As we see it, Adelson has managed to take home the most audacious souvenir of the city, as suggested by his decision to call his reconstructed city The Venetian. His use of the definite article gave his project a proprietorial claim to the city that appears to be missing among the other luxury casinos built about the same time: Bellagio, Mandalay Bay, and Paris, for example. "The Venetian" suggests that here is a summation of all that is "essentially" Venice. The structure goes beyond MacCannell's notion of the souvenir as a marker for a site that must otherwise be left behind for other tourists to reuse, since Adelson has made another Venice: even if he continually stresses that The Venetian was created as an act of homage to the prototypical Venice, in his assertions is also the implicit intention of rendering the original version irrelevant, or at least superfluous. As he himself once put it, during construction, "We're actually rebuilding Venice, authentically. I'll be standing in front of The Venetian Hotel Casino Resort in Las Vegas and saying 'This is Venice.'" Sure enough, just a few months later, waving his arms before reporters in front of the nearly finished Venetian,

Adelson exclaimed, "You feel that you're standing in the middle of St. Mark's Square, don't you? You're in Italy, in Venice!!"[46]

One could, of course, simply dismiss Sheldon Adelson as delusional: rich, certainly, but still basically insane. One could also dismiss The Venetian, as did the (then) mayor of Venice, Massimo Cacciari, as "the dregs of art [used] to make money; a circus tent of bad taste, created [for] shamelessly exploiting the image of Venice." Still, The Venetian continues to exist (and grow!) as a transported, reinvented distillation of the Most Romantic City in the World, carrying with it some profound and disturbing implications. In a sense, Adelson has rendered Venice irrelevant, no longer necessary to visit, because his "Venice" is more accessible to American tourists in their own country, and with roomy, high-quality hotel rooms. His vainglorious claim that "we're actually *rebuilding* Venice" reveals the particular and fascinating mindset behind the project, one very different from that behind the rebuilding of the collapsed Campanile of San Marco or structurally unsound palaces in Venice itself. The Venetian is not a *re*building or a *re*construction at all, but an entirely new creation expressed in the (literally) superficial forms of the original. Adelson and his team did not take anything physical from Venice for this construction, nor did they try to imitate the city's original building techniques or materials. Here everything has been made from entirely modern materials—steel, fiberglass, artificial stonework, and Styrofoam. The result is an image of Venice, the titrated visual essence of some of its key sights, just sufficient to represent Venice. Image is fundamental to Adelson, who, when standing in Piazza San Marco, once asserted to an interviewer, "Trust me: it will appear very, very, very authentic. To the eye it's going to look as authentic as it does here."

The eye is the key sense for The Venetian (see figure 27), and only the visual validates the authenticity that Adelson has so single-mindedly (and expensively) pursued. Perhaps this is because the other sensations that would make this place truly Venice—like sea breeze, cold stone, or church bells—cannot be so easily faked; in any case, this fixation on appearance has also led the project into some interesting (dare we say?) blind alleys. In arguing over the proper color to paint the windows of his Ducal Palace, for example, Adelson had recourse at one point to a coffee-table book of Venetian views:

Is that the book? Hah? It all depends on the view of the lights. It depends where the sun is. I don't care if you use chartreuse as long as it looks like what it looks like in Venice. And don't pick it up and reinterpret it, interpret it exactly the way it's supposed to be. Don't change history. I want it to look exactly what it looks like in the book.

FIGURE 27. "The Venetian": Ducal Palace, Torre dell'Orologio, and gondolas. (Photo by Quintin Lindsmith.)

Despite Adelson's assertion, "history" was of course not in his book, which showed only a representation of the original Ducal Palace, caught by the photographer on a particular day and in a particular play of light and shadow. Nevertheless, in this basic way, The Venetian steps right out of time, and how it was *seen* when it was photographed became the authentication of how it must be seen henceforth, frozen in the Ducal Palace of The Venetian. In reality (dare we use the word?), only the lights of Las Vegas, whether blaring desert sun or neon cacophony of the night, actually illuminate the "authentic" light of Venice that Adelson has had painted on the facade.

The proper colors and the shapes of shadows were constant concerns throughout the design and building phase of The Venetian, partly because of Adelson's obsessive desire to make the new look old, though not in the way that new brick was smoked to age it for Venice's rebuilt Campanile. Though the planners were relentless in producing suitably weathered-looking stonework on, for example, the column capitals of Adelson's Ducal Palace, background elements beyond human reach were most sketchily done, mostly not in stone at all, but rather in Styrofoam.

Since in The Venetian many of the buildings, as well as the "Grand Canal," are actually indoors, their features are bathed in light from an artificial sky, resplendent with Tiepolo colors of vivid blues set off by stacks of gold-and rose-tinged clouds—where there is no change, no day or night, and no weather. Such an illusion was easy for skilled artists—after all, Tiepolo did the same with his own ceiling frescoes—but Adelson's team ran into more difficulties in creating their own Grand Canal. Despite repeated efforts to paint the *bottom* of the channel to make it look the same color as the original canal, it has proved impossible to gain the same effect, since the water used at The Venetian is necessarily clear and chlorinated (and fresh instead of salt) rather than, as in the real thing, full of suspended material. Filling up his own Grand Canal with the same sort of pungent waters that one finds in Venice apparently struck even Adelson as push-ing authenticity a bit too far. His solution was to proclaim the experience of a gondola ride along his Grand Canal as actually *better* than on the one in Venice, and the Frommer's guide to Las Vegas willingly agrees: "All that's missing is the smell from the canals, but we are happy to let that one slide."[47]

Indeed, the desire to offer the public (but also, evidently, himself) a "new, improved" Venice manifests itself throughout Adelson's creation. His Rialto Bridge features a people mover rather than stairs: not that many would regret missing the chance to pause and gaze dreamily from the balustrade anyway, since Adelson's bridge spans a six-lane access road rather than the Grand Canal. As far as the romance goes, Frommer's com-ments, "Down the middle runs a canal, complete with singing gondo-liers. The 10-minute ride costs about $8, which seems a bit steep, but trust us, it's a *lot* more in the real Venice." Actually, the price is only slightly higher in Venice proper, but there's more, as other admirers have pointed out: "You'd be hard pressed to find a singing gondolier in Venice today—in fact, only a few hundred of them remain in Venice, modern victims of water taxis and buses *(il vaporetti [sic])*. But you'll find them here at The Venetian, with their trademark tight black pants, striped shirts and crim-son sashes."[48]

These are not, of course, real gondoliers. They are actors recruited for the job, given Italian stage names, and told always to speak to their pas-sengers with something like an Italian (though not a Venetian) accent. Although they have been taught some of the movements appropriate to rowing their craft, this is only to cloak the essential fakeness of their ac-tivities. In fact, the gondolas—which are considerably shorter than the real thing and built more like a *sandolo*—are driven by small electric motors activated by a concealed button under the gondolier's foot. Such replicas may seem like a mockery of the original, but, once again, they may also be taken (as indeed they often are by many users) as a distinct improvement. The gondoliers at The Venetian work for (and answer to) Las Vegas Sands, Inc., not for themselves or for a cooperative, as they do in Venice. As a result, they are not only always singing (they have to), but they will also not overcharge, make rude jokes about their passengers in dialect, or talk on their cell phones while on the job. They are, in fact, almost as perfect in doing what they do as the robot figures that cavort through Disney's "Pirates of the Caribbean" ride.

Jean Baudrillard was thinking of Disneyland when puzzling over the postmodern differences between representation and simulation, each of which he posited as arising out of "the entangled orders of simulacra." Thus, as he put it:

Representation starts from the principle that the sign and the real are equivalent (even if this equivalence is Utopian, it is a fundamental axiom). Conversely, simulation starts from the Utopia of this principle of equivalence, *from the radical negation of the sign as value,* from the sign as reversion and the death sentence of every reference. Whereas representation tries to absorb simulation by interpreting it as false representation, simulation envelops the whole edifice of representation as itself a simulacrum.[49]

Such distinctions matter with The Venetian, where Adelson's aim—or at least his claim—was to rebuild the Venice of Italy in Las Vegas in such a way that he could claim that "this *is* Venice." The Venetian was not to be simply modeled on Venice or a model of Venice but was to be an equivalent of Venice. His claim here is one of representation, that the sign (that is, The Venetian) and the real (Venice) are, in fact, equivalent. In order to achieve such representation, everything in The Venetian has to be authentic, equivalent to the original in Venice. And, indeed, throughout the building process in Las Vegas there was a continual referral back to the original Ducal Palace or Rialto Bridge as the necessary validation for the new structure: in the *process* of construction, the new depended on the original for its emerging existence. Venice had to exist for The Venetian to exist: the latter could make sense only as a representation of the original. After a few years of dealing with both together, Adelson seems to have learned to effortlessly conflate the two Venices, in his head and in conversation: "Every day a good fifty thousand people visit our Casino, strolling between [our] Rialto and San Marco. [They are] people who are left ecstatic before your [*sic*] monuments, dumbfounded by the details of our reconstructions."[50]

At this stage, Baudrillard's point about dissimulation can be applied to representation: namely, that "dissimulating leaves the reality principle intact: the difference is always clear, it is only masked." Once it has been completed, however, The Venetian then achieves a quite different status, that of simulacrum. Now it has become Venice in its own right and, in an important sense, no longer needs Venice. The linkages that Baudrillard has identified between religious iconography and the simulacra one finds in such amusement parks as Disneyland can just as readily be applied to Venice: "What if [Venice] can be simulated, that is to say, reduced to the signs which attest to [its] existence? Then the whole system becomes weightless; it is no longer anything but a giant simulacrum: not unreal but a simulacrum, never again exchanging for what is real, but exchanging in itself, in an uninterrupted circuit without reference or circumference." Whereas representation leaves the relationship between the sign and the real intact, "simulation threatens the difference between 'true' and 'false,' between 'real' and 'imaginary.'"[51] Once constructed, The Venetian floats free from its anchor in Venice to become a place unto itself, with its own integrity and identity. So we find that publicity material for The Venetian often refers to Venice itself, but as a means of further validating the experiences offered by the casino complex, rather than as an encouragement for the visitor to seek out the real city. The imaginary or imagined Venice hovers as a backdrop, an inchoate presence expropriated for what is on offer right now, which is, of course, The Venetian. In each of the three thousand suites, pamphlets and brochures, along with a short video, all sing the praises of the original Venice, whose thousand-year history has been brassily taken over and welded to this profit-driven enterprise, without (as far as we know) paying a cent for the privilege. The introduction to the hotel website proclaims and neatly sums up the appropriation:

THE ROMANCE OF VENICE! WELCOME TO THE VENETIAN!

Indeed, the visitor is rarely asked to admire the fakery that went into making The Venetian or to marvel at how closely it approximates Venice, but rather is urged to enjoy the complex on its own terms of originality and splendor. How this interplay—and the potential conflict it implies—is resolved to the ultimate benefit of The Venetian is demonstrated in the corporation's blurb for the gondola ride:

No trip to Venice—or to The Venetian—would be complete without a graceful and romantic glide down the Grand Canal in an authentic Venetian gondola. Float beneath bridges, beside cafes, under balconies and through the vibrant Venetian streetscape as your singing gondolier sweeps you down the Grand Canal for a "thrill ride" like no other. A relaxing and romantic ride that reveals all the charm, excitement and passion that is Venice — and The Venetian.[52]

Once the authenticating link between Venice, gondolas, and The Venetian is made, this becomes strictly a description of a boat ride in a hotel casino: the "bridges" and "streetscape" mentioned here have nothing to do with Venice. The streetscapes are, indeed, simply stores in a mall, but the iconic stature of these too has been assiduously cultivated through linkage to Venice itself, by reproducing certain of the (unacknowledged) architectonic elements that actually occur in storefronts along the Mercerie and elsewhere in Venice. There is also a (rather odd) attempt to strike a Renaissance air by calling the whole area "Grand Canal Shoppes." A few stores, moreover, seek a Venetian connection: the Lido Beach Shop, which "take[s] its name from the trendy beach area of Venice"; Il Prato, providing "a respected source of collectible masks and fine paper goods"; Ripa de *[sic]* Monti, a "Venetian-themed retail venue . . . [for] imported Venetian glass and collectibles"; and, of course, the Emporio d' *[sic]* Gondola, offering (what else?) "gondola-related merchandise from logo t-shirts and sweatshirts to pens and pencils." Perhaps none of these shops (shoppes?) so comfortably absorbs something of Venice to create something "Venetian" as the jewelry store called Ca' d'Oro, which in a single sentence neatly appropriates the original: "Known for generations as Venice's 'Palace of Gold,' Ca' d'Oro features traditional fine Italian jewelry, created with unsurpassed craftsmanship and elegant designs, that will surely set a new American standard." From such promotional literature the casual reader/shopper might easily conclude that the Ca' d'Oro was to Venetians something like the House of Sofas to modern shoppers — a place to buy gold jewelry, in other words, rather than a palace commissioned by Marino Contarini in 1420 whose facade was covered in gold leaf. As we can see, Adelson and his team have not let their rhetoric of authenticity lead them into "museumification." Their Venice need not be approached with the sort of high-cultural respect that many tourists feel is required by the real place; nor do visitors have to avail themselves of the mediation of guides, guidebooks, maps, or other explanatory texts. The Venetian is instead a ludic or fantasy space (one appropriately enough enclosed in the wider, ludic space of Las Vegas), a completely postmodern construction that blends and then fractures all distinctions between high and popular culture. Although the images it offers are expressed in and sanctified by their recurring references to high European culture, The Venetian itself has no need for anything other than the visual surface of this culture, and it neither seeks meaning from it nor offers any interpretation of it.[53]

Yet, if these Venetian-themed buildings, artwork, and waterway may seem to be mere facades, the pure surfaces appropriate to a stage set, The Venetian is still composed of real buildings, just like the water is real water (though the sky is not sky)—all solidly and three-dimensionally present in space. One has to keep in mind that behind this Ducal Palace, Rialto Bridge, and Campanile—and dwarfing them all—are Adelson's three-thousand-suite hotel (largest in the world), casino, and convention center. It is to serve these that The Venetian, as a new Venice, exists—an interesting reversal of what might seem the usual relationship between location and lodgings, since hotels *ought* to develop where a tourist attraction exists, to serve the site. Adelson expects that people will come to gaze on his Venetian, and, having done so, they will then make him rich(er) by staying to consume his creation in other ways: eating and drinking, shopping, gambling, and sleeping there. On reflection, however, we realize that this is very much the position taken by those in Venice's own hotel and restaurant sector when they complain how the *turisti mordi e fuggi* simply come to gaze for free on the sights of "their" Venice, without paying to stay the night, eat, drink, shop, and sleep over.[54]

The resemblance here is not between imitator and ideal, however, but between equals—or soon will be, as it enters a process of what might be called "emergent authenticity." The Venetian shows promise of becoming a tourist attraction in its own right. It is already bringing visitors in for reasons beyond its functionality as a convention center or gambling hall, and before long it will no doubt supersede its derivative connections to Venice proper. This is only reasonable, in this postmodern world of free-floating signifiers—especially in Las Vegas, that most postmodern of cities. Visitors will not come to Adelson's creation just to admire its architectural allusions to Venice any more than they go to Caesar's Palace or the Luxor to experience ancient Rome or pharaonic Egypt. Instead they will come to see The Venetian, a famous landmark in its own right. How this will all develop in terms of bricks and mortar (and Styrofoam) is not yet completely clear. According to *Nuova Venezia,* Adelson's plans two years after opening called for expanding "his" Venice with replicas of the Arsenal, the Scala del Bovolo, Ca' Vendramin Calergi (thus creating, as the article pointed out, "a Casino that pays homage to another Casino"), and Palazzo Salviati (with its nineteenth-century pseudo-Gothic mosaics). These or other additions may come to pass, but as of mid-2003 Adelson contented himself with (merely!) adding on a new tower, called "Venezia," containing 1,013 additional guest suites. Though not especially Venice-themed, the structure does boast a "Renaissance Wedding Chapel" that claims to "encompass the romanticism of Venice" and offers itself somewhat funereally as "an enchanted edifice to seal the special moment of eternity."[55]

Having built himself another Venice, Sheldon Adelson is now ready to dispense with the original, as not quite a real place any longer: "I see [the original] Venice with different eyes. Without offending anyone, I think that Venice isn't theirs. . . . it doesn't only belong to The Venetians or to the Italians . . . but to the whole world. It is a jewel box of various unique and magic treasures that must be appreciated and enjoyed by every-body."[56] Adelson is, of course, just repeating the mantra that late Victorian fans of Venice found so useful in staking their own claims to the city— that Venice no longer belongs to the Venetians but rather to the world, which is to say, to us. The threat to Venetian sovereignty over their own city is clear in such an assertion—what would it imply to claim that New York belongs to the world, not to New Yorkers? Yet Adelson, by evoking jewel boxes and magic treasures, also underscores, perhaps unwittingly, another implication in this mantra: that a city that does not belong to its inhabitants is no city at all; without its vital human element, Venice is indeed what he claims—a toy town, a heap of baubles, magic or otherwise. That Adelson, of all people, should think of Venice this way is hardly surprising: with The Venetian, he can reasonably claim to have created a city now more real than what the original has become. With its hotel, mall, restaurants, spas, casino, and resident staff of thousands, The Venetian is indeed arguably a genuine town in its own right, going about the postmodern world's most real business, that of selling image for profit.

AFTERWORD

Chi ciapa schei xe contento

In mid-July 2003, after an absence of several years, we returned to Venice for a short visit. While there, we had lunch with an old friend, a professor at the University of Venice whose observations have appeared several times in this work. This time, he noted how increasingly difficult it was for Venetians to make their way around within the triangle of the city marked out by San Marco, the Rialto, and the Accademia. He called this area Venice's "Bermuda Triangle," which caused us to laugh, since earlier in this book we had referred to it as the city's Bermuda-Shorts Triangle. His triangle was a rather more anguished and personal one than what we had intended, however: for him the center of Venice had become a place where the tourist density was such that "once you go in, you feel you may never make it back out again."

It was, in a way, a sad commentary on how degraded the heart and core of this most touristed of all cities has become, and we had plenty of chances to experience it in all of its pathos during a single week in July. Every part of the Triangle seemed to be more crammed with tourists than in the past. Most crowded of all, of course, was Piazza San Marco, which had long lines for the Basilica and the Campanile, hundreds of half-clad visitors slouching on the steps of the Procuratie and the Piazzetta dei Leoncini, thousands in a carpet stretching out to the Molo, and hundreds more struggling back and forth over the Ponte della Paglia. The Rialto too was jammed, all its approaches more obstructed than ever with fleets of souvenir carts and the side walkways on the bridge filled several deep with tourists elbowing for a view through the diesel fumes of the masses of *vaporetti*, taxis, and *moto topi* plowing along beneath them. Worst of all,

though, was the petrifying density along the connector alleys: the Mercerie, the Calli della Mandola, dell'Ovo, dei Fabbri, della Madonetta, and
all the others were packed all day with foreigners who were ambling, eating, shopping, or just standing in the way. All this at a time when the local papers were running articles—with headlines like "Tourism: The
Barometer Predicts Storms"—that noted an 8 percent drop in visitors compared to the previous year. What they were talking about, of course, was
tourism measured in hotel registers, which had certainly taken a dive since
September 11, 2001, and the ensuing world recession.[1] Yet as those crowds
testified, Venice as a destination for day-trippers, for the *mordi e fuggi*
tourists, is, if anything, more compelling than ever. By all accounts—and
not just our friend's—every year the foreign presence is greater, lasts more
months, and covers more of the city. Every year, Venice becomes more
like an amusement park and less livable for the Venetians.

The signs of this dissolution have gone on throughout the writing up
of our research and show little sign of abatement; indeed, even the current *giunta,* vigorously led by Paolo Costa, often seems less to govern according to any predetermined program than to spend its time racing to
cope with one unexpected disaster after another. These have taken many
forms, though almost always carrying with them the same subtext: the
resource that is the historic center of Venice continues to be ruthlessly
exploited for profit, even as it is being allowed to disintegrate before the
eyes of the uncaring, or at least uncomprehending, world. In the late summer of 2001, for example, the locally based insurance firm Assicurazioni
Generali, which had been headquartered in the Procuratie Vecchie at San

Throughout this book we have tried to maintain a nonprescriptive
stance on Venice's many problems, on tourism to the city, and on the future choices that might await the Venetians. This has seemed appropriate for a work that has aspired to be part history and part ethnography
but in no way polemical or political. It is, of course, into the political—
the often highly polarized and fragmented world of Italian politics—that
anyone proposing solutions to Venice's woes would have to go, something we, as outsiders, would never dare to do or even to contemplate
doing. Perhaps we are restrained even more by the knowledge that both
the current and the previous *giunte,* or city councils, were led by profes-
sors in the humanities and social sciences—not so different from us, in
other words—men who have held positions at the University of Venice,
who know the city intimately, and who have still found it well nigh impossible to reverse (or even slow down) the continuing dissolution of
Venetian society.

Marco from 1832 until the late 1980s, suddenly announced that it was planning to turn the two thousand or so square meters that it still owned in the historic building into "superdeluxe" condos. Priced out at around twenty-two thousand dollars per square meter, these would unquestionably be targeted at wealthy foreigners, since "probably Venetians wouldn't like to live precisely in Piazza San Marco, invaded [as it is] by throngs of tourists, constipated by sickly pigeons, and perpetually pervaded by little orchestral concerts." The *giunta* was caught completely off guard by this proposal and, convinced that "seeing the [Procuratie] transformed into apartments for VIPs from around the world would in some way signify the definitive surrender of Venice to its touristic transformation," has had to scramble to find a way to block the project, if indeed it can legally be blocked at all. "Watch out for these speculators," warned the *Gazzettino*. "They could transform Venice into 'Veniceland,' a city made-to-order for tourists, especially the elite ones."[2]

Despite administration efforts, those who want to buy up underused or unused properties in Venice and convert them into condos, hotels, or bed-and-breakfasts to further feed (and feed off of) the city's tourist boom are still largely free to do so. The city has had a few successes in turning back attempted takeovers of its famous structures—evidently condos will not be coming to the Procuratie any time soon, nor will the Railway Palace at Santa Lucia be converted to a hotel—but the trend seems to be clearly toward large-scale development.[3] As we have seen in the course of this book, the Venetian market for lodgings, whether in the form of luxury-hotel suites or rental rooms, is poised on the edge of a boom, one that promises to push the city much further into its "touristic transformation." The signs of this are everywhere in town, especially in the form of the big construction cranes that dominate the Venetian skyline—"an absolute forest of cranes" at the moment, poking up from the islands of San Servolo, San Clemente, Sacca Sessola, San Giorgio Maggiore, Sant'Elena, and all along the Giudecca, as well as in the historic center itself, at Sant'Anna and San Lorenzo in Castello, at Calle Vallaresso near San Marco, and at Campo Manin and Campo San Bartolomeo near Rialto. Some of them are involved in building on behalf of the University of Venice or for the local administration, but the greater part—at least three-quarters of the sites—is tourist oriented: hotels, condo conversions, megastores, conference centers.[4] By itself the city government simply cannot compete in such a overheated marketplace, as the mayor had to admit when he seemingly threw in the towel in the competition over who would buy the historic Fondaco dei Tedeschi, now that the post office is selling out its place there and moving to the mainland. With the post office asking about $25 million for the structure, Costa said, the city was "priced out of the market"; the result is that "the front runners, also in this case, are the colossal hotel chains or investment groups that could transform the ample internal courtyard and the loggias into a 'hip' commercial center, a reworking, Lagoon-style, of the mega-store [concept], like London's Harrod's or Berlin's Kadewe."[5]

Even as this building boom for tourists and wealthy foreigners goes on, other, less favored parts of Venice have continued falling to pieces. Every month, it seems, another house has to be evacuated somewhere in the periphery, usually rendered unsafe by the continual beating of the *moto ondoso* thrown up by the power boats, its residents forced to move to a hotel or to the mainland.[6] One day in the midsummer of 2001, moreover, four of the historic covered docks along the north side of the Venetian Arsenal simply collapsed. There had been plans afoot to restore this part of the old shipyards, much of which is in private hands, and in fact the removal of old detritus was already underway at the time of the collapse. Initial hopes that the lost section could be rebuilt according to the favorite Venetian mantra, *com'era e dov'era,* have been frustrated, since most of the marble and brick making up the structures has crumbled to dust, such that, according to the experts, this historic piece of Venice is "'by now irredeemably lost.'"[7]

Meanwhile, the city's big cultural events continue to lurch forward, year by year, ever more caught up in the contradiction between being legitimate celebrations of Venetian culture or merely shills to bring in yet more tourists. In 2001, both the Vogalonga and the Redentore were called off at one point, as the organizers and local government found themselves overwhelmed by technical and administrative difficulties and (it would seem) simply exhausted by the unrewarding work such spectacles demand. Only after a storm of protest—by the Venetians themselves, but also by the hotel and restaurant owners—were these two holidays saved that year. That one edition of the Redentore cost the city over half a million dollars, needed to put together a votive bridge from scratch in order "to guarantee to the city the scenario from forever"—*dov'era e com'era,* in other words.[8] Even so, when the event finally reached its costly climax, with the fireworks on the night of July 15, the Venetians who went out to the Bacino in their own boats found that, this time more than ever, their view was largely blocked by big *lancioni di granturismo,* "those that host 100 to 150 people, often coming in from outside [the Lagoon]."[9]

The Carnival also has continued to grow, even after the collapse of

American tourism in Europe after September 11, 2001. The celebration of 2002, though somewhat underpopulated in its first days, actually set new records of attendance for the tumultuous final weekend: as many as 260,000 maskers, photographers, gawkers, and pickpockets showed up for the final weekend—the "record of records," as the papers called it.[10] Yet it all played out, as always, amid a rising tide of complaints and protests from Venetians about its gigantic, loud, and consumeristic character, about their own irrelevance to the whole affair, and about "these hordes of miserable maskers looking for a festival. Once again the big bluff has been pulled off, to the joy of those few who have earned (a lot) and to the desperation of the many who have endured it."[11]

As the old-timers put it, *chi ciapa schei xe contento, i altri no:* "[those] who make the bucks are happy, the others aren't."[12] One thing that has become increasingly obvious over the course of this book, however, is that those who are making the bucks are in fact very often the Venetians themselves. As tempting as many locals may find it to blame outsiders for exploiting their city, much if not most of the selling out of Venice that makes this possible is still being done by Venetians. It is Venetians, as we have seen, who since the days of the Serenissima have been the most eager to sell off their unique city to outsiders, to fill up its streets with tourists, convert its buildings for tourist use, and pollute its waterways in the hope of attracting just a few more paying foreigners. It is Venetians who are, for the most part, the ones who run their boats too fast or in the wrong places, creating ever more *moto ondoso;* they are also the owners and operators of more than a few of the mask stores and touristware shops that clog their city. It is also Venetians—or perhaps Muranese— who are said to be involved in the various mafioso-tinged attempts to monopolize tourist flows from the parking island of Tronchetto to the glass factories on Murano, and thence to the favored restaurants in the historic center. There is, indeed, some indication that certain Venetians are the ones who are aiding and protecting the *vu' cumprà,* allowing these hapless illegals to continue their difficult and often unwelcomed work on the city streets.

Under the circumstances, then, it is hard to sympathize with the Venetians, or rather it is hard to know which Venetians one ought to sympathize with. Certainly those who are struggling to defend their city and its culture, even at this late date and with so little possibility of success, deserve respect, even if in their zeal many of them have proved as vigorously determined to resist changes that would benefit the lives of residents as changes that just bring in more tourism.[13] On the other hand, in a certain grudging way, one has to admire those who are resolutely selling off bits of their city, whether as second homes for the rich, new hotels, or souvenir shops for the masses, since they at least are asserting some sort of ownership over the place: they can sell only what is theirs, after all, even if this last, in a sense, defiant act ultimately will leave them with nothing whatsoever. Knowing how, for so many years, outsiders from Rome, London, New York, and everywhere else in the world have told the Venetians how to run their city (and that it is not really their city anyway), it becomes easier to understand, if not really accept, the destructive (and self-destructive) cynicism with which so many locals treat this place.

How Venice will end up remains anyone's guess. The current administration recently brought out a "Venice card," a sort of debit card that is intended to regulate tourism to the city and to provide Venetian residents with services at differentiated costs. There are also plans—at fairly advanced stages—to build a fourth bridge across the Grand Canal at Piazzale Roma and to put in a *sublagunare,* an underwater tram line, to connect the city at its Arsenal end to the Marco Polo airport. Such schemes (and there are others, less developed) are seen as vital if Venice is ever to have hope of becoming a real city again, allowing businesses—especially those in the universally beloved (and courted) information-technology sector—to set up shop there. Trams, tunnels, and bridges will also serve the tourism sector, however, and it is a safe bet that each of these plans, once (or if) realized, will be accused on some future day of having contributed to the Death of Venice through Tourism.

Meanwhile, though the Venetians are no longer leaving at the rate they were for decades—indeed, the decline seemed to have actually stopped briefly in 2000—most of them are not especially optimistic about hanging on in the face of a tourist flood that shows no sign of abating in their lifetimes. Such is the emotive power of Venice that foreigners will keep coming, despite the high prices, the crowds (of themselves), and the many discomforts: "It's so romantic!" they told us over and over again in English, French, Spanish, German, and Italian. Even if Venice does strike many as dirty, dilapidated, or smelly, it still fits the images they have nourished, sometimes their entire lives, of what romance means: "It has always been my dream since I was little," one elderly Austrian woman exclaimed to us, laughing and then misciting for us: "'To see Venice and die!'" And the Venetians? They carry on, working their way through the crowds and writing their letters to the papers, asking, "How far can we go?" before "the canals are completely jammed, and one is able to cross the Grand Canal jumping between the passing boats"? How does one survive in a city that has given itself completely over to tourism?

Poor Venice! Who knows if one day you will see some people in a cage in Piazza San Marco, with a sign saying, "Genuine Venetians." It could happen, then, that a tourist will pass over to the resident some leftovers from the sack lunch furnished with his admission pass to the living museum.[14]

Notes

List of Abbreviations

The following abbreviations are used in the notes:

ASV   Archivio di Stato di Venezia, at the Frari, Venice

GZ   *Il Gazzettino,* published daily in Venice

GZV   *La Gazzetta di Venezia,* precursor to *Il Gazzettino*

LT   *London Times*

MPD   Mss., *provinenze diverse,* Museo Correr, Venice

NV   *La Nuova Venezia,* published daily in Venice

NYT   *New York Times*

VMC   Venice, Museo Correr

Introduction. The City Built on the Sea

1. Jong, p. 250.

2. The film's original title is *Italiensk for Begyndere,* directed by Lone Scherfig (Denmark, 2001).

3. As McCarthy notes, "It has been part museum, part amusement park, living off the entrance fees of tourists, ever since the early eighteenth century when its former sources of revenue ran dry." McCarthy's *Venice Observed,* quoted in Specter, p. 40.

4. *GZV,* 19 July 1896.

5. Speaking in terms of "presences": two tourists coming to Venice for one day represent two presences, as does one staying for two days. In 1997, just under 11 million visitors came to Venice and left the same day; another 1.4 million stayed

there for just over two days on average, giving an additional 3 million presences, for a total of 14 million: see Isman, pp. 88–92; also *NV,* 14 March 2002.

6. Nearly three times the average found in Salzburg, which boasts the Continent's next highest visitor-resident ratio: Isman, p. 92.

7. *GZ,* 13 August 2000.

8. James (1959), p. 290.

9. From the video *Super Cities: Venice.*

10. Muschamp and Sottsass, p. 208.

Chapter 1. Pilgrims' Rest

1. Indeed, see Redford, p. 14, who states that "the Grand Tour is not the Grand Tour unless it includes [among other things] . . . a young, British male patrician (that is, a member of the aristocracy or the gentry)."

2. These foreign colonies included Germans, Turks, Dalmatians, Greeks, Albanians, Armenians, and Jews, plus Italians from Lucca, Florence, Milan, Bergamo, Friuli, and Padua, among other towns. See Costantini, pp. 881–86; and Fedalto, p. 499.

3. Urry (1990), pp. 1, 3.

4. Casola, p. 153. On pilgrims as tourists, a well-used trope among social scientists, though less so among historians, see Turner and Turner, esp. pp. 1–39.

5. Turner and Turner, p. 20; also Graburn.

6. See Newett's introduction to Casola, esp. pp. 23–113; also Ashtor, pp. 197–223; Tucci (1980), pp. 348–53; and Costantini, pp. 886–87.

7. Casola, pp. 124–25, evidently was referring both to the common talk among contemporary travelers and to written descriptions by earlier pilgrims—such as that of fellow Milanese Santo Brasca, who may in turn have been following the *Itinerario* of twenty-two years earlier by Gabriele Capodilista: Brasca/Capodilista, pp. 32–33.

8. See Wey, p. 4–7; von Harff, pp. 69–71; and Brasca/Capodilista, pp. 128–29.

9. Brasca/Capodilista, p. 48; von Harff, pp. 54–55; Casola, p. 125; Fabri, p. 110.

10. Casola, pp. 125–32, 137.

11. Casola, p. 124; Ashtor, pp. 205–12; Tafur, pp. 33, 47; Newett says that by 1451 the fall galleys had been abandoned completely, largely due to declining demand: Casola (intro.), pp. 77–78.

12. See Turner and Turner, pp. 6–8.

13. Fabri, pp. 84, 107; Casola, pp. 337–42.

14. Keeping in mind that a skilled Venetian worker earned around forty ducats a year in 1500: Lane (1934), p. 177, n. 6, and pp. 178–79, 251–52; Lane (1973), pp. 46, 63; Tucci (1985), pp. 64–65.

15. Interestingly, those fifteenth-century pilgrims who did go to Florence dispensed with the city in just a few paragraphs, staying only briefly in their apparent eagerness to get to Rome. Von Harff, pp. 12–13, called it "very pleasant," and Tafur, pp. 227–28, "one of the most wonderful in Christendom."

16. Casola, p. 129.

17. Von Harff, p. 51; Casola, pp. 129–33.

18. Brasca/Capodilista, p. 49.

19. Translating so many relics from Constantinople to Venice simply replaced one traditional pilgrimage destination with another; see Geary, pp. 98–128; Muir (1981), pp. 96, 207; and Lane (1973), pp. 41–42, 104, 106, 394.

20. Simon Fitz-Simon, quoted in Parks, vol. 1, pp. 579–81; later visitors would add the names of Saints Helen, Massimo, Paul the Martyr, Barbara, Giovanni the Duke, and Eustachio; see also Guylforde, p. 8.

21. Casola, p. 324, had made his vow when seeking safety from a storm at sea during his return voyage from the Holy Land; see also Tafur, p. 27, for a similar vow.

22. Fabri, p. 110. Fabri went all over Venice and out to Murano to see the shrines appropriate to travelers—to San Raffaello, San Michele, San Cristoforo, and finally Santa Marta.

23. Frescobaldi, p. 7, visited 11 major saints' relics, along with "a great piece of the Wood of the Cross," and 198 complete bodies of the infants killed by order of Herod.

24. Wey, pp. 89–90; Brasca/Capodilista, pp. 50–51.

25. Fabri, pp. 93–94, also carried the jewels of some of his travel companions. Brasca/Capodilista, pp. 50–51, had also "seen and touched" all of the many relics on his list; see also Geary, pp. 32–35.

26. Evelyn, pp. 200–201. The treasury was also said to contain "the very ring which St. Mark wore on his thumb" (Leo of Rozmital, p. 156), a painting of the Virgin, and the original Gospel of Saint Mark, both done in the saint's own hand. Misson, pp. 175–76, claimed that by the 1670s the latter was "so worn, torn, defaced, and rotten with Moisture, and other Injuries of Time . . . [that] it is a hard matter to discern anything in it," even if it was written in Latin or Greek.

27. Barbatre, p. 98; Casola, p. 126; von Harff, pp. 52, 54–55.

28. Fabri, pp. 94–95; Tafur, pp. 157–58; Casola, p. 127.

29. Fabri, pp. 83–84.

30. Muir (1981), pp. 119–34, 223–30; Evelyn, p. 198.

31. De Voisins, p. 18.

32. Da Sanseverino, pp. 17–18; Tafur, pp. 158–59; Fabri, p. 99; Brasca/Capodilista, p. 49.

33. Casola, pp. 146–53, the source of this quote, was the first to mention actually marching in the procession, although Brasca/Capodilista, p. 50; Fabri, pp. 108–9; and Barbatre, pp. 106–7, had described it in detail.

34. Guylforde, p. 11; also Muir (1981), pp. 189–211.

35. Newett's introduction to Casola, p. 49.

36. See Greenwood, pp. 171–85; on Greenwood's position, see Boissevain (1996), pp. 1–26; and Selwyn, pp. 1–31.

37. On pairing off slaves and indigents with senators, see Davis (2000), pp. 473–78; see also Newett's introduction to Casola, p. 113.

38. ASV, *Cattaveri, busta* 2, *registro* 4, 22 March 1387, 20 May 1401, 9 March

1429 (cited by Newett). Tucci (1985), p. 63, notes that *tolomazi* was a Venetian corruption of the German *Dolmetscher*, or guide, and that *messeta*, or agent, was also used for this profession. Vera Costantini has pointed out to us that *Dolmetscher* (and thus *tolomazi*) was ultimately Turkish in origin, however, and that Venetians used the term more in the sense of "interpreter" than "guide."

39. This was their first mention in the deliberations of the Senate: see Newett's introduction to Casola, pp. 24–49.

40. The *tolomazi* were permitted a 5 percent commission for booking their clients' passage to the East: ASV, *Cattaveri, registro* 4, 22 March 1387, 28 June 1448, 14 January 1455; Newett's introduction to Casola, pp. 41, 73.

41. Costantini, pp. 891–92, termed the conditions in many of the *osterie pubbliche* as "terrifying"; on the *dazio*, see ibid., p. 890.

42. Zaniboni, p. 57; Fabri, p. 79.

43. Fabri, p. 80, also wrote of the Saint George's famous "big black dog," who, when Germans came to the inn, "showed how pleased he was by wagging his tale," but who also raged, "barking loudly, leap[ing] furiously upon . . . Italians or Lombards, Gauls, Frenchmen, Slavonians, Greeks, or men of any country except Germany."

44. Casola, p. 123; Brasca/Capodilista, p. 48; von Harff, pp. 50–51, also had fellow Germans from the Fondaco as his guides about the city.

45. Fabri, pp. 105–6; on the White Lion and the Savage Man, see Padoan Urban (1989).

46. VMC, MPD, 396c/II c.524, *Forestieri a Venezia nel sec. XVI; osterie e albergatori,* mid-sixteenth century (undated).

47. ASV, *Inquisitori di Stato, busta* 760; the *albergatori* have recently begun to resurface in Venice, in the form of the "bed and breakfast": see *NV,* 3 February 2000.

48. Casola, pp. 141–42; see Crouzet-Pavan, p. 256, for a partial census of the city in 1509.

49. Fabri's inn of Saint George was located just behind the Fondaco dei Tedeschi. Evidently the only churches his group walked to were San Marco, San Bartolomeo (the parish church of the Fondaco), San Salvador, Santa Maria dei Miracoli, and Santi Giovanni e Paolo: Fabri, pp. 84–110; also Tafur, p. 167.

50. Coryat, p. 314.

51. Lassels, p. 226; Lassels visited Venice several times between 1635 and 1665.

52. Du Mont, p. 395, who reported that gondoliers hired out for between a half ducat to a ducat and a quarter per day; the base daily wages of shipbuilders at this time were between a fifth (for apprentices) and a third of a ducat (for masters); see Davis (1991), pp. 28–30.

53. Casola, pp. 111–13. But cf. Gailhard, p. 119, who claimed in the 1660s that the Cattaveri were still available to adjudicate disputes between pilgrims to the Holy Land and the sea captains who carried them.

54. Although both Casola, pp. 144–45, and von Harff, pp. 64–65, discussed the dress and appearance of such women—who were almost certainly street whores—they never named them as such.

55. See Coryat, pp. 325–26, 365; Montaigne journeyed in Italy from 1580 to 1581.

Chapter 2. Strumpets and Trumps

1. These options originally offered by Palmer, in Pine-Coffin, pp. 4–7.

2. Redford, pp. 12–15.

3. Sainte-Marie, p. 19; also Miller, p. 190.

4. According to Balfour, pp. 214–15; also Evelyn, pp. 210–12; Moryson (1971), pp. 74–76.

5. Chard, pp. 78–83.

6. Keysler, p. 153.

7. In particular, see Howell, Gailhard, and St. Didier; Venetian government is also treated extensively in Ray, pp. 132–67; but see also Eglin, esp. chaps. 1–2.

8. Misson, p. 195; Northall, pp. 431–32.

9. So says Keysler, p. 156; also Eglin, pp. 80–84.

10. Quote from von Archenholtz, p. 46. De Blainville, pp. 270–71, 281–82, wrote at length on begging nobles.

11. According to Sperling, pp. 26–29, 115–16, 246–48, tab. A2, from the 1550s to 1700, at least half of all noblewomen in Venice were cloistered nuns.

12. Ayscough, p. 199; see also Jeffereys, p. 89.

13. Quote from Stevens, p. 343; see also Nugent, pp. 44–45.

14. Including such "natural Curiosities [as] Minerals, Fossils, and Petrifactions," Keysler, pp. 180–82. Both the Ducal Palace and the Arsenal had weapons rooms, "fill'd with curious engines of death," that attracted military buffs then much as they do today: Davis (1997b), pp. 80–81.

15. Barbatre, p. 98; von Harff, p. 52; Moryson (1971), pp. 80–81; Evelyn, pp. 200–201; d'Anglure, pp. 32–34; only d'Anglure wrote of the large tooth.

16. Misson, pp. 175–76; also Balfour, p. 220.

17. Misson, p. 182.

18. See, for example, Evelyn, p. 202, for a somewhat vague observation on the Council chamber: "On the roof are the famous Acts of the Republic, painted by several excellent masters."

19. Keysler, pp. 167–69, 186–204; also Lalande, pp. 130–34 (where Venetian art is treated painter by painter, instead of in specific examples); Cochin, pp. 152–60.

20. Or just leave it out of their narrative. Nugent, Stevens, Ayscough, von Archenholtz, and Ann Miller all fall into this category. On the continual need of Grand Tourists for new discoveries, see Chard, pp. 1–5, 18–22.

21. St. Didier, p. 55; de Blainville, p. 288.

22. Misson, p. 205.

23. Von Archenholtz, p. 37; also Misson, p. 199; St. Didier, p. 96.

24. Quote from Keysler, p. 157; see also von Archenholtz, pp. 36–37.

25. De Blainville, p. 291; Keysler, p. 156.

26. Pollnitz, p. 411; du Val, pp. 73–74.

27. Du Val, p. 74; Reresby, pp. 58–59.

28. Moryson (1903), p. 457; Reresby, p. 58.

29. Quote from Skippon, p. 520; see also De Blainville, p. 287.

30. See de Lalande, p. 104; Ayscough, pp. 201–2.

31. Du Bocage, p. 140; also de Lalande, p. 104. On gondolas, see Keysler, p. 153.

32. Von Archenholtz, pp. 35–36; Ayscough, p. 204.

33. Quote from Misson, p. 203; see also Reresby, p. 59; Broderick, p. 333.

34. De Blainville, p. 279; Reresby, p. 59; *The Curious Traveller,* p. 275.

35. De Blainville, pp. 370, 373.

36. Ayscough, p. 201; see also Drummond, p. 65.

37. Misson, p. 199; see also de Blainville, p. 288; von Archenholtz, pp. 29–30, 37.

38. Gailhard, p. 146.

39. Skippon, p. 520; Evelyn, p. 228.

40. Evelyn, p. 229; also Gailhard, pp. 146–47; St. Didier, p. 62.

41. "Four or Five hundred Pistoles with the Charges of the Journey," according to St. Didier, p. 62. According to von Archenholtz, p. 31, this meant that impresarios could never afford to engage more than one singer for any given role, and "of course the least indisposition of one of the principals, or any other little accident may ruin the whole body of those heroes, and frustrate all their hopes."

42. "Bless you and bless your father who made you!" St. Didier, p. 63.

43. De Blainville, pp. 374–75, defended the comparatively poor showing of Venetian opera around 1700: "I grant . . . that their Machines are pitiful, especially if we compare them with those of the Opera of *Paris.* But this is not surprising, when it is considered, that in *France* it is a great Monarch who defrays the whole Expence, whereas at *Venice* it is only supported by private persons."

44. St. Didier, p. 61. It was a complaint echoed by Misson, p. 199: "The Ornaments and Recreations of these here fall extremely short of the others [especially at Paris], the Habits are poor, there are no Dances, and commonly no Machines, nor any Illuminations, only some Candles here and there, which deserve not to be mentioned." See also von Archenholtz, p. 31, a century later, who asserted that "the serious and comic operas . . . are of no moment to him who has frequented the theatres of London, Paris, Vienna; nay, even those of Naples, Rome, Turin, and Florence."

45. St. Didier, p. 58.

46. Keysler, p. 155; St. Didier, pp. 58–60. Far from being in mask, the noble dealers were often in their official dress, with "a great peruke and [senatorial] robes"; de Lalande, p. 105 and note; Misson, pp. 201–2.

47. Lalande, p. 105; De Blainville, p. 361; also Nugent, p. 90: "Noblemen keep the bank, and fools lose their money."

48. Nugent, p. 90; St. Didier, p. 59, observed that "the Silence here observ'd is much greater than that in the Churches."

49. Skippon, p. 520.

50. Coryat, p. 403.

51. Moryson (1903), p. 411.

52. St. Didier, pp. 47–54.

53. Moryson (1903), p. 412; Miller, p. 211; also Keysler, p. 155.

54. Misson, p. 197.

55. Quote from Moryson (1903), p. 467; see also Coryat, p. 311.

56. Quote from St. Didier, p. 52; Gailhard, p. 126, wrote of prostitutes coming to Venice "from other parts of Italy, and several of Candia and other Greek women"; on branding, see Keysler, p. 155; also Eglin, pp. 87–89.

57. Coryat, p. 406.

58. St. Didier, p. 54, with a leering modesty rather typical of the genre, put it: "I shall rather leave it to be imagin'd than to express [it] here."

59. Keysler, p. 156.

60. De Blainville, p. 289; also Keysler, p. 156.

61. Pollnitz, p. 411.

62. Pine-Coffin, p. 55.

63. St. Didier, p. 95; Coryat, pp. 405–6; see also Nugent, pp. 87–88.

64. Keysler, pp. 154–55.

65. De Caraccioli, pp. 76–77, noted that "a fortnight's stay at *Venice* [is] enough for any one, who has not a passion for either women or gambling," saying of himself that "they wanted to draw our Philosopher into some amorous intrigues . . . but Reason, though a friend of the fair sex, avoids adventures." Also Keysler, p. 155.

66. Lithgow, p. 24; Gailhard, p. 125; St. Didier, pp. 53–54; Misson, pp. 198–99.

67. *The Curious Traveller,* p. 276; also Pollnitz, p. 411; Howell (1645), pp. 64–65, opposes the concept of Venice the maiden (unconquered) city with its many courtesans; yet he then goes on to talk of himself being "a guest to this hospitable maid a good while yet."

68. Evelyn, p. 204; de Blainville, p. 289; see also Montaigne, p. 92, who noted that "the crowd of foreigners seemed to him [one of] the most noticeable features."

69. Pratt, pp. 2–10.

70. St. Didier, pp. 83–84.

71. Du Mont, pp. 401–2, goes on to observe that "though it will not cost you a Sigh to gain your Mistress's Heart, you must pay dear for the Enjoyment of her Person: for you cannot purchase a handsome Maiden-Girl under 150 Crowns. . . . As for the Maiden-head, if the Mother promise to warrant it, you may depend upon her word." See also Redford, pp. 57–58.

72. Du Mont, pp. 403–4.

73. Misson, p. 197; St. Didier, p. 96, notes how "this Festival of *Ascension* seems rather a *Carnival* that draws hither again this time a great number of Strangers, who agreeable pass away these few Days of the finest Season of the year."

74. St. Didier, pp. 55–58.

75. Coryat, pp. 401, 408.

76. Mundy, p. 98; Misson, p. 98.

77. St. Didier, p. 53; Moryson (1903), p. 411; Coryat, p. 403.

78. Misson, p. 199.

79. Von Archenholtz, pp. 27–30.

80. *NV,* 6 October 2000.

Chapter 3. The Heart of the Matter

1. *Knopf City Guides: Venice,* p. 220; *NV,* 21 August 2001.

2. *Fodor's: Exploring Venice,* p. 130.

3. Von Archenholtz, p. 35.

4. Tafur, p. 165.

5. Ibid., p. 164; Casola, p. 128.

6. Coryat, pp. 314–18, 327, 332–33.

7. Cosgrove, p. 150.

8. According to de Caylus, p. 79, the location of the Broglio shifted according to the weather.

9. *NYT,* 6 September 1863; quotes from du Bocage, p. 145; and Miller, p. 189.

10. Du Bocage, p. 149.

11. "They soon became wearied with the sage discourse of their teacher," Ayscough, pp. 205–6, continued, "and left him in order to pay their respects to the comedia and the ridutto."

12. For a 1727 procession that had to be rerouted out of the Piazza altogether because the Piazza was filled with the booths and displays of the Ascension festival, see Davis (2000), pp. 471–74 and n. 67; also du Bocage, p. 149.

13. Tafur, p. 164.

14. Piozzi, pp. 88, 97; de Blainville, p. 286.

15. Shelley, vol. 2, p. 244. Some commercial activity merely found itself a new niche, however: Carter, p. 412, reported in 1826 that "three sides in the basement [of the Campanile] are lined with paltry retail shops, and in the fourth, or front, is a sort of temple [i.e., the Loggetta]."

16. *NYT,* 6 September 1863; Hewins, p. 66; Carter, p. 422.

17. *NYT,* 28 August 1881; *GZV,* 10 November 1881.

18. *NYT,* 11 November 1881; Twain, p. 183; James (1959), pp. 290–91; James (1984), pp. 97–98.

19. *NYT,* 28 August 1881.

20. *GZ,* 17 July 2001.

21. *GZ,* 13 January 2000.

22. Despite claims by the café owners that "a great many [Venetians come here] . . . because they know they will be specially treated"; *NV,* 1 September 2000.

23. James quoted in Garrett, p. 53; Balzac quoted in *Knopf City Guides: Venice,* p. 245.

24. Quote from Frye, p. 340; see also Pemble, esp. pp. 30–49.

25. Del Negro, p. 199.

26. Ibid., p. 202.

27. Shields, p. 65.

28. MacCannell, pp. 42–43.

29. Miller, p. 183; Ayscough, p. 201; Beckford, p. 265 (cited in Pemble, p. 118); Shelley, vol. 2, p. 240.

30. As John Moore, p. 69, put it: "We acquire an early partiality for Rome, by reading the classics, and the history of the ancient republic. . . . [But] Venice claims no importance from ancient history, and boasts no connection with the Roman republic . . . and whatever its annals offer worthy of the attention of mankind is independent of the prejudice we feel in favour of the Roman name."

31. See GZ, 14 June 2000, for such a package given to sixty local managers of Chrysler of Mexico.

32. Coryat, p. 315.

33. GZ, 28 April 2000; NV, 25 April 2000.

34. GZ, 9 May 2000.

35. GZ, 25 April 2000.

36. Di Monte and Scaramuzzi, p. 212, tab. 1.

37. Ibid., pp. 192, 215. A more conservative estimate for 1993, based a low estimate of 6.3 million visitors to the city, indicated that a sixth visited the Ducal Palace and another eighth or so (obviously with considerable overlap) saw some of the other two dozen museums in the city.

38. Scaramuzzi, p. 6.

39. As Coryat, p. 326, put it, "From every side of which square gallery you have the fairest and goodliest prospect that is (I thinke) in all the world. For therehence may you see the whole model and forme of the citie sub uno intuito, a sight that doth in my opinion farre surpasse all the shewes under the cope of heaven."

40. Fodor's: Exploring Venice, p. 130.

41. James (1959), pp. 290–92.

42. On the attempts of vendors to work their way into the Piazza, see GZ, 23 July, 8 August 1998.

43. James (1959), p. 295.

44. Fodor's: Exploring Venice, p. 67, acknowledges the passive quality of the experience, pointing out that "the order in which people are herded around the basilica changes from time to time."

45. GZ, 14 July 2000; NV, 26 August 2000.

46. Moore, pp. 59–60.

47. GZ, 22 October 2000; NV, 13 August 2000.

48. NV, 15 April 2000.

49. Fodor's: Exploring Venice, p. 131; NYT, 8 January 1880, 23 August 1925, 17 May 1964.

50. Quote in NV, 4 June 2003; see also GZ, 20 and 21 May 2000, 10 April 2001; NV, 14 February 2001.

51. GZ, 29 December 2001.

52. NV, 12 August 2001.

Chapter 4. Lost in the Labyrinth

1. Muir and Weissman, p. 93.

2. *Rialto* is said to have derived from *rivo alto,* or "high bank," referring to the slightly raised mudflat where Venice's first church of San Giacometto was supposedly constructed in 521: see Lane (1973), pp. 4–5; also *NYT,* 18 December 1983.

3. Attendances as of 1998: see Isman, pp. 154–68.

4. See *Fodor's: Exploring Venice,* p. 24.

5. *NYT,* 1 June 1952.

6. *NYT,* 30 April 1961.

7. Isman, p. 92, citing Perego and Sbetti.

8. *Fodor's: Exploring Venice,* p. 236.

9. See van der Borg, pp. 160–63.

10. De Grancourt, p. 136.

11. *NYT,* 9 December 1979, 25 October 1981 (source of quote), 31 July 1988, 23 May 1999.

12. These are San Sebastiano, Redentore, San Stae, Frari, San Polo, San Giacomo dall'Orio, Miracoli, Santa Maria Zobenigo, Santo Stefano, Santa Maria Formosa, San Pietro di Castello, Sant'Alvise, and Madonna dell'Orto. See the CHORUS website; also *GZ,* 10 February 1998.

13. *GZ,* 28 June 2001; *NV,* 3 April 2000.

14. *NYT,* 12 June 1988.

15. See *NV,* 22 September 2000.

16. Thomas Cook/Passport, p. 26.

17. Cooper, p. 279; also Young, p. 219: "As to streets, properly so called, there is nothing similar to them in the world; twelve feet is a broad one; I measured the breadth of many that were only four and five."

18. On Venetian bridges, see Davis (1994), esp. pp. 14–19.

19. Bromley, p. 226; de Grancourt, pp. 134, 136; Cooper, p. 281; Frye, p. 340.

20. Goethe, pp. 60, 63.

21. One can be found in *Colors,* pp. 24–25.

22. *Knopf City Guides: Venice,* p. 392.

23. *GZ,* 10 August 1980.

24. Quote from *NV,* 26 April 2000; see also *GZ,* 2 August 1980.

25. *NYT,* 23 March 1968.

26. *NYT,* 30 April 1961.

27. Evelyn, p. 198; Veryard, p. 127, noted that there was "neither Coach, Cart, nor so much as an Horse used, or seen, in the whole City."

28. *NYT,* 6 September 1863.

29. *Laura McKenzie's Travel Tips;* Radisson Seven Seas ad, spring 2001.

30. *GZ,* 8 July, 8 October 2000; *NV,* 6 July 2000.

31. Muschamp and Sottsass, pp. 168–69.

32. James (1959), p. 290.

33. In the mid-1970s UNESCO divided Venice into a Zone A (essentially Scaramuzzi's inner two zones), where historical buildings so dominated that no new

construction would henceforth be permitted, and Zone B (her outer zone), where newer structures predominated and new construction would be allowed: see *NYT,* 20 October 1974.

34. See, for example, under listings for shops in Baedeker (1930), p. 345, also pp. 368, 383.

35. Isman, pp. 113–14; also *NV,* 17 June 2000; *GZ,* 31 July 2000.

36. *GZ,* 22 December 1999, 24 July 2001.

37. *NV,* 20 September 2001, 15 March 2002.

38. *NV,* 27 March 2001.

39. Some claim that accents are noticeably different in, for example, Santa Marta and Secco Marina.

40. *NYT,* 30 April 1961; see also 16 March 1959: "It is calculated that 20,000 people live in rooms that daylight never enters."

41. As of 1991, residents of Sant'Elena, Giudecca, and Sacca Fisola were on the average half as likely to have college degrees as those of the inner *sestieri* of the city; those of Sacca Fisola were less than a third as likely to have graduated from high school as those in San Marco: Pedenzini and Scaramuzzi, p. 135.

42. Many residents of these areas do their shopping on the mainland, taking a *vaporetto* to Piazzale Roma and then a free shuttle bus to one of the great "hypermarkets" in Porto Marghera: *GZ,* 16 July 1998.

43. *GZ,* 4 February 1995.

44. Thomas, 17 January 1992.

45. *GZ,* 26 July 1998, 21 February 2001; *NV,* 29 March 2000.

46. *NV,* 21 June 2001.

47. For example, Plante.

48. Quotes from *NV,* 30 October 2002, 14 and 21 May 2003. In May 2003, *La Nuova* announced that "the last bakery of the area" had just closed down in San Vio, that part of the International Zone near the Salute; joining it in oblivion by the end of 2003 were said to be the last butcher and greengrocer: *NV,* 18 May 2003.

49. *NYT,* 23 February 1969.

Chapter 5. Contested Ground

1. See di Monte and Scaramuzzi, p. 199; Isman, pp. 88–90; UNESCO website: "Threats: Transformation of Venice into a Museum-City."

2. *Fodor's: Exploring Venice,* pp. 8, 10.

3. James (1959), p. 296; Miller, pp. 181–82.

4. *NYT,* 15 June 1930.

5. On the aesthetic debate over putting more benches into the city, see *NV,* 15 June 2001.

6. *GZ,* 4 August 1998.

7. *GZ,* 12 August 1982.

8. *GZ,* 19 July 1998.

9. Isman, pp. 88–92.

10. Van der Borg, pp. 160–63.

11. *GZ*, 9 March 2000.

12. Van der Borg, p. 163.

13. *GZ*, 7 August 1998.

14. *NV*, 25 August 2000.

15. Permitted as of 1998 to operate without formal licensing, bed-and-break-fasts in Venice have gone through a boom period, with more than one thousand beds put on the market just in 1999: see *NV*, 3 February, 19 May 2000.

16. *GZ*, 18 July, 17 and 21 August 1980; more recently, see *GZ*, 9 August 2001.

17. *GZ*, 17 August 1980.

18. *NV*, 22 July 2001.

19. "Certainly, the vendors are a problem that does not involve only Venice, but only in this city has this assumed levels that are absolutely intolerable": *GZ*, 5 April 2001.

20. For a rare (Chinese) female, see *NV*, 17 May 2001.

21. For a rare example of a Senegalese selling with a license, see *NV*, 7 September 2000.

22. *NV*, 15 December 2000.

23. *NV*, 18 and 29 May 2001; *GZ*, 10 August 2000.

24. *GZ*, 12 August 2000. Police have determined that many vendors also work in factories or in the more legitimate areas of the tourist economy (in hotels and restaurants, primarily) in addition to their time as *vu' cumprà;* see *NV*, 31 May 2000, 29 April 2001.

25. For Venetians defending the *vu' cumprà*, see *NV*, 8 March 2000; *GZ*, 21 May 2000.

26. Quote from *NV*, 31 May 2000; see also *NV*, 23 June 2000; *GZ*, 24 July 2000. For the most recent estimate of *vu' cumprà* numbers, see *NV*, 23 October 2002.

27. *NV*, 31 May 2000; *GZ*, 11 February, 29 March, 24 April 2001.

28. *GZ*, 21 and 24 May 2000, 8 May 2001.

29. Quote from *GZ*, 8 August 1998; see also *GZ*, 26 April 2001; *NV*, 20 May 2000.

30. *GZ*, 1 September 2000, 6 May 2001.

31. Quote from *GZ*, 10 May 2001; see also *GZ*, 19 and 20 May 2000, 24 March 2001. Police in Venice give higher priority to pickpockets, and will often abandon a roundup of *vu' cumprà* if they receive a call that a gang of *borseggatori* has been sighted.

32. *GZ*, 7 April 2001; *NV*, 14 June 2001.

33. *GZ*, 10 August 2000; *NV*, 26 July 2000.

34. Quote from *NV*, 24 May 2000; see *GZ*, 2 July 2001, for "the parish priest of the church of the Scalzi, who laments because the doorway [to the church] is blocked by the sheets of the vendors."

35. *LT*, 24 August 1860; *GZV*, 12 June, 4 September 1876.

36. Fabri, p. 93.

37. *GZ*, 11 May 2001.

38. *GZ*, 8 August 1998.

39. Quote from *NV*, 27 June 2001; see also *GZ*, 29 April 2001.

40. *GZ*, 19 May 2000.

41. *NV*, 17 May 2000; *GZ*, 29 July 1998.

42. *NV*, 5 July 2001; *GZ*, 25 May 2000, 6 March 2001.

43. *GZ*, 24 March, 10 May 2001; *NV*, 15 July, 6 and 7 September 2000.

44. *GZ*, 8 August 1982, 22 and 25 July 1998, 11 January 2000, 12 December 2001; *NYT*, 14 June 1989.

45. *NV*, 25 May 2001; *GZ*, 27 April, 16 May 2001.

46. *NV*, 17 December 1999, for a shopkeeper who stressed that he had been evicted "not because of market constraints, but because . . . 'with tourism one can earn more'" selling tourist goods.

47. *NYT*, 31 October 1971.

48. Though such shops are sometimes called *negozi di specialità veneziane*, giving a sense that they offer local handicrafts, few locals indeed follow these traditional crafts. Most of these "stores of Venetian specialties" get their wares from elsewhere in Italy or east Asia: *GZ*, 14 August 1983; *Colors*, pp. 40–41, 45, 50–51, 82–83.

49. Just between 1995 and 1998, Venetian ice-cream shops blossomed by over 30 percent, from 42 to 56, or roughly 10 times what the industry itself considers an "optimal" residential ratio: *GZ*, 25 July 1998. The Veneto region also boasts more than twice as many pizzerias as the Campania (where Naples is located), with 229 in Venice and Mestre alone: *GZ*, 19 April 2000.

50. See *NYT*, 22 October 1935; also Lane (1973), pp. 12–13; Muir (1981), pp. 3–6.

51. Miller, p. 190.

52. "*Campi* and *calli* that make the most delicate and most sought-after city in the world resemble a gigantic outdoor restaurant": *NV*, 30 May 2001. In Venice dining hours are much longer than in most Italian cities, to accommodate foreigners' various preferred eating times: lunch runs from 11:30 A.M. until after 2 P.M., dinner from 5 until 10 P.M.; many restaurants simply stay open continuously through the afternoon.

53. There is no easy English definition for this word, which refers both to the "tax" paid for the space and to the space itself.

54. *GZ*, 27 June 2001.

55. *GZ*, 25 July 1998; on Campo Santa Margherita in particular, see also Kahn, pp. 119–25.

56. *GZ*, 19 November 1999, 23 May 2000; *NV*, 10 June 2003.

57. *GZ*, 9 May 2000.

58. Quote from *NV*, 29 June 2001; see also *NV*, 30 May 2001.

59. *GZ*, 28 April 2000.

60. *NV*, 2 and 31 March, 3 September, 13 August 2001.

61. Upon inspection by the police, several of the bars were found to have removed the required volume limitation devices from their amplifiers: *GZ*, 30 May 2001.

62. *GZ,* 12 and 19 May 2001; *NV,* 13 May 2001.

63. See *GZ,* 12 August 1982.

64. Several Venetians have told us that virtually nothing is still in use of the huge lexicon of Venetian words that 150 years ago enabled Giuseppe Boerio to make his *Dizionario del dialetto veneziano* a volume of over eight hundred pages.

65. *GZ,* 20 January 2000, 21 May 2001.

66. *NV,* 3 and 10 October 2000.

67. *NV,* 4 April 2000.

68. See *GZ,* 4 August 1982.

69. Recently, during such special events as Carnival and the Redentore, dozens of portable toilets have been brought into the San Marco area. The rest of the tourist core of Venice is not much improved from the 1980s, however, and late on Carnival and summer nights many visitors continue "to piss wherever they happen [to be]": *GZ,* 6 March 2000.

Chapter 6. The Floating Signifier

1. Depending on who is counting: *Fodor's: Exploring Venice,* p. 18, says "118 islets and 170 canals"; but the *NYT,* 17 May 1964, came up with 177 canals.

2. To be precise, there were 6,169 registered mooring spots along Venice's canals in 1999, plus an additional 5,761 boat slips and dry-docks around the city's periphery; thanks to Isabella Scaramuzzi of COSES for this information.

3. Taking one's private craft on the Grand Canal is supposedly forbidden by local law, though clearly some boaters get away with it: *NV,* 27 August 2000.

4. Moryson (1903), p. 77; Reresby, p. 60.

5. Some visitors' estimates and the dates they were made (which may precede publication dates) are as follows:

Barbartre (1480): 15,000–20,000

von Harff (1497): 50,000

Wm. Thomas (1549): 12,000

Sansovino (1576): 9,000–10,000

Moryson (1595): 800–1,000

Harleian (1605): 8,000

Coryat (1608): 10,000

Lassels (1630s): 20,000

Raymond (1647): 40,000

Reresby (1656): 80,000

Mortoft (1659): 8,000–9,000

Skippon (1663): 9,000–10,000

Balfour (1668): 14,000–15,000

Gailhard (1668): 15,000

Veryard (1682): 12,000–14,000

Bromley (1691): 20,000

du Mont (1691): 25,000

de Fer (1694): 4,000–5,000 (at the Festa della Sensa)

Chiswell (1696): several thousand for hire

de Blainville (1707): 10,000–20,000

Chancel (1714): 25,000

Ray (1738): 8,000–15,000

Nugent (1755): 14,000

Stevens (1756): 15,000

Northall (1766): 5,000

6. Romano, pp. 360–61; Coryat, p. 314; Balfour, p. 217; du Bocage, p. 136; Raymond, pp. 198–99. Du Mont, p. 395, claimed that "every *Gondola* requires four Men," but he may have been talking of a backup crew.

7. Donatelli, pp. 34–37.

8. Raymond, p. 198; Coryat, p. 314.

9. Gailhard, p. 10. In Marino Sanudo's time there were thirty-seven *traghetti* in Venice, grouped in three different types: *da viaggio,* for entering and leaving the city; *da bagatin,* for crossing the Grand Canal; and *per guardagnar,* for hire — usually by the day. Each *traghetto* operated as a sort of independent corporation, electing new members, policing infractions, and fighting turf wars with its neighbors: Romano, p. 368.

10. Quote from de Blainville, pp. 269–70; see also Lassels, p. 226.

11. De Blainville, p. 270.

12. See James (1959), p. 289; Pemble, pp. 23–24.

13. Twain, pp. 170–71; James (1959), p. 288. Cooper, p. 285, noted, "The livery of a private gondolier used to be a flowered jacket and cap; and a few such are still to be seen on the canals."

14. James (1959), pp. 299–300. For James, the gondoliers were "the children of Venice, . . . associated with its idiosyncrasy, with its essence, with its silence, with its melancholy."

15. Although the Austrians had built the railroad bridge connecting Venice with the mainland in 1846, the line to Milan was not finished until 1857, and the final link to the north — the Mount Cenis tunnel — was completed only in 1871: see Pemble, p. 15; quote from *NYT,* 6 September 1863.

16. Baedeker (1877), p. 207, emphasis in original; James (1959), pp. 292, 297.

17. The daily rate (for ten hours) was 6 lire, or $1.20: Baedeker (1903), p. 261.

18. By Baedeker (1895), p. 235, one gondolier was considered sufficient; cf. Baedeker (1877), pp. 205–6.

19. *NYT,* 28 August 1881. Venice experienced its first Cook's tour in 1867; by the 1880s, Cook's was running as many sixty visitors at a time to the city: Pemble, pp. 15, 175.

20. *NYT,* 11 May 1885.

21. Baedeker (1877), p. 205.

22. First quote from the *London Times,* quoted in *NYT,* 23 October 1885; second and third quotes from *London Standard,* quoted in *NYT,* 8 November 1881 and 23 October 1885.

23. *London Daily News* quoted in *NYT,* 5 December 1881; also *NYT,* 24 October 1896.

24. *NYT,* 5 December 1881, 1 August 1909.

25. By 1930, when even the shortest gondola ride cost eight lire (forty-two cents), a *vaporetto* trip along the entire length of the Grand Canal cost the equivalent of less than a nickel: Baedeker (1930), pp. 343–44.

26. For the story of the strike of 2 to 4 November 1881, the demands of the gondoliers, and the results, see *GZV,* 2–4, 10, and 16 November 1881; and *L'illustrazione italiana,* p. 335.

27. Quote from *NYT,* 21 July 1909; see also 1 August 1909.

28. Provoking a good deal of nervous laughter among gondoliers: *GZ,* 7 January 2000.

29. Quote from *NYT,* 10 May 1925; see also *NYT,* 7 September 1924, and 31 May 1925.

30. Baedeker (1930), p. 343, gives the tariff for one to three passengers, "from any point in the town, including the railway station, to any other point of the town, including the islands of St. Michael, St. George, or St. Helena," as eight lire, or forty-two cents, the same as the rate for one to three people for half an hour.

31. St. Didier, pp. 55–60.

32. The *ferro* was actually a counterweight to the gondolier and a warning indicator for low-lying bridges, but, engraved with the arms of the owner's family, it also served for some aggressive patrician display: see St. Didier, pp. 55–56 (source of quote in this paragraph); Northall, p. 428.

33. Moryson (1971), pp. 77–78; du Mont, p. 395; Romano, p. 360; Northall, p. 428; *NYT,* 6 June 1982.

34. Donatelli, pp. 98–102. Until this time, it appears that as long as gondolas were outfitted with a *felze,* they maintained the same configuration of seats that was described by St. Didier, pp. 57–58, two hundred years earlier: "At the lower end of [the *felze*], they place cross-ways a Board covered with black leather that serves for a seat; as likewise another Bench handsomely covered against the Demi-circle behind, which serves as a Back-board to a couple of Persons that may conveniently sit here, as the most Honourable Place: They have likewise a Bench on each side of the first, that will hold Four People."

35. In all editions of Baedeker's *Northern Italy* between 1877 and 1930 readers were given the Venetian phrase for "remove the *felze*" *(cavar il felze),* presumably because many sightseers found the device an annoyance; after 1930, it would seem, the gondoliers got rid of it altogether, for it was mentioned no longer. *NYT,* 31 May 1925, 21 July 1930, 6 June 1982.

36. St. Didier, p. 55.

37. Cosulich, pp. 113–15, 136, 138, 194.

38. *NYT,* 6 August 1966. Presently, the uniform is fixed in law by communal decree 238, of 22–23 December 1994.

39. *NV,* 12 September 2000.

40. *GZ,* 10 June, 29 August 2000; *NV,* 26 July, 2 and 3 September 2000.

41. "Una certa eleganza di voga; dona natura come bel remiere": quotes from the BBC show *The Last King of the Gondoliers.*

42. *Fodor's: Exploring Venice,* pp. 116–17.

43. Quote from de Blainville, p. 269; Cooper, p. 285, noted, "It requires practice to keep the oar in its place, as I know by experience, having tried to row myself, with very little success"; see also Broderick, p. 313.

44. Assuming the dollar of 1999 was worth about twenty cents in 1950 terms, and that the ratio of lire to dollar has varied between 500:1 and 2,000:1. See *NYT,* 1 June 1952, 22 July 1962, 17 May 1965, 3 December 1979, 12 June 1988; *Knopf City Guides: Venice,* p. 390; and *Last King of the Gondoliers.*

45. *NYT,* 31 October 1999.

46. Cosulich, p. 136.

47. James (1959), p. 299.

48. Thus, Gianpaolo D'Este, licensed at the Molo *traghetto* and a champion oarsman who often competes in the Venetian regattas, notes, "My gondolier friends let me take time off [to practice] even during the high seasons, because they want our group to win": *Last King of the Gondoliers.*

49. Though gondoliers are free to swap turns, to gain more consecutive days off. The typical schedule is four days on and one off in the summer, two and two in the winter. One of the most common forms of punishment a co-op inflicts on errant members is to reduce the hours they are allowed to work (known as *levar la volta*), thus cutting their share of the monthly take.

50. Admittedly, we have never seen more than about fifteen gondolas *in carovana,* but we were told that groups of 150 or more passengers do appear sometimes, forcing the gondoliers to call in their reserves and substitutes, or even those from neighboring *stazi.*

51. *NYT,* 22 May 1968.

52. For the territories of the ten existing *stazi* (located at and known as Danieli, Molo, Dogana, Trinità, Santa Maria del Giglio, San Tomà, San Beneto, Carbon, Santa Sofia, and Ferrovia), as described by the *mariegole* of 21 April 1995, see Zanelli, pp. 90–93.

53. The president of the Ente Gondola is not a gondolier himself, but generally a Venetian of some social standing, appointed by the mayor of Venice to act as a liaison between the city administration, the gondoliers, and the public: see, for example, *GZ,* 10 June 2001. On the attempt by angry gondoliers to have the president fired and replaced, see *NV,* 15 May 2003.

54. The *ganzer* is traditionally an older gondolier who stays ashore, docks the gondolas, and helps passengers in and out; currently, he also handles the bookings, rather like a restaurant maitre d'. See also *NYT,* 7 September 1924.

55. MacCannell, p. 42.

56. Ibid., p. 23.

57. Ibid., pp. 109–33.

58. Keeping in mind that, since most of the singers are non-Venetian, they themselves usually do not know the old dialect tunes.

59. *NYT,* 21 February 1896.

60. Quote from *NYT,* 1 August 1909; Donatelli, p. 41.

61. A sight that has stopped us in our tracks several times. Since a wedding gondola with two gondoliers costs over a thousand dollars for an afternoon's hire, few Venetians engage one anymore for the traditional ride from the bride's house to the parish church: if one sees an obviously Venetian bride and groom going about in this way, we were once told, very likely he is a gondolier who received the excursion as a gift from others in his co-op.

Chapter 7. Behind the Stage

1. On the building of Venice on these mudflats, see Crouzet-Pavan, pp. 57–216.

2. Coryat, p. 396; also Davis (1991), pp. 103–4.

3. Casola, pp. 129–32; Coryat, pp. 395–96.

4. *NYT,* 2 July 1880.

5. Du Bocage, pp. 148–49; Moryson (1903), p. 467; *GZ,* 25 February 2001.

6. Sperling, esp. 18–71 and *passim.*

7. Du Val, pp. 28, 30, 42, 46, 51, 54, 65.

8. St. Didier, p. 8.

9. Evelyn, p. 234; Skippon, p. 518; also Balfour, pp. 223–25.

10. On the origins of glass making in Venice and the industry's move to the islands of Murano to reduce the risk of fire in Venice proper, see Lane (1973), pp. 157–60.

11. Coryat, pp. 327, 387; "A True Description," p. 80; Moryson, p. 89; Raymond, p. 199.

12. De Fer, p. 62.

13. For eighteenth-century perceptions that Muranese glass had been overtaken by foreign competitors, see de Caylus, p. 125; Miller, pp. 206–7. John Moore, pp. 28–30, asserted in the 1780s that "the great manufactures of looking-glasses, are the only inducements which strangers have to visit this place."

14. Only foreigners traveling about in the gondola of their own ambassador would be waved on: *Venice under the Yoke,* pp. 221–24 (referring to the 1780s); Chancel, pp. 100–1; Keysler, p. 153.

15. First two quotes from Shelley, vol. 1, p. 328; see also Cooper, pp. 289, 291.

16. There now remain only two island monasteries: San Francesco del Deserto, of the Franciscans, and San Lazzaro degl'Armeni, home of Mechitarist Armenian priests: *GZ,* 11 September 2000.

17. Shelley, vol. 1, p. 329; *LT,* 14 January 1840. The process has continued: it has been estimated that at the fall of the Republic, the Lagoon's *barene,* or mudflats, amounted to around 149 square kilometers of the Lagoon's total surface area of

580. In the year 2000, they had been reduced to just 48 square kilometers: *GZ,* 7 August, 30 October 2000.

18. "Of the throngs who visit Florence, Rome, and Naples, even, there are many who never stray off toward the 'city in the sea,' which [has] . . . certainly not become a necessary stopping place in a brief European tour"; *NYT,* 28 August 1881.

19. *NYT,* 2 July 1880; James (1959), p. 312, also spoke of Burano as "celebrated for the beauty of its women and the rapacity of its children . . . [who] assail you for coppers, and in their desire to be satisfied pursue your gondola into the sea."

20. See *NYT,* 2 August 1863, for a description of Lido as "a long, narrow island between the lagoons and the sea, with a village at either end and bath-houses on the beach, which is everywhere faced with forts."

21. For pleasures, Misson, p. 183, had noted in the 1680s: "There you find Shell-fish, and the Walk is very diverting when the Weather is Calm": also *NYT,* 2 August, 6 September 1863; Carter, p. 418.

22. *NYT,* 24 October 1897.

23. James (1959), p. 312.

24. Vanzan Marchini, pp. 76–82; *GZ,* 29 July 2000.

25. A number of these early organizations were foreign-based: see Vanzan Marchini, pp. 94–96; also *GZ,* 5 July 2000.

26. See *NYT,* 2 August 1863, 26 July 1880. *NYT,* 24 October 1897, sought to dispel the "theory that Venice is very hot and unhealthy in July and August [which] keeps away the all-pervading tourist during these months."

27. *NYT,* 21 November 1909.

28. Vanzan Marchini, pp. 104–8.

29. *NYT,* 21 November 1909, 30 August 1925.

30. *NYT,* 17 September 1922.

31. *NYT,* 3 April 1960, 29 April 1962.

32. *NV,* 6 August, 24 November 2000.

33. Di Monte and Scaramuzzi, pp. 145, 169; also *GZ,* 25 April, 12 July 2000.

34. Marco Polo Airport has been repeatedly upgraded, and as of 2001 was ranked as Italy's third largest in terms of air traffic, mostly of tourists. See *NYT,* 3 April 1960.

35. Reports of ticket sales from the Cavallino landing at Punta Sabbione to San Marco indicated that, already by 1980, there were around eight thousand day-trippers per day to the city in the main summer months, with holiday peaks of over ten thousand; twenty years later, these numbers might double: *GZ,* 12 and 17 August 1980, 15 August 1982; *NV,* 10 August, 1 September 2000.

36. See *GZ,* 18 July 1980, which noted that unseasonably cool July weather had emptied the beaches, but "in recompense, [there is] an enormous flow of excursionists into the city"; also *GZ,* 10 August, 1 September 2000.

37. See *GZV,* 19 September 1881, when, for the Regatta that year, "it is calculated . . . that 40,000 foreigners have come to Venice during these four or five days, not counting those who leave in the evening of the same day they arrived [i.e., do not spend the night]."

38. *GZ,* 18 July 1980.

39. Cited in a letter he wrote to the *GZ*, 24 July 1998, and cited again in *GZ*, 19 August 2001.

40. Quote from *GZ*, 1 August 1998; see also *Colors*, p. 49.

41. *NYT*, 12 June 1986.

42. See *GZ*, 18 August 1980, 1 August 1998.

43. *NV*, 3 November 2000, 1 June 2003.

44. *NV*, 27 August 2000.

45. *NV*, 15 February 2001.

46. Pemble, p. 129; Baedeker (1895), pp. 289–90.

47. See Cosulich, pp. 196–99. According to Moore, p. 30, in the eighteenth century the Muranese were adept in turning out "an infinite quantity of glass trinkets (margaritini, as they are called) of all shapes and colours."

48. *NV*, 30 December 1999, 11 October 2000.

49. *NV*, 12 October 2000.

50. *Knopf City Guides: Venice*, p. 76.

51. *NV*, 11 and 12 October 2000.

52. According to some, the cartel has also protected its interests through threats and violence: *NV*, 12 and 13 October 2000; *GZ*, 20 February 2001.

53. Pedenzini and Scaramuzzi, pp. 124–36, tabs. 27, 31–33.

54. *GZ*, 5 December 1998, 17 December 1999; *NV*, 20 June 2000.

55. *GZ*, 26 July, 1 September, 28 October 2000; *NV*, 26 November 2000.

56. *GZ*, 15 January, 5 April 2000, 14 January, 3 March 2001.

57. *GZ*, 5 December 1999, 26 February, 10 May 2000.

58. Quotes in *NYT*, 31 July 1988; see also *GZ*, 11 May, 1 September, 28 October 2000; *NV*, 3 February 2001.

59. *NYT*, 18 December 1983.

60. Indeed, San Secondo is known locally as Rat Island: *NV*, 30 December 1999.

61. The city council's solution to dealing with this unwanted gift was to pretend the offer had never been made: *GZ*, 14 and 21 August 2000.

62. *NV*, 28 June 2000.

63. Italia Nostra, 4–15.

64. *NV*, 18 February, 29 May 2000, 14 January, 7 February 2001. On the Lazzaretto Vecchio, however, plans are underway to construct an archaeological museum by 2007, an attraction that will certainly merit its own *vaporetto* stop: *NV*, 8 May 2003.

65. *GZ*, 7 February 2000; *NV*, 7 February 2000.

66. Venice, unlike Florence or Rome, has never hosted more than one or two foreign universities, though not for want of trying: *NYT*, 25 June 1978; *GZ*, 28 September 2000.

67. Craven amounts to barely half a hectare, boasting a "little nineteenth-century fort in decent condition"; altogether it will cost another million to stabilize and restore: *NV*, 28 June 2000, 7 February, 22 March 2001.

68. Currently the total hotel capacity, of all categories, in the historic center is around fifteen thousand beds. It will take around a hundred million dollars to

convert the old hospital facilities to luxury hotels: *GZ*, 7 April, 15 July 2000; *NV*, 15 July 2000. For the results, see *NYT*, 14 December 2003.

69. Quote from *GZ*, 30 December 1999; see also *NV*, 30 December 1999.

70. Some of which has been carried out with convict labor: *NV*, 21 September 2000; *GZ*, 25 September 2000.

71. *GZ*, 1 October, 7 November 2000; in the early 1980s, Torcello was said to have "about 100" residents: *NYT*, 18 December 1983.

72. *GZ*, 24 May 2000; *NV*, 30 June 2000; Isman, p. 104; Di Monte and Scaramuzzi, p. 212.

73. *GZ*, 24 May 2000; *NV*, 30 May 2000.

74. Quotes from *GZ*, 9 June 2000; *NV*, 15 August 2000.

75. *NV*, 20 September 2000; *GZ*, 26 October 2000.

76. *NV*, 30 May, 1 and 21 June, 15 and 22 August, 24 and 28 October 2000.

77. Federica Millich and Isabella Scaramuzzi, COSES/1999 Doc.199; *NV*, 22 June 2001, 7 June 2003.

78. *NV*, 28 February 2001.

79. *NYT*, 7 June 1912.

80. *GZ*, 8 January, 7 April, 10 August 2000; *NV*, 26 April, 26 June 2000.

81. *NV*, 21 June 2000.

Chapter 8. Dangerous Waters

1. *LA Weekly*, 19–25 November 1999.

2. Gailhard, p. 12.

3. Quotes from Nugent, p. 47; de Caylus, p. 123; de Blainville, p. 274; see also *The Polite Traveller*, p. 44; Moore, p. 40; Keysler, p. 153; Shelley, vol. 1, p. 330.

4. De Fer, p. 48.

5. Coryat, p. 313.

6. Von Archenholtz, pp. 83–84; also All Aboard website.

7. Tafur, p. 167; Warcupp, p. 273.

8. It would be linked up to the network that currently serves Lido (soon to be connected with Pellestrina) and runs around thirty-five meters below sea level: *NV*, 23 June 2000; *GZ*, 29 March 2001.

9. Figures for 1998: Isman, pp. 84–88.

10. *NV*, 11 February 2001.

11. *GZ*, 1 August 1980.

12. "Venice and Tourism," case 79 in Trade and Environment Database website.

13. *NV*, 24 March 2001; *GZ*, 24 March 2001; also "Safeguarding Venice."

14. Gailhard, p. 12.

15. From the minutes of the Venetian city council, quoted in Vanzan Marchini, pp. 68–70. There were also to be a theater, ballroom, restaurants, gardens, and upward of five hundred rooms "for every class of person."

16. Vanzan Marchini, p. 71. The Venetian city council, on the other hand, en-

thusiastically approved the project; only the Austrians apparently expressed any doubts about the impact of such a monstrous structure on Venice's scenic ambience.

17. Baedeker (1877), p. 207, assured his readers that "baths of every description . . . are situated between the Riva degli Schiavoni and the Isola S. Giorgio"; by the 1895 edition, p. 236, however, readers were advised that "the excellent Lido Sea-Baths are much pleasanter"; thereafter the floating baths were never mentioned. See also Vanzan Marchini, pp. 53–56, 72.

18. Cited in Vanzan Marchini, p. 53.

19. The author offered this account to readers of the *New York Times* as "a very novel and amusing episode of Venetian life": *NYT*, 1 November 1880.

20. Quote from ibid. On four drunken Russians going for a swim in the internal canal of San Giuseppe, see *GZ*, 9 August, 3 October 2000; also *NV*, 17 July, 23 August 2001.

21. Baedeker (1870), p. 187, offered only the caution that "the proper time for bathing is when the tide commences to rise; at low tide the water is shallow and muddy."

22. *NV*, 4 April, 29 December 2000, 3 February 2001; *GZ*, 12 September 2000.

23. The mosquito plague of the late 1980s is graphically featured in *Will Venice Survive Its Rescue?* See also "Safeguarding Venice," pp. 26–28; *NV*, 12 September 2000; *GZ*, 17 May 2000.

24. See Zompini, drawing 19.

25. So said von Archenholtz, p. 84.

26. Shelley first visited the city in 1816 after one such lapse and complained that "the canals are becoming choked up, and the stench from them at low water is often dreadful"; she praised Venice's Austrian occupiers fifteen years later for having imposed "very severe penalties [against] throwing refuse into the canals": vol. 1, p. 330, and vol. 2, p. 244.

27. *NV*, 29 March 2001.

28. *NV*, 25 July, 7 October 2000; *GZ*, 12 September 2000.

29. *GZ*, 27 November 2000, 19 November 2002.

30. *GZ*, 11 December 1999, 17 May, 21 November 2000; *NV*, 13 March 2001.

31. *GZ*, 11 July, 7 November 2000; *NV*, 20 September 2000.

32. *NV*, 25 January 2001.

33. *NV*, 22 November 2000.

34. *GZ*, 13 January 2000.

35. Helped along by the siren calls of various travel writers: *NYT*, 9 December 1979, 25 October 1981, 2 October 1988.

36. *GZ*, 6 September 1998, 7 April, 11 July 2000; *NV*, 1 October 2000.

37. *NV*, 19 January 2003.

38. *GZ*, 13 October 2000.

39. *GZ*, 7 November 2000.

40. See *Will Venice Survive Its Rescue?*; also *NYT*, 31 January 1989.

41. *NV*, 17 June 2001, 15 May 2003.

42. In Antonio Canaletto's *View of Piazza San Marco*, at the National Gallery

in Washington, D.C., the panorama shows that in the 1730s one entered the Basilica from the Piazza by ascending a one-meter bank of steps; nowadays, the steps into the narthex go *down* about a meter.

43. *GZ,* 17 November 1999.

44. *GZ,* 29 March 2001.

45. *GZ,* 12 January 2001; *NV,* 25 January 2001, 25 April 2003.

46. *NV,* 30 May 2001.

47. *Lancioni,* workboats, and taxis currently account for around 55 percent of all boat traffic in the city, the rest divided among the *vaporetti* (13 percent), public service vessels (10 percent), and private pleasure boats (also 10 percent); a mere 4 percent of the boats one sees in Venice these days are still rowed in the traditional style. *NV,* 31 May, 20 September 2000; *GZ,* 27 August 2000.

48. Only in the late evening does the Grand Canal regain anything of its former tranquillity and romance. See *NV,* 18 June 2000 (source of quote), 30 May 2001.

49. *NV,* 24 July 2001; *GZ,* 15 June 2001.

50. Quote from *GZ,* 4 August 2000; see also *GZ,* 21 February 2001.

51. *NV,* 26 April, 7 May 2003.

52. As counted by COSES on 21 April 2000. This represented an increase of 20 percent over a similar survey conducted on 4 September 1998: see *NV,* 31 May, 20 September 2000.

53. *L'illustrazione italiana,* p. 335.

54. *NYT,* 23 October 1885, 21 February 1896.

55. *NV,* 11 March 2001.

56. Quote from *NV,* 22 March 2001; see also *NV,* 21 March 2001; *GZ,* 21 March 2001.

57. *NV,* 6 April 2001; *GZ,* 6 April 2001.

58. *GZ,* 28 April 2000.

59. Quote from *NV,* 1 October 2000; see also *NV,* 20 September 2000.

60. *NV,* 18 June 2000.

61. *GZ,* 26 April 2000.

62. Quote from *GZ,* 20 September 2000; see also *NV,* 1 October 2000.

63. *GZ,* 6 August 2001.

64. Princess Cruise Lines' *Grand Princess,* currently the largest of these ships, weighs in at 109,000 metric tons, with a draft of 26 feet and a motor-propeller system capable of 24 knots; parts of the channel the *Princess* has to traverse to reach its berth at the port are barely 40 feet deep: see *Grand Princess* website.

65. *NV,* 11 January 2002.

66. *NV,* 23 August 2001; *GZ,* 23 August 2001.

67. *GZ,* 15 June 2000.

68. *NV,* 1 October 2000.

69. Quote from *GZ,* 16 December 1999; see also *GZ,* 8 August 2000.

70. *GZ,* 11 May 2000; *NV,* 11 May, 9 July, 17 December 2000. The steady growth of cruise passengers through Venice received something of a setback after 11 September 2001, however.

71. *GZ,* 13 October 2000.

72. Quote from *NV,* 2 December 2000; see also *NV,* 18 March 2001.

73. The *Grand Princess* was already surpassed, in November 1999, by Royal Caribbean's *Voyager of the Seas,* at 142,000 metric tons, 1,020-foot length, and 3,114-passenger capacity; the *Voyager* was joined by its 138,000-ton sister ship, *Adventure of the Seas,* in November 2001: see the Royal Caribbean website. For the moment, fortunately for the Venetians, neither of these behemoths sail outside the Caribbean.

74. Frater; see also *GZ,* 19 June 2001.

75. The Campanile of San Marco is 315 feet (95 meters) high.

76. All quotes are from the Radisson Seven Seas Cruises ad run in spring 2001 in various American magazines; see Radisson website.

77. Quoted in *NV,* 23 May 2003. According to *GZ,* 30 November 2002, the noise from these ships can exceed the legally permitted nighttime limits by six or seven decibels, a level "that would be acceptable in an industrial zone."

Chapter 9. Restoration Comedies

1. *NYT,* 27 April 1961, May 1965, 3 May 1970, 21 December 1975, 22 June 1980; *GZ,* 7 February 2000, 5 April 2001; Pemble, pp. 146–47.

2. Shelley, vol. 1, pp. 329–30; Carter, pp. 405–7; Frye, p. 341; also Pemble, pp. 124–30.

3. Quoted in Pemble, pp. 131–34.

4. Ibid., pp. 128–29, 145–47.

5. For Ruskin's position on the Basilica, see *NYT,* 9 February 1880; Pemble, pp. 145–49, quote on p. 149.

6. Quote from *NYT,* 18 June 1880; on Saccardo, see Pemble, pp. 147–53.

7. Horatio Brown quoted in Pemble, pp. 154–55.

8. The adjoining Procuratie Vecchie were also reworked in the eighteenth century, altering the structures' original proportions designed by Jacopo Sansovino: see *GZ,* 26 August 2001.

9. Thus, restoration (funded by Piaget) of the mechanism in the Torre dell' Orologio has raised a great outcry—especially in Britain, spearheaded by the *Horological Journal*—for having replaced some of the nineteenth-century parts and shortening the pendulum: see *NV,* 18 October 2000.

10. *Athenaeum,* 2 July 1881: "It is astounding that a city the municipality of which acknowledges that it depends entirely on tourists should . . . take every occasion to destroy the ancient monuments which are its attraction." Quoted in Pemble, p. 145; also Links, p. 230.

11. *GZV,* 19 July 1896; James (1959), p. 295.

12. Von Archenholtz, pp. 27–30.

13. From the *London Standard,* quoted in *NYT,* 23 October 1881.

14. *LT,* 18 November 1879. The dean of Christ Church Oxford asserted a particularly British claim to Venetian art, declaring that "Venice had been made sacred by Byron's poetry"; quoted in Pemble, p. 151.

15. James (1959), p. 294.

16. Edward Hutton, quoted in Pemble, pp. 151, 156–57.

17. *NV,* 22 March 2000. The present-day electric elevator, which can carry up to twenty-five people at two meters a second, was not installed until 1950.

18. *NYT,* 27 March, 9 October 1903; Pemble, p. 157.

19. For example, *Knopf City Guides: Venice,* p. 248; *Fodor's: Exploring Venice,* p. 84.

20. The burst of such articles in just the *New York Times* alone is impressive: 26 July, 4 and 7 August, 7 October 1902, 10 January 1903, 13 December 1904; see also *LT,* 10, 11, and 13 January 1903.

21. One Italian American called such a charitable subscription "a gratuitous insult." The Italian state later agreed to accept donations, however. New York's subscription list could boast such artistic luminaries as Louis Comfort Tiffany and the sculptor Augustus Saint-Gaudens: *NYT,* 19 July, 4 October 1902; Pemble, pp. 156–57.

22. Pemble, pp. 179–80, 187–91.

23. The Special Law initially made $500 million available for the city: *NYT,* 21 December 1975, 17 March 1976; *GZ,* 14 February, 6 and 14 April, 18 July 2000.

24. As of 2000, eleven of these private committees were Italian. The rest were from Australia, Austria, Denmark, France, Germany, Great Britain, the Netherlands, Sweden, Switzerland, and the United States, along with five organizations that are international in nature; several of the foreign committees are based in Venice: UNESCO website: "International Solidarity: Mobilizing Private Committees."

25. Venice in Peril's thirty-five patrons include no fewer than twenty-two titles, including twelve peers; many of the rest are British academics.

26. Save Venice Inc. grew out of the Venice Committee of the International Fund for Monuments, which was begun soon after the great *acqua alta* of 1966. Besides Save Venice, the American committees include the America-Italy Society of Philadelphia and Venice Heritage, Inc.

27. From the Save Venice Inc. *1999 Regatta Week Gala* souvenir brochure, p. 9.

28. *NYT,* 19 May 1974.

29. *NYT,* 11 October 1998.

30. The Treasure Hunt has since been expanded to two days. On Save Venice programs more generally, see the Save Venice website.

31. Quotes in *NYT,* 11 October 1998.

32. To insure that enough attractive small-ticket items are available, according to Conn, the organization sometimes asks the Superintendencies to add more of these to their rolls. For the current wish list, see the Save Venice website.

33. Most notably in the recent restoration work at the Badoer-Giustinian Chapel in San Francesco della Vigna, dedicated to the memory of Save Venice's cofounders John and Betty McAndrew.

34. See also the booklet *Save Venice, 1968–1998.*

35. *NYT,* 19 May 1974.

36. From the Save Venice website, "The Venice Collection," and the brochure *The Serenissima Society,* sent to members by Save Venice Inc. in 2002.

37. For more on this and other donors to Venice in Peril, see the committee's website.

38. See the UNESCO website: "International Private Committees for the Safeguarding of Venice," "Results: Restoring Buildings and Works of Art," and "Results: International Solidarity"; also membership correspondence from Save Venice Inc., November 1998.

39. First quote from *The Serenissima Society* brochure; second and third quotes from UNESCO website.

40. From *The Serenissimia Society*.

41. UNESCO website: "International Private Committees: Projects Completed in 1998"; and Friends of Venice website, newsletter of January 1997. On Ghetto restorations, see Venetian Ghetto website, "The Jewish Community Today, 2000," which lists no fewer than twelve past projects (several more are currently underway) in the two-hectare space of the Ghetto.

42. For restoration as a form of archaeology in Venice, see UNESCO website: "Results: Restoring Buildings and Works of Art: The Facade of Santa Maria del Giglio"; and *NV*, 10 May 2001. Save Venice has been involved in the "cleaning and consolidation" of several rediscovered works of art, in particular from the ex-church of Sant'Andrea della Zirada. See also similar comments from Venice in Peril, in its online newsletter "Friends of Venice," April 1997.

43. *NYT*, 3 May 1970.

44. *NYT*, 20 October 1974.

45. Quote from *NYT*, 3 June 1974; see also *GZ*, 20 October 2000, 12 February, 23 March 2001.

46. Quote from UNESCO website. It is doubtful that locals took much comfort from UNESCO's rather Pollyanna observation that this was "a telling example of how intimate a role art and history play in the fabric of the city": UNESCO website, "Results: International Private Committees for the Safeguarding of Venice, Projects Completed in 1999." The arch is sculpted with the figure of an angel: see Fuga and Vianello, p. 63.

47. *NV*, 11 February, 4 March 2002. It should be noted that the Comité Français purchased the statue in previous agreement with the city government, which obviously failed to realize that putting up such a work (some claim it is only a copy) in the middle of Venice would arouse so much resistance.

48. See Muir (1989), pp. 25-40.

49. Indeed, "reciting the rosary in the evening in front of the little chapel *[capitello]* is an old popular tradition that still lives on": *NV*, 16 June 2000. The article also notes that the site, "two steps from via Garibaldi, [was] recently restored thanks to the sensitivity of . . . an inhabitant of the area, who, with the help of his friend . . . fixed up and painted the little altar *[altarino]* and the enclosure *[sacello]* of the *capitello*."

50. Quoted in *GZ*, 18 July 2000.

51. Quotes from the Save Venice newsletter, autumn 1997; also see UNESCO website, "Results: Restoring Buildings and Works of Art—Santa Maria dei Miracoli."

52. Garrett, p. 140.

53. *NV,* 11 March 2001.

54. *Save Venice, 1968–1998,* p. 42. On its website, Save Venice also stresses that "the sheet music for Wagner's wedding march on the organ . . . reveals the church's present day popularity . . . for marriages." *GZ,* 27 May 2001; also Weddings in Venice website.

55. Quotes from the Save Venice newsletter, autumn 1997; see also *NYT,* 30 April 1961; and *NYT,* 18 December 1983, which numbers the Miracoli among "these little-visited monuments [that] are worth a detour."

56. Norwegian Cruise Lines website: "Shore Excursions, Venice, Italy: Private Concert"; see also CHORUS website.

57. Quotes from *GZ,* 24 October 2002; and *NV,* 1 November 2002.

58. Quotes from *GZ,* 30 December 1999, 14 January 2000.

59. UNESCO website: "Threats."

60. On these VIP projects, see Friends of Venice website newsletter of December 1996; January, April, and September 1997; and April 2001. The VIP's housing project on the Calle delle Beccarie in Cannaregio was completed by mid-2003, according to its website; see also UNESCO website, "International Private Committees: Projects Completed in 1999."

61. Friends of Venice website newsletter, April 2001, January 1997.

62. Examples of the smallest and largest versions of the Mangia Onda were delivered to the ACTV for testing in January 1999 and November 2000: see Save Venice newsletter, autumn 1999 and autumn 2000; also *GZ,* 9 August and 5 October 2000.

63. Heartening at least for the spirit such efforts display. In fact, when such wave-free boat-buses do finally appear on Venice's canals, they will most likely be of a design and specification worked out by Italian engineers working directly for the ACTV: see *NV,* 20 September 2001.

64. *NV,* 6 July 2001. For some examples of degraded housing, see *NV,* 15 June, 11 July 2001, 29 May 2003; *GZ,* 5, 7, and 31 July 2001.

65. *NV,* 6 July 2001.

Chapter 10. Ships and Fools

1. Baedeker (1895), p. 239; *NYT,* 15 June 1930.

2. Credit goes to Lauro Bergamo, Carlo Gottardi, and Toni Rosa Salva, who modeled the Vogalonga on the Swedish Vasaloppet (a mass cross-country skiing event): *NV,* 11 May 2000.

3. *GZ,* 9 May 1983.

4. Ibid.

5. Named Uka Rapu, she came from Rapa Nui (Easter Island): *NV,* 24 May 2003.

6. See *Fodor's: Exploring Venice,* p. 23; *Knopf City Guides: Venice,* p. 55.

7. *GZ,* 27 August 1998, quotation from 12 May 2000.

8. Quote from *NV,* 11 May 2000; see also *GZ,* 20 March 2001.

9. Quotes from *GZ,* 21 and 29 May 2001.

10. During which some 50,000 out of a population of around 175,000 died: *GZ,* 11 July 2000.

11. See *GZ,* 11 July 2000.

12. The bridge set up for the Festa del Salute only crosses the Grand Canal and measures about sixty-two meters: *NV,* 14 July, 22 November 2000.

13. *GZ,* 15 July 1983.

14. *GZ,* 18 July 1983.

15. *GZ,* 15 July 1983.

16. Baedeker (1930), p. 346; *GZV,* 16 July 1876; also *GZV,* 19 July 1896, which noted that over fifty-five hundred tourists came to the festival in 1896 in specially chartered trains; see also *NYT,* 21 July 1930; *GZ,* 12 July 2000.

17. *NV,* 16 July 2000.

18. First quote from *GZ,* 5 August 1980; second quote from *GZ,* 18 July 1983; see also *GZ,* 19 July 1982.

19. *GZ,* 16, 17, 18, 19, and 20 July 1989.

20. *GZ,* 18 July, 5 August 1980, 18 July 1982.

21. *NV,* 16 July 2001; earlier quotes from *NV,* 17 July 2000; *GZ,* 17 July 2000.

22. Quote from *GZ,* 16 July 1998; see also *GZ,* 20 July 1998.

23. With the explanation that the *Princess,* which was scheduled to set off to sea the following morning, would not otherwise have been able to reach the sea from its normal berth behind Piazzale Roma, because of the *ponte votivo* blocking the Giudecca Canal.

24. *GZ,* 20 July 1998.

25. Quote from *GZ,* 17 July 1998; see also *GZ,* 16 July 1998; and see *GZ,* 13 July 2000 for a similar problem with the equally large *Costa Atlantica.*

26. *NV,* 8 June 2000.

27. *NV,* 14 July 2000; *GZ,* 20 November 2000.

28. *GZV,* 10 March 1886.

29. Baedeker (1895), p. 238; second quote from *NYT,* 9 December 1979; see also *NYT,* 10 April 1910.

30. On local revivals of the Carnival, see *GZ,* 21 and 25 February 1979, 16 February 1980. The Burano Carnival was reborn in 1968.

31. *GZ,* 8 February 1979.

32. *GZ,* 23 and 28 February 1979.

33. Since these first years, the Carnival of Venice has been staged through the combined efforts of as many as six of these festive *scuole* (usually) in concert with one of a succession of semipublic boards whose members are drawn from the city council, the APT, and the organization of hoteliers. In 2001, the latest incarnation of this entity was the Consorzio Carnivale: *GZ,* 9 February 2001.

34. *GZ,* 12 February 1980.

35. Quote from *GZ,* 18 February 1980; see also *GZ,* 12 February 1980.

36. *GZ,* 15 and 17 February 1980.

37. *GZ,* 18 and 20 February 1980.

38. *GZ,* 20 February 1980.

39. *GZ,* 17, 18, 19, 21, and 22 February 1980, 13 February 1983, 18 February 1985.

40. *GZ,* 15 and 23 February 1980.

41. Quote from *GZ,* 24 February 1980; see also *GZ,* 19 February 1980.

42. *GZ,* 23 February 1982, 13 February 1983, 5 March 1984, 18 February 1985.

43. *GZ,* 1 March 1984, 17 February 1985: the morning after the Ombralonga in 1985, the *Gazzettino* joked, one could take part in the *farmalonga*—a tour of the city's pharmacies, looking for headache pills.

44. *GZ,* 1 March 2000; for 2003, the theme was "Federico Fellini": *NV,* 21 November 2002.

45. The original Festival of the Twelve Marys dated back to at least 1039 and was banned in 1379: Muir (1981), pp. 135–56. In recognition of both the festival's sacred origins and the season, the contestants are clad in designer medieval gowns rather than bathing suits. On the modern version, see Isman, p. 126; *GZ,* 26 February 2000, 9 and 17 February 2001; *NV,* 26 February 2000.

46. *GZ,* 19 February 1980.

47. In 1987, these included television and radio crews from "America, Japan, Canada, Italy, Spain, France . . . and the BBC": *GZ,* 5 March 1987.

48. Best known among them have been Katia Ricciarelli (1996) and Ornella Muti (1997): *GZ,* 14 February 1996. By 2000, a second, subsidiary *madrina* was being chosen to act as a kind of duenna for the Marys Festival: *GZ,* 9 February 2001.

49. This first "Flight of an Angel" was performed by a non-Venetian, a woman from Ascoli Piceno in southern Italy: *GZ,* 9 and 19 February 2001; *NV,* 19 February 2001. In 2003 it was done by the Venetian-born fencing champion Frida Scarpa: *GZ,* 24 February 2003.

50. *NV,* 18 February 2001.

51. *GZ,* 3 February 1988.

52. *GZ,* 19 February 1995.

53. *GZ,* 25 February 2000.

54. Among those willing to do so was the Italian food conglomerate Alivar, followed by the local bank, the Cassa di Risparmio di Venezia, the Italian State Railways, and La Gioisa, producers of wines: *GZ,* 18 March 1984, 8 February 1995.

55. *GZ,* 1 February 1988, 5 February 1995.

56. In 2001, the city supplied 590 million lire (around $300,000), while corporate sponsors paid about $750,000 collectively: *GZ,* 16 February 2001.

57. *GZ,* 15 January 2001, 22 February 1996.

58. *GZ,* 4 and 5 March 1987.

59. *GZ,* 14 February 1988.

60. The council made wearing a mask illegal—although not, the police specified, actually "criminal"; apparently no one was made to take off his or her disguise: *GZ,* 3 and 4 February 1991, 13 February 1993.

61. *GZ,* 22 February 1996, 13 February 1997, 26 February 1998, 9 March 2000, 14 February 2002.

62. Garbage collection figures for 2002 and 2003 can be found in *GZ,* 3 March 2003, along with some estimates of the number of visitors in 2003. These latter

were reduced by about 10 percent from the previous year, apparently due to the tensions that preceded the American invasion of Iraq.

63. *NV,* 6 March 2000.

64. *GZ,* 20 February 1996.

65. *GZ,* 13 February 1994.

66. *GZ,* 22 February 1998.

67. *GZ,* 22 and 23 February 1982. Especially at risk were the Porta della Carta, the statue of the four "Tetrarchs" nearby, the arcades of the Ducal Palace and the Procuratie, and the walls of the Basilica of San Marco itself: *GZ,* 21 February 1985, 7 March 1987.

68. *GZ,* 22 February 1982, 13 March 2000.

69. *GZ,* 3 March 1987.

70. *NV,* 6 and 13 March 2000.

71. Quote from *GZ,* 10 February 1983; see also 19 February 1985.

72. *GZ,* 3 March 2000.

73. In the city's 2001 Carnival budget of around $300,000, about $100,000 went to the neighborhoods: *NV,* 8 February 2001.

74. *GZ,* 9 February 2001.

75. *GZ,* 12 February 1994.

76. *GZ,* 27 February 1983.

77. *GZ,* 15 February 1983, 6 February 1986, 8 February 2001.

78. *NV,* 9 March 2000; *GZ,* 9 March 2000.

79. *GZ,* 12 February 1986.

80. See *NYT,* 11 February 2001.

81. See Davis (1991), p. 45; Davis (1997a), pp. 275–90; Bertelli, 29–32; *GZ,* 11 July 2000.

82. The Carnival's costumed *vogada* (which has over the years expanded to over a hundred participating boats) is not, of course, competitive: *GZ,* 23 February 1979, 17 February 1980.

83. *GZ,* 20 February 1980.

84. *GZ,* 8 March 1987.

85. *GZ,* 11 February 1983.

86. *GZ,* 1 March 1995.

87. *GZ,* 16 February 1983, 15 and 17 February 1985, 2 March 1995. The Carnival program for 2001 mentioned performers from India, Argentina, Bosnia, Greece, Austria, Brazil, Cuba, Japan, and Ivory Coast, among others: *GZ,* 9 February 2001.

88. *GZ,* 6 March 1987.

89. *GZ,* 18 February 1985.

Chapter 11. Taking It All Home

1. MacCannell, p. 158.

2. Sontag, pp. 9–10.

3. Quoted in Chalfen, p. 101.

4. Urry (1990), pp. 3, 11–14.

5. MacCannell, p. 19.

6. Goethe, p. 59.

7. Ayscough, p. 199. See also Young, p. 218: "The city, in general, has some beautiful features, but does not equal the idea I had formed of it, from the pictures of Canaletti." Nearly a century later, Mark Twain, p. 171, waggishly complained how "cherished dreams of Venice have been blighted forever" once he actually had seen the place.

8. *NYT,* 28 August 1881.

9. Osborne, p. 79.

10. Albers and James, p. 136.

11. Osborne, pp. 79, 82.

12. Ibid., p. 71.

13. Ibid., p. 72; emphasis ours.

14. Graburn, pp. 21–36.

15. On the speed of such responses, see Markwell, pp. 131–55.

16. Which would require the city's thirteen million annual visitors to take fewer than eight pictures each on average.

17. Quote from *GZ,* 26 February 2000; see also *NV,* 26 February 2000.

18. Stewart, p. 138.

19. *GZ,* 19 August 2001.

20. See especially the illustrations in Cosulich, pp. 135–38, 151, 158–59, 161; Vanzan Marchini, pp. 92–93; Pemble, pp. 113–39; Carter, pp. 405–7.

21. If they are not available, such sunsets and crescent moons will be Photoshopped in: *Colors,* pp. 54–55, 62.

22. *San Francisco Examiner,* 26 December 1999.

23. But see, for example, Edwards, pp. 197–221; Marwick, pp. 417–38.

24. Stewart, p. 135.

25. MacCannell, p. 147.

26. See, for example, Greenwood, pp. 171–85.

27. MacCannell, p. 157.

28. Goethe, p. 56; Cosulich, pp. 186–88.

29. Quotes from Fabri, p. 98; Casola, p. 142; see also Barbatre, pp. 104, 109; Brasca/Capodilista, p. 50; von Harff, p. 66.

30. Quotes from "A True Description," p. 80; Lassels, p. 260, noted that "sometimes to shew their art, they make here pretty things," and went on to mention the castle, ship, and organ.

31. So wrote Evelyn, p. 237, without noting whether his "Curiosities" ever arrived intact.

32. Goethe, p. 74.

33. As Stewart, pp. 148–49, has suggested.

34. *NV,* 24 August 2001.

35. *NYT,* 2 July 1880; Baedeker (1895), p. 290, (1913), p. 415.

36. For example, *Knopf City Guides: Venice,* pp. 60–61 and 370–73; and *Fodor's:*

Exploring Venice, pp. 72–73; both extol the history of lace making on the island, without presenting lace itself as the same sort of accessible touristware as glass: thus, see *Knopf City Guides: Venice*, pp. 404–5, and *Fodor's: Exploring Venice*, p. 235, which deal with souvenir shopping.

37. An arrangement similar to that employed by the prostitutes in Amsterdam.

38. Thus, Pollnitz, pp. 411–12.

39. Stewart, p. 135.

40. Pemble, pp. 27–28; *NV*, 21 December 2000, 1 February and 24 August 2001; *GZ*, 22 December 1999.

41. Quote from *LT*, 30 May 1844; other, varying descriptions in *LT*, 22 August 1843, 1 June, 11 July 1844.

42. *LT*, 7 June 1892.

43. See *Colors*, pp. 14–19.

44. The corporate presence behind The Venetian is Las Vegas Sands, Inc., of which Sheldon Adelson is chairman of the board. Much of the story that follows, along with many of the attendant quotes, is from the BBC documentary *Viva Las Venice!* Venetians like to refer to Adelson as "Las Venetian": *NV*, 4 April 2000.

45. For details, see The Venetian website.

46. Quoted in Booth; see also *Viva Las Venice!*

47. *Frommer's 2000: Las Vegas*, p. 80.

48. Ibid., p. 81; see also Fine.

49. Baudrillard, pp. 170–71; emphasis in original.

50. Quoted in *NV*, 4 April 2000.

51. Baudrillard, pp. 168, 170.

52. From The Venetian website.

53. Quotes from the bulletin/insert "The Grand Canal Shoppes," provided by The Venetian in its publicity pack.

54. *NV*, 29 July 2000.

55. See *NV*, 4 April 2000; and The Venetian website, "Venezia, Opening June 2003."

56. Quoted in *NV*, 4 April 2000; and *GZ*, 5 April 2000.

Afterword. *Chi ciapa schei xe contento*

1. *NV*, 12 July 2003.

2. Quotes from *GZ*, 26 August, 1 September 2001; see also *NV*, 27 August 2001; *GZ*, 28 August 2001.

3. *GZ*, 26 June 2000; *NV*, 24 October 2000, 17 February 2001.

4. Quote from *GZ*, 27 May 2001; see also *GZ*, 25 August 2001.

5. *GZ*, 1 September 2001.

6. *NV*, 11 July 2001, 29 May 2003.

7. Quote from *NV*, 27 August 2001; see also *GZ*, 28 August 2001.

8. The Italian Army declined to put up its own bridge in 2001: see *NV*, 6 July 2001.

9. *NV,* 16 July 2001.

10. The loss of the Americans was apparently made up by a large influx of French: *NV,* 10 and 11 February 2002.

11. *GZ,* 28 February 2001, 11 February 2002.

12. *GZ,* 31 May 2001.

13. Most famously Piero Pazzi, tireless crusader against what he calls the *brutture,* the ugliness, of modernity in Venice, including visible air conditioners, power cables, TV aerials, and other commonplaces of modern life: see *GZ,* 27 June 2001.

14. Quotes from *GZ,* 6 and 24 August 2001.

Bibliography

This bibliography is divided into seven sections, in the following order: Primary Narrative Sources, Guidebooks, Secondary Sources, Newspaper and Magazine Articles, Films and Television Programs, Websites, and Archival Sources.

Primary Narrative Sources

Ayscough, George. *Letters from an Officer in the Guards to His Friend in England: Containing some accounts of France and Italy.* London: T. Cadell, 1778.

Balfour, Sir Andrew. *Letters Written to a Friend.* Edinburgh: n.p., 1700.

Barbatre, Pierre. "Le Voyage de Pierre Barbatre a Jérusalem en 1480." In Pierre Tucoo-Chala and Noël Pinzuti, eds., *Annuaire-Bulletin de la Société de l'Histoire de France* (1972–73): 73–111.

Beckford, William. *Vathek and European Travels.* London: Richard Bentley, 1891.

Brasca, Santo. *Viaggio in Terrasanta di Santo Brasca, 1480, con l'Itinerario di Gabriele Capodilista, 1458.* Ed. Anna Laura Momigliano Lepschy. Milan: Longanesi, 1966.

Broderick, Thomas. *The Travels of Thomas Broderick, Esq.* London: M. Cooper, 1754.

Bromley, William. *Remarks made in Travels through France & Italy.* London: Thomas Bassett, 1693.

Carter, Nathaniel Hazeltine. *Letters from Europe, comprising the Journal of a Tour.* Vol. 2 of 2. New York: G. C. Carvill, 1829.

Casola, Canon Pietro. *Canon Pietro Casola's Pilgrimage to Jerusalem, in the Year 1494.* Ed. Margaret Newett. Manchester, U.K.: Manchester University Press, 1907.

Chancel, A. D. *A New Journey over Europe; from France thro' Savoy, etc.* London: John Harding, 1714.

Cochin, N. *Voyage d'Italie, ou recueil de notes.* Vol. 3 of 3. Paris: Charles Jombert, 1769.

Cooper, James Fenimore. *Gleanings in Europe: Italy.* Ed. J. Conron and C. Denne. 1838. Reprint, Albany: State University of New York Press, 1981.

Coryat, Thomas. *Coryat's Crudities Hastily gobbled up in five Moneths travells in France, Savoy, Italy, etc.* Vol. 1 of 2. London, 1611. Reprint, Glasgow: James MacLehose, 1905.

The Curious Traveller, Being a Choice Collection of Very Remarkable Histories, Voyages, Travels, etc. London: J. Rowland, 1742.

d'Anglure, Baron. *Le Saint Voyage de Jérusalem par le Baron d'Anglure, 1395.* Paris: Bibliothèque Catholique, 1858.

da Sanseverino, Roberto. *Viaggio in Terra Santa fatto e descritto per Roberto da Sanseverino.* Ed. Gioacchino Maruffi. Bologna: Romagnoli dall'Acqua, 1888.

de Blainville, Sieur. *Travels through Holland, Germany, Switzerland, and Other Parts of Europe, but Especially Italy.* Trans. George Turnbull. London: W. Strahan, 1707.

de Caraccioli, Louis Antoine. *The Travels of Reason in Europe.* London: Charnley and Vesey, 1778.

de Caylus, Anne Claude Philippe, Comte. *Voyage d'Italie, 1714–1715.* Ed. Amilda A. Pons. Paris: Fischbacher, 1914.

de Fer, Nicolas. *Voyages and Travels over all Europe.* Vol. 3 of 3. London: H. R. Hodes, 1694.

de Grancourt, Bergeret. *Voyage d'Italie, 1773–1774.* Ed. Jacques Wilhelm. Paris: M. de Romilly, 1948.

de Voisins, Philippe. *Voyage à Jérusalem de Philippe de Voisins, Seigneur de Montaut.* Ed. Ph. Tamizey de Larroque. Paris: Honoré Champion, 1883.

Drummond, James, Fourth Earl of Perth. *Letters from James Earl of Perth, Lord Chancellor of Scotland, &c. to his sister, etc.* London: Camden Society, 1845.

Du Bocage, Anne Marie. *Lettres sur l'Angleterre, la Hollande et l'Italie.* 1762. English trans., *Letters Concerning England, Holland and Italy.* London: Edward and Charles Dilly, 1770.

Du Mont, Jean, Baron de Carlscroon. *Nouveau voyage du Levant.* The Hague, 1694. English trans., *A New Voyage to the Levant,* 2nd. ed. London: M. Gillyflower, 1696.

du Val, Jean-Baptiste. *Les Remarques triennales: Un Français en Italie.* Ed. F.-G. Pariset. Algiers: Boughet, 1955.

Evelyn, John. *Diary and Correspondence.* Vol. 1 of 4. 1818. Reprint, London: Hurst and Blackett, 1854.

Fabri, Felix. *The Wanderings of Felix Fabri.* Pt. 1. Ed. and trans. Aubrey Stewart. Library of the Palestine Pilgrims' Text Society, vol. 7 of 12. London: AMS Press, 1887–97.

Franco, Giacomo. *Habiti d'huomeni et donne venetiane con la processione della serenissima signoria et altri particolari cioè trionfi feste ceremonie publiche della nobilissima città di Venetia.* Venice: n.p., 1610. Reprint, Venice: F. Ongania, 1878.

Frescobaldi, Lionardo. *Viaggio in Terrasanta.* Novara: De Agostini, 1961.

Frye, William. *After Waterloo: Reminiscences of European Travel, 1815–1819.* Ed. Salomon Reinach. London: William Heinemann, 1908.

Gailhard, Jean. *The Present State of the Republick of Venice.* London: John Starkey, 1669.

Goethe, Johann Wolfgang von. *Italienische Reise.* Stuttgart, 1816–17. Trans. Robert Heitner, *Italian Journey.* New York: Suhrkamp, 1989.

Guylforde, Richard. *This is the begynnynge, and contynuaunce of the Pylgrymage of Sir Richard Guylforde Knygth, & controuler unto our late soueraygne lorde kynge Henry the vij.* London: Richard Pynson, 1511.

Hewins, Amasa. *A Boston Portrait-Painter Visits Italy.* Ed. Francis Allen. Boston: Boston Athenaeum, 1931.

Howell, James, S.P.Q.V. *Epsitolae Ho-Elianae.* London: Humphrey Moseley, 1645.

———. *A Survey of the Signorie of Venice.* London: Richard Lowndes, 1651.

James, Henry. *The Aspern Papers and The Turn of the Screw.* Ed. and intro. Anthony Curtis. London: Penguin, 1984.

———. *Italian Hours.* New York: Grove, 1959.

Jeffereys, David. *A Journal from London to Rome, by way of Paris, Lyons, Turin, Florence, &c., and from Rome back to London, by way of Loretto, Venice (giving a particular Account of the Government of that Republic), Milan, Strasburg, &c.* London: W. Owen, 1755.

Keysler, Johann Georg. *Neueste Reise durch Deutschland, Böhmen, Ungarn, die Schweitz, Italien, und Lothringen.* Hanover: n.p., 1751. English trans., *Travels through Germany, Hungary, Bohemia, Switzerland, Italy, and Lorrain.* Vol. 4 of 4. London: A. Linde, 1758.

Lalande, Jérôme Le Français de. *Voyage d'un François en Italie, Fait dans les Années 1765 & 1766.* Vol. 8 of 8. Paris: Yverdon, 1770.

Lassels, Richard. *An Italian Voyage, or a Compleat Journey through Italy.* 2nd ed. London: Richard Wellington, 1698.

Lithgow, William. *A Most Delectable, And True Discourse.* London: Nicholas Okes, 1609.

Miller, Anna. *Letters from Italy, Describing the Manners, Customs, Antiquities, Paintings, &c. of that Country.* Vol. 3 of 3. London: Edward and Charles Dilly, 1776.

Misson, François. *Nouveau voyage d'Italie.* The Hague: n.p., 1691. English trans., *A New Voyage to Italy with Curious Observations on several other Countries.* Vol. 1 of 2. London: R. Bently, 1699.

Montaigne, Michel Eyquem de. *The Diary of Montaigne's Journey to Italy in 1580 and 1581.* Ed. and trans. E. J. Trenchmann. New York: Harcourt, Brace, 1929.

Moore, John. *A View of Society and Manners in Italy.* Vol. 1 of 2. London: W. Strahan, 1781.

Mortoft, Francis. *Francis Mortoft: His Book. Being His Travels through France and Italy 1658–1659.* Ed. Malcolm Letts. Hakluyt Society ser. 2, vol. 57. London: Hakluyt Society, 1925.

Moryson, Fynes. *An Itinerary Written by Fynes Moryson, Gent.* London: John Beale, 1617. Reprint, Amsterdam: n.p., 1971.

———. *Shakespeare's Europe.* Unpublished chapters of Fynes Moryson's *Itinerary.* London: Sherratt and Hughes, 1903.

Mundy, Peter. *Travels in Europe and Asia, 1608–1667.* Vol. 1 of 4. Journey in 1620. Cambridge: Hakluyt Society, 1907.

Northall, John. *Travels through Italy.* London: S. Hooper, 1766.

Nugent, Thomas. *The Grand Tour; Or, A Journey through the Netherlands, Germany, Italy and France.* London: D. Browne, 1749.

Palmer, Thomas. *An Essay on the meanes how to make our travailes into forraine countries the more profitable and honourable.* London: Mathew Lownes, 1606.

Piozzi (née Thrale), Hester Lynch. *Observations and Reflections made in the course of a Journey through France, Italy, and Germany.* London: A. Strahan, 1789. Reprint, ed. Herbert Barrows, Ann Arbor: University of Michigan Press, 1967.

The Polite Traveller: Being a Modern View of Part of Italy, Spain, Portugal, and Africa. London: John Fielding, 1783.

Pollnitz, Karl Ludwig von. *Mémoires de Charles-Louis Baron de Pöllnitz.* Amsterdam, 1737. Trans. Stephen Whatley, *The Memoirs of Charles-Lewis, Baron of Pollnitz.* Vol. 2 of 2. London: D. Browne, 1737.

Ray, John. *Travels through the Low Countries, Germany, Italy and France.* Vol. 1 of 2. London: J. Walthoe, 1738.

Raymond, John. *An Itinerary Contayning a Voyage Made through Italy, in the Year 1646, and 1647.* London: Humphrey Moseley, 1648.

Reresby, John. *Memoires and Travels.* London: Kegan, 1904.

Rozmital, Leo of. *The Travels Leo of Rozmital through Germany, Flanders, England, France, Spain, Portugal, and Italy, 1465–1467.* Ed. and trans. Malcolm Letts. Cambridge: Cambridge University Press, 1957.

St. Didier, Alexandre Toussaint Limojon de. *La Ville et la République de Venise.* Paris, 1680. Trans. Francis Terne, *The City and Republick of Venice.* London: Charles Brome, 1699.

Sainte-Marie, Charles. *Nouveau voyage de Grèce.* Anonymous English trans., *A New Journey through Greece, Ægypt, Palestine, Italy, Swisserland, Alsatia and the Netherlands.* London: J. Batley, 1725.

Sansovino, Francesco. *Venetia, città nobilissima et singolare, descritta in XIII libri.* Venice: I. Sansovino, 1581.

Shelley, Frances. *The Diary of Frances Lady Shelley, 1818–1873.* Ed. Richard Edgcumbe. Vol. 1, 1787–1817. Vol. 2, 1818–73. New York: Scribner's, 1913.

Skippon, Philip. *An Account of a Journey made thro' part of the Low-Countries, etc.* In A. Churchill and J. Churchill, eds., *A Collection of Voyages and Travels,* 6 vols., vol. 6, pp. 359–736. Journey in 1663. London: Churchill and Churchill, 1732.

Stevens, Sacheverell. *Miscellaneous Remarks Made on the Spot, in a late Seven Years Tour through France, Italy, etc.* London: S. Hooper, 1756.

Tafur, Pero. *Pero Tafur: Travels and Adventures, 1435–1439.* Ed. and trans. Malcolm Letts. New York: Harper and Brothers, 1926.

"A True Description of What Is Most Worthy to Be Seen in Italy, &c." Ca. 1605. In *Harleian Miscellany,* vol. 12, pp. 73–130. London: Robert Dutton, 1811.

Twain, Mark. *The Innocents Abroad.* New York: Harper and Row, 1989.

Venice under the Yoke of France and Austria. Anonymous ("A Lady of Rank"). Vol. 1 of 2. London: G. and W. B. Whittaker, 1824.

Veryard, Ellis. *An Account of Divers Choice Remarks, as well Geographical, as Historical, Political, Mathematical, Physical, and Moral.* London: Smith and Walford, 1701.

von Archenholtz, Johann Wilhelm. *England und Italien.* 1785. Trans. Joseph Trapp, *A Picture of Italy.* Vol. 1 of 2. London: G. G. J. and J. Robinson, 1791.

von Harff, Arnold. *The Pilgrimage of Arnold von Harff, Knight.* Ed. and trans. Malcolm Letts. London: Hakluyt Society, 1946.

Warcupp, Edmund. *Italy, in its Original Glory, Ruine, and Revival.* London: S. Griffin, 1660.

Wey, William. *The Itineraries of William Wey, Fellow of Eton College. To Jerusalem, A.D. 1458 and A.D. 1462; and to Saint James of Compostella, A.D. 1456.* London: J. B. Nichols and Sons, 1857.

Young, Arthur. *Travels during the Years 1787, 1788, and 1789.* London: W. Richardson, 1792.

Zompini, Gaetano. *Le arti che vanno per via nella città di Venezia.* Facsimile reprint of 1785 ed. Venice: Filippi Editore, 1968.

Guidebooks

Baedeker. *Northern Italy.* London: T. Fisher Unwin, 1870, 1877, 1895, 1903, 1913, 1930.

Fodor's: Exploring Venice. New York: Fodor's, 1998.

Frommer's 2000: Las Vegas. New York: Frommer's, 2000.

Fuga, Guido, and Lele Vianello. *Corto sconto: Itinerari fantastici e nascosti di Corto Maltese a Venezia.* Rome: Lizard Edizioni, 1998.

Garrett, Martin. *Venice.* New York: Interlink Books, 2001.

Kahn, Robert, ed. *City Secrets: Florence, Venice, and the Towns of Italy.* New York: Little Bookroom, 2000.

Knopf City Guides: Venice. New York: Alfred A. Knopf, 1993.

Links, J. G. *Canaletto.* Oxford: Phaidon, 1994.

Thomas Cook/Passport. *Illustrated Travel Guide to Venice.* Chicago: Thomas Cook, 1998.

Secondary Sources

Albers, P., and W. James. "Travel Photography: A Methodological Approach." *Annals of Tourism Research* 15 (1988): 134–58.

Ashtor, Eliyahu. "Venezia e il pellegrinaggio in Terrasanta nel basso medioevo." *Archivio storico italiano* 143 (1985): 197–223.

Baudrillard, Jean. *Jean Baudrillard: Selected Writings.* Ed. Mark Poster. Stanford, Calif.: Stanford University Press, 1988.

Bellavitis, Giorgio, and Giandomenico Romanelli. *La città nella storia d'Italia: Venezia.* Bari: Laterza, 1985.

Bertelli, Stefania. *Il Carnevale di Venezia nel Settecento.* Rome: Jouvence, 1993.

Boissevain, Jeremy, ed. *Coping with Tourists: European Reactions to Mass Tourism.* Oxford: Berghahn, 1996.

——, ed. *Revitalizing European Rituals.* London: Routledge, 1992.

Boorstin, Daniel. *The Image: A Guide to Pseudo-Events in America.* New York: Harper and Row, 1964.

Brewster, Harry. *Out of Florence.* London: Radcliffe, 2000.

Chalfen, Richard. *Snapshot Versions of Life.* Bowling Green, Ohio: Bowling Green State University Press, 1987.

Chaney, Edward. *The Evolution of the Grand Tour: Anglo-Italian Cultural Relations since the Renaissance.* London: Frank Cass, 1998.

Chard, Chloe. *Pleasure and Guilt on the Grand Tour: Travel Writing and Imaginative Geography, 1600–1830.* Manchester, U.K.: Manchester University Press, 1999.

Clifford, James. *Routes: Travel and Translation in the Late Twentieth Century.* Cambridge, Mass.: Harvard University Press, 1997.

Cosgrove, Dennis. "The Myth and the Stones of Venice: An Historical Geography of a Symbolic Landscape." *Journal of Historical Geography* 8 (1982): 145–69.

Costantini, Massimo. "Le strutture dell'ospitalità." In Alberto Tenenti and Ugo Tucci, eds., *Storia di Venezia: Dalle origini alla caduta della Serenissima,* 18 vols., vol. 5, *Il Rinascimento: Società ed economia,* pp. 881–911. Rome: Treccani, 1996.

Cosulich, Alberto. *Viaggi e turismo a Venezia.* Venice: Edizione "I Sette," 1990.

Crouzet-Pavan, Elisabeth. "Police des mœurs: Société et politique à Venise à la fin du moyen âge." *Revue Historique* 264 (1980): 241–88.

——. *"Sopra la acque salse": Espaces, pouvoir, et société à Venise, à la fin du moyen âge.* Rome: École Française de Rome, 1992.

Davis, Robert C. "The Geography of Gender in the Renaissance." In Judith Brown and Robert C. Davis, *Gender and Society in Renaissance Italy,* pp. 19–38. Harlow, U.K.: Longman, 1998.

——. *Shipbuilders of the Venetian Arsenal: Workers and Workplace in the Preindustrial City.* Baltimore: Johns Hopkins University Press, 1991.

——. "Slave Redemption in Venice, 1585–1797." In John Martin and Dennis Romano, eds., *Venice Reconsidered: The History and Civilization of an Italian City-State, 1297–1797,* pp. 454–87. Baltimore: Johns Hopkins University Press, 2000.

——. "The Trouble with Bulls: How to Stage a *Caccia dei Tori* in Early-Modern Venice." *Histoire Sociale/Social History* (1997a): 275–90.

——. "Venetian Shipbuilders and the Fountain of Wine." *Past & Present* 156 (1997b): 56–86.

——. *The War of the Fists: Popular Culture and Public Violence in Late Renaissance Venice.* Oxford: Oxford University Press, 1994.

Del Negro, Giovanna. "'Our Little Paris': An Ethnography of the *Passeggiata* in Central Italy." In Mario Aste, ed., *Industry, Technology, Labor, and the Italian American Communities,* pp. 198–209. Staten Island, N.Y.: The Association, 1997.

Di Monte, Giuseppina, and Isabella Scaramuzzi, eds. *Una provincia ospitale: Itinerari di ricerca sul sistema turistico veneziano.* Bologna: Il Mulino, 1996.

Donatelli, Carlo. *The Gondola: An Extraordinary Naval Architecture*. Venice: Arsenale Editore, 1994.

Edwards, Elizabeth. "Postcards: Greetings from Another World." In Tom Selwyn, ed., *The Tourist Image: Myths and Myth in Making Tourism*, pp. 197–221. Chichester, U.K.: John Wiley, 1996.

Eglin, John. *Venice Transfigured: The Myth of Venice in British Culture, 1660–1797*. Basingstoke, U.K.: Palgrave, 2001.

Featherstone, Mike. *Consumer Culture and Postmodernism*. London: Newbury Park, 1991.

——. *Undoing Culture: Globalization, Postmodernism, and Identity*. London: Sage, 1995.

Fedalto, Giorgio. "Stranieri a Venezia e a Padova." In Girolamo Arnaldi and Manlio Pastore Stocchi, eds., *Storia della cultura veneta*, 5 vols., vol. 3, pt. 1, *Dal primo '400 al Concilio di Trento*, pp. 499–535. Vicenza: Neri Pozza, 1980.

Geary, Patrick J. *Furta Sacra: Thefts of Relics in the Central Middle Ages*. Princeton, N.J.: Princeton University Press, 1990.

Graburn, H. H. "Tourism: The Sacred Journey." In Valene L. Smith, ed., *Hosts and Guests: The Anthropology of Tourism*, pp. 21–35. Philadelphia: University of Pennsylvania Press, 1989.

Greenwood, Davydd. "Culture by the Pound: An Anthropological Perspective on Tourism as Cultural Commodification." In Valene L. Smith, ed., *Hosts and Guests: The Anthropology of Tourism*. Philadelphia: University of Pennsylvania Press, 1989.

L'illustrazione italiana. No. 47, 20 November. Milan: Garzanti, 1881.

Isman, Fabio. *Venezia: La fabbrica della cultura*. Venice: Marsilio, 2000.

Lane, Frederic C. *Venetian Ships and Shipbuilders of the Renaissance*. Baltimore: Johns Hopkins University Press, 1934.

——. *Venice: A Maritime Republic*. Baltimore: Johns Hopkins University Press, 1973.

MacCalancy, Jeremy, and Robert Parkin. "Revitalisation of Continuity in European Ritual? The Case of San Bessù." *Journal of the Royal Anthropological Institute* 3 (1997): 61–78.

MacCannell, Dean. *The Tourist: A New Theory of the Leisure Class*. 3rd. ed. Berkeley: University of California Press, 1999.

Markwell, Kevin. "Dimensions of Photography in a Nature-Based Tour." *Annals of Tourism Research* 24 (1997): 131–55.

Marwick, Marion. "Postcards from Malta: Image, Consumption, Context." *Annals of Tourism Research* 28 (2001): 417–38.

Muir, Edward. *Civic Ritual in Renaissance Venice*. Princeton, N.J.: Princeton University Press, 1981.

——. "The Virgin on the Street Corner: The Place of the Sacred in Italian Cities." In Steven Ozment, ed., *Religion and Culture in the Renaissance and Reformation*, pp. 25–40. Kirksville, Mo.: Sixteenth Century Journal Publications, 1989.

Muir, Edward, and Ronald Weissman. "Social and Symbolic Places in Renaissance Venice and Florence." In John A. Agnew and James S. Duncan, eds., *The*

Power of Place: Bringing Together Geographical and Sociological Imaginations, pp. 81–103. Boston: Unwin Hyman, 1989.

Ortalli, Gherardo. "Il degrado e il saccheggio." In Italia Nostra, *San Clemente: Storie veneziane di civiltà e inciviltà,* pp. 4–15. Venice: Centro Studi Ricerche Ligabue, 1995.

Osborne, Peter. *Travelling Light: Photography, Travel, and Visual Culture.* Manchester, U.K.: Manchester University Press, 2000.

Padoan Urban, Lina. *Locande a Venezia, dal XIII al XIX secolo.* Venice: Centro Internazionale della Grafica, 1989.

Parks, George B. *The English Traveler to Italy.* 2 vols. Stanford, Calif.: Stanford University Press, 1954.

Pedenzini, Cristiana, and Isabella Scaramuzzi, eds. *Commercio e città: Un laboratorio per il piano commerciale di Venezia.* Bologna: Il Mulino, 1997.

Pemble, John. *Venice Rediscovered.* Oxford: Oxford University Press, 1995.

Perego, Francesco, and Francesco Sbetti. *Vivere a Venezia.* Venice: n.p., 1998.

Pine-Coffin, R. S. *Bibliography of British and American Travel in Italy to 1860.* Florence: Leo Olschki, 1974.

Pratt, Mary Louise. *Imperial Eyes: Travel Writing and Transculturation.* London: Routledge, 1992.

Reader, I., and T. Walter, eds. *Pilgrimage in Popular Culture.* London: Macmillan, 1993.

Redford, Bruce. *Venice and the Grand Tour.* New Haven, Conn.: Yale University Press, 1996.

Rizzo, Tiziano. *I ponti di Venezia.* Rome: Newton Compton, 1983.

Romano, Dennis. "The Gondola as a Marker of Station in Venetian Society." *Renaissance Studies* 8 (1994): 359–74.

"Safeguarding Venice." *Chemical and Engineering News* 78 (2000): 23–31.

Save Venice, 1968–1998: Venetian Treasures Restored and Preserved. Venice: Grafiche Veneziane, 1998.

Save Venice Inc. *1999 Regatta Week Gala.* Brochure. New York: Save Venice Inc., 1999.

Scaramuzzi, Isabella. "Memories and Suggestions." In Consorzio per lo Sviluppo Economico e Sociale della Provincia di Venezia (COSES), *Group on Tourism Sustainability.* Study document 90. Venice: COSES, 1997.

Selwyn, Tom, ed. *The Tourist Image: Myths and Myth Making in Tourism.* Chichester, U.K.: John Wiley, 1996.

Shields, Rob. "Fancy Footwork: Walter Benjamin's Notes on *Flânerie.*" In Keith Tester, ed., *The "Flâneur,"* pp. 61–80. London: Routledge, 1994.

Smith, Valene L., ed. *Hosts and Guests: The Anthropology of Tourism.* Philadelphia: University of Pennsylvania Press, 1989.

Sontag, Susan. *On Photography.* London: Farrar, Straus, and Giroux, 1977.

Sperling, Jutta. *Convents and the Body Politic in Late Renaissance Venice.* Chicago: University of Chicago Press, 1999.

Stewart, Susan. *On Longing: Narratives of the Miniature, the Gigantic, the Souvenir, the Collection.* Baltimore: Johns Hopkins University Press, 1984.

Tucci, Ugo. "Mercanti, viaggiatori, pellegrini nel Quattrocento." In Girolamo Arnaldi and Manlio Pastore Stocchi, eds., *Storia della cultura veneta*, 5 vols., vol. 3, pt. 2, *Dal primo '400 al Concilio di Trento*, pp. 348–53. Vicenza: Neri Pozza, 1980.

———. "I servizi marittimi veneziani per il pellegrinaggio in Terrasanta nel Medioevo." *Studi Veneziani* 9 (1985): 43–66.

Turner, Victor, and Edith Turner. *Image and Pilgrimage in Christian Culture*. New York: Columbia University Press, 1978.

Urry, John. *Consuming Places*. London: Routledge, 1995.

———. *The Tourist Gaze: Leisure and Travel in Contemporary Societies*. London: Sage, 1990.

van der Borg, Jan. *Tourism and Urban Development*. Amsterdam: Thesis Publishers, 1991.

Vanzan Marchini, Nelli-Elena. *Venezia: I piaceri dell'acqua*. Venice: Arsenale Editore, 1997.

Zanelli, Guglielmo. *Traghetti veneziani: La gondola al servizio della città*. Venice: Il Cardo, 1997.

Zaniboni, E. *Alberghi italiani e viaggiatori stranieri (sec. XIII–XVIII)*. Naples: Detkin and Rocholl, 1921.

Newspaper and Magazine Articles

Booth, Cathy. "In with the New." *Time*, 26 October 1998.

Colors 33 (August–September 1999). The issue title is *Venice: How Much?* Published by Benetton Group SpA, Ponanzo, Italy.

Fine, Adam. "The New Renaissance." *Casino Player*, August 1999.

Frater, Alexander. "Mother Ship." *Observer*, "Life" section, 13 September 1998.

Jong, Erica. "My Italy." *Travel and Leisure*, September 1996.

Muschamp, Herbert, and Ettore Sottsass. "Venice Rising." *House and Garden*, April 1990.

Plante, David. "Alone with Venice in Winter." *New York Times Magazine*, 2 October 1988.

Specter, Michael. "A Sinking Feeling." *New Yorker*, 12 July 1999.

Thomas, Stephen. "Just One Gelato." *Times Educational Supplement*, 17 January 1992.

Films and Television Programs

Italiensk for Begyndere. Directed by Lone Scherfig. Denmark, 2001. Film.

The Last King of the Gondoliers. BBC, 1998. Television broadcast.

Laura McKenzie's Travel Tips: Venice. Los Angeles: Republic Pictures Home Video, 1987. Video recording.

Super Cities: Venice. San Ramon, Calif.: International Video Network, 1994. Video recording.

Viva Las Venice! Produced and directed by Simon Dickson. BBC/TLC, 1999. Tele-vision documentary.
Will Venice Survive Its Rescue? BBC/NOVA, 1989. Television broadcast.

Websites

All Aboard: www.all-aboard.com/stories/venice3.htm.
American University, Trade and Environment Database: www.american.edu/ted/venice.htm.
CHORUS (L'Associazione delle Chiese di Venezia): www.chorus-ve.org.
Grand Princess: www.sealetter.com/Jul-98/newgp.html.
Norwegian Cruise Lines: www.ncl.com.
Radisson Seven Seas: www.rssc.com.
Royal Caribbean International: www.rccl.com.
Save Venice: www.savevenice.org.
UNESCO: www.unesco.int/culture/heritage/tangible/venice.
The Venetian: www.venetian.com.
Venetian Ghetto: www.venice-ghetto.com.
Venice in Peril: www.veniceinperil.org and www.friendsofvenice.org.
Weddings in Venice: www.venetohouse.com/guest/albatravel.htm.

Archival Sources

Archivio di Stato di Venezia, Frari (abbreviated in notes as ASV). *Cattaveri* and *Inquisitori di Stato.*
Museo Correr, Venice (abbreviated in notes as VMC). Mss. *provenienze diverse* (various origins) (abbreviated in notes as MPD).

Index

role in, 60–62, 308nn8,11,12; street
vendors of, 62, 118
Serenissima Society (Save Venice Inc.
program), 224–25
sewage system: *acque alte*'s impact on, 193;
canal dredging's support of, 188–89,
322n26; of hotels/restaurants, 184–
85; problematic solution for, 206–7;
pump system proposal for, 183–84,
321n8; tourists' reaction to, 182–83.
See also pollution
sex trade. *See* prostitution
Sforza, Francesco, 22
Shelley, Frances, 68, 164, 322n26
Shelley, Percy Bysshe, 164, 272
shopping. *See* retail sector
shrines *(capitelli),* 229, 326n49
sightseeing. *See* tourist gaze
simulation, representation vs., 288–89
Skippon, Philip, 38, 43, 163
Società dei Bagni, 166
Society for the Preservation of Ancient
Buildings, 214, 221
Sontag, Susan, 261
Sottomarina (beach town), 168
sottoporteghi (underpasses), 85, *86,* 89
souvenirs: Burano lacework as, 279–80,
331n36; generic type of, 274; homeward orientation of, 261; importation
of, into Venice, 172, 278, 313n48; as
markers of place, 274–75, 281; masks
as, 120, 280–81; Murano glass as, 171,
276–77, 320n47, 331n30; nostalgic
worth of, 273; of pilgrim-tourists,
19–20; sign/site linkage of, 262;
staged production of, 278–79; *vu'
cumprà*'s sale of, 118. *See also*
photographs, tourist; postcards
Special Law on Venice (1973), 220, 325n23
statistics: on *acque alte,* 190, 191, 192; on
annual tourists to Venice, 3–4, 105,
109, 301n5; on annual tourists worldwide, 1; on beach city day-trippers,
319nn35,37; on Campanile tourists,
71; on canal muck, 189; on canals, 134,
314nn1–2; on Carnival attendance,
250, 253, 297, 333n10; on Carnival
garbage, 254, 329n62; on Carnival
profits, 256–57; on Carnival sponsorship, 252, 329n56; on cruise ships' size,
204, 205, 324n73; on daily boat trips,
199, 323n52; on Ducal Palace tourists,

70, 71, 309n37; on fast-food outlets,
120, 313n49; on foreigners' gambling
losses, 42–43; on Galleria dell'Accademia tourists, 71, 82; on galley of Jaffa
bookings, 15; on gondola ridership,
152–53; on gondolas, 27, 136, 314n5;
on gondoliers, 27, 142; on hotel accommodations, 103, 112, 312n15; on
hotel bookings, 184; on Lagoon
passengers, 179; on lifting project,
195–96; on Murano glass industry,
171, 174; on Museo Correr tourists,
71; on Pala d'Oro tourists, 70; on
pedestrian traffic, 84; on Periphery
residents, 311nn40–41; on Piazza San
Marco tourists, 55, 84; on pigeons,
77; on Port of Venice traffic, 203–4;
on prostitutes during Serenissima, 43,
44; on restoration funding, 221, 225;
on Torcello tourists, 178; on university students, 102; on Venetian population, 3, 88–89, 109, 298; on *vu'
cumprà,* 115
Statmania (Carnival contest), 252
St. Didier, Alexandre, 44; on foreigners
at Carnival, 35; on gondolas, 145,
316n34; on Lagoon settlements,
163; on prostitution, 45, 48–50,
307nn56,58; on *ridotti,* 42, 306n48;
on Sensa, 307n73
Stewart, Susan, 270, 273, 281
Stiching Nederlands Venetië Comité
(Dutch Private Committee), 228
streets, of Venice: Calle dei Fabbri, 83,
294; Calle della Madonetta, 294;
Calle della Mandola, 294; Calle
dell'Ovo, 294; Calle di Mezzo, 111;
Calle Larga di San Marco, 265; Calle
Saoneri, 85, *86,* 111; Calle Vallaresso,
41, 151, 295; Lista di Spagna, 83; Merceria di San Zulian, 83; Ruga Vecchia,
83; Salizada di San Giovanni Grisostomo, 111; Salizada di San Lio, 83;
Secco Marina, 101; Sottoportego
del Luganegher, 85; Strada Nova,
83, 111
street vendors. See *vu' cumprà*
sublagunare (underwater tram line), 298
Superintendencies of Art and Architecture, projects list of, 220–21, 223,
325n32
syphilis, 45, 307n56

Text:	10/13 Galliard
Display:	Galliard
Cartographer:	Bill Nelson
Compositor:	Integrated Composition Systems
Printer and Binder:	Thomson-Shore, Inc.